THE LOCOMOTIVES

OF

THE STOCKTON & DARLINGTON RAILWAY

T. R. Pearce, I.Eng.,
A.M.I.Mar.E.

This book is dedicated to the memory of Ken Hoole

Published by the Historical Model Railway Society.

Typeset by the Author and Penshurst Press London Ltd.

ISBN 0 902835 14 9

Historical Model Railway Society

The Historical Model Railway Society was founded in 1950 to foster the research and recording of information relating to the railways of the British Isles and to make it widely available not only to those engaged in building accurate and authentic models but also to all with an interest in Britain's railway heritage. Recognition of the Society's valuable contribution in this educational field came in 1977 when charitable status was granted to it.

The Society has an active publications policy which is a key way of ensuring that the fruits of members' research reach a wider audience. It also co-operates with like-minded organisations in order to make their members' work more generally available. The Society has published a number of books which at first concentrated on railway liveries but which have latterly been expanded to cover the infrastucture as well. Other books, such as this, have examined a particular aspect of a railway in depth. All railway history is considered as falling within the Society's province, from the earliest years to the present day. The British Railways era has a notably strong following within its membership.

Members of the Society receive a high quality quarterly Journal and a bi-monthly newsletter, as well as discounts on HMRS publications and on the HMRS range of transfers. The Society also has extensive collections of drawings and photographs, and a large library. A system of Company Stewards, each specialising in a particular company, forms a clearing house for the exchange of information and a point of contact for enquiries. Local Area Groups meet regularly and talks are given in many parts of the country.

Further information and a prospectus/-membership application form can be obtained from:

The Membership Secretary,
Historical Model Railway Society,
c/o Chaworth House,
15 Byron Gardens,
SOUTHWELL,
Nottinghamshire NG25 0DW.

Penshurst Press | LONDON

Author's Foreword

I wrote this book because it didn't exist. Really, I suppose, Rail 150 started things off, but not to any great extent other than a renewal of interest in railways, and an awareness of the Stockton & Darlington Railway that had not been there before.

A few years later, I bought some postcards of S. & D.R. engines at an exhibition at Sunderland, a Tanfield Show, if I remember. These were obviously copies, and, as photographs, not too good, but again, they interested me, and I wanted to identify them, not having bothered to ask the chap selling them. And here I ran into difficulty. Yes, engines of the Stockton & Darlington Railway were illustrated in various books, if one knew which, but not many of them, either books or engines. I did have Robert Young's "Timothy Hackworth & the Locomotive", but had not then read it thoroughly.

Someone suggested that I contacted Ken Hoole, so I did. He was an approachable man in that way. I rang him up and asked, and he was helpful. Try the "Locomotive" magazine back in the 1920's, he said. R. H. Inness had written a whole series of articles on this subject, and Ken considered that this series was the definitive work on the matter. So, where to find them? Darlington Library Local History Section was the obvious starting point, and sure enough, they had copies, not the originals but early photo-copies, so old that they had been made by a wet solution negative/positive reflex method, and many of the positives were missing. They supplied me with Xerox copies of the copies and I used a mirror. They also showed me lists of locomotives made by several people long ago, but none of them complete.

So I began to make up my own list of engines, and went back to Ken with it. I got criticism, and more suggestions. So it went on, and gradually I learned to find my way through the maze, and to recognize types and classes; and I also managed to identify the postcards in the process. By then I had found and bought more such postcards, and was beginning to know where to look for other information.

I learned that a Mr. Theodore West had presented a paper on this subject to the Cleveland Institution of Engineers in 1886, and I was able to get a copy of this. I then realized that I had seen some of his sheets of sketches in the Hackworth Museum at Shildon in 1975, and had been very interested in them without recognizing their significance.

As always, one thing led to another. I found that the National Railway Museum had a better set of the "Locomotive" magazine, and got myself a reader's card and a full set of photo-copies. In the process I noticed advertisements in these magazines for books by Dendy-Marshall that had an obvious bearing on the matter. The N.R.M. had them. The information in them was fascinating and very enlightening, and also referred to others, such as E. A. Forward and his articles in the "Engineer" magazine. So it went on, and on, and on.

Every time that I thought that the work was finished fresh data has come to light, and I have had to set to again. These delays have enabled me to add much material, and, I think, to produce a better book. I must apologize to my readers for my many inconsistencies, grammatical and otherwise, and especially for my apparent inability to decide on the gender of a locomotive! I can only blame the many sources used and their differences.

Now there is a book on the Locomotives of the Stockton & Darlington Railway, where there was a gap before. To my regret, Ken Hoole never saw it, even in its first state, but I discussed it with him quite often, and got a lot of help from him. I like to think that he would have approved of it; and I hope that my readers do, as well.

I have been told that it is no longer fashionable to put forward alternative postulates. I consider this to be nonsense, and where more than one possibility exists I shall state all the facts at my disposal, so that my reader can decide for himself which is correct, if he so wishes. I can also point out that there is much research remaining to be done, especially from the Kew P.R.O. records, should anyone care to try!

This book contains no general index. I make no apology for this; it would have added considerably to both cost and time. Appendix IV includes a page and figure index of the locomotives

T.R.P./12-2-94
Nunthorpe.

Introduction : Tony Hall-Patch

There have been many locomotive histories published over the years which have been based on secondary sources for reference at the best and on myths which have grown in strength at each repeating at worst.

Tom Pearce has refreshingly sought out and studied primary sources in the main in his efforts to produce this erudite volume.

It is surprising that the Stockton & Darlington Railway Locomotives were not fully chronicled at the time, such was the novelty of this, the first public railway in the world to use steam locomotives.

Even the primary sources are sometimes open to interpretation and where this has been necessary the Author's engineering background and extensive knowledge of the subject have been used to best effect.

No complete, contemporary, list exists today and Tom Pearce's work not only remedies this but enhances it with detailed descriptions wherever this has been possible.

A. Hall-Patch, C.Eng., BSc.Eng., M.I.Mech.E., A.M.I.Mar.E.,
Senior Curator of Rail Transport,
Science Museum, London,
(Retired)

Contents

Timothy Hackworth

William Bouch

George Stephenson

The Protagonists

Chapter 1. Background

The People

In the beginning (of the Stockton & Darlington Railway, of course) was Stephenson. No matter how loyal and enthusiastic a supporter or admirer one may be of the other great engineering pioneers of the time, Cap'n Dick Trevithick, Matthew Murray, Timothy Hackworth, I. K. Brunel and Daniel Gooch, to name but a few, George Stephenson's charisma, persuasiveness, persistence and sheer ability in his own field overshadow them all. Of this there can be no doubt. And it was Stephenson's character and Stephenson's locomotives that made it possible for the Stockton & Darlington Railway to open when it did and to operate with steam traction. He was certainly not the great engineer that he wished to be known as, but he got things done! One has to admit that Edward Pease had a bit to do with it as well!

Nevertheless, the line's continued success was due to Timothy Hackworth, whose mechanical aptitude and refusal to be beaten by circumstances enabled him to solve problem after problem, and to provide a solid basis for the line to continue as well as it did after he retired to work for himself.

This book is about engines, not about people, but the machines reflected their designers' characters in many ways, and so I provide a brief background to Timothy Hackworth and his successor before starting on the main subject. For those who would like to know more, Robert Young's "Timothy Hackworth and the Locomotive" will provide much information about him, and so will the Timothy Hackworth Museum in Soho Cottage at Shildon. He was born at Wylam, where his father John was foreman blacksmith at the colliery, at the end of 1786. His family were religious folk, and he followed them in this. He was apprenticed to this same trade of blacksmith at the age of 14, under his father, who died when Timothy was 16. In his turn, Timothy became foreman in 1807 after completion of his apprenticeship at the age of 21, which indicates his prowess and standing. He was certainly concerned to some extent in the manufacture of the early Wylam locomotives with Blackett, Hedley and Forster. He would have been well acquainted there with George Stephenson, a Wylam man, although then employed at Killingworth.

Hackworth married in 1813, and became a Methodist at about the same time. He lost his job in 1815/16, at least partly due to his religious principles, and went to work at Walbottle Colliery for the next eight years. It seems that after his departure from Wylam no further major developments took place with the locomotives there, although track deficiencies and changes led to modifications to wheel arrangements. No doubt there was little need for further development, since they continued to do their job for many years.

In 1824, during the absence of both Stephensons, Robert in South America and George on surveying work and preparation for the newly proposed Liverpool & Manchester Railway, and others, Hackworth went to work for the Stephensons at their new South Street Works in Newcastle-upon-Tyne, in a temporary capacity, and although little is known definitely about this employment it is probable that he was concerned with the construction, and possibly the design, of Darlington Engine No.1, later to be called "Locomotion". Philip Brooks and others say that he was in charge of the South and Forth Street Works, at George Stephenson's request, and that he was also offered a partnership, which he declined. He would have had little enough capital to invest in the business.

He then worked on his own behalf for a short time; Layson says that he worked also for the Tyne Iron Co. on making boilers. But he was soon offered by George Stephenson a position as Resident Engineer (described in the Committee minutes as "Superintendent of the permanent and locomotive engines", although there were no locomotives for another five months) to the Stockton & Darlington Railway, which he accepted in May, 1825, at a salary of £150 p.a. plus free house and coal. He held this position, as a direct employee of the railway until 1833, and then as a major contractor, until May, 1840, when he retired in order to look after his own business in New Shildon. More of this later.

At the time George Stephenson was also being given a very hard passage as a witness at the debate on the first Liverpool & Manchester Railway Bill, which failed. However, he was still the official consultant engineer for the Stockton & Darlington Railway, and it is probable that he was responsible for the specification that accompanied the enquiry to Forth Street for the first two locomotives. It is also probable that he was responsible as well for at least some of the design of No.1 engine (see Chapter 3).

At the formal opening of the railway Timothy

Fig.1 Forth Street Works; South Street just round the corner.

Hackworth acted as guard and traffic manager for the inaugural train, while George Stephenson was on the engine, two of his brothers helping him.

The exact sequence of the following events, and of Timothy Hackworth's precise status at times, has not been clear, and there are many minor discrepancies and contradictions in the various accounts, but the recent appearance of the hand-written notes of Mr. E. Craven and a typed resumé by Ken Hoole has helped considerably. Hackworth was employed directly by the Railway until late 1833, when it was decided to change the system. The sub-committee minutes first mention the new system on 14th. March 1834, when agreement was reached with Wm. Lister to take three engines, No.12 "Majestic", No.7 "Rocket" and No.3 "Black Diamond" for 12 months, operate, maintain and repair them at an agreed figure and to pay the Railway 5% on a value to be agreed for them. Next were Kitchings (29th. March 1834) taking three more engines, and these contracts were to begin on 1st. May 1834. The remaining twelve engines were to be operated by Hackworth, but the valuations are not mentioned until 7th. August 1835, when both Hackworth and Kitching got an additional new engine each. Kitching also got No.2 "Hope" to rebuild at the same time. Other contracts were soon negotiated with Hackworth and Wm. Lister for operating "coaching" engines and with Neasham & Welch for repairs. There were others, including a Mr. Harrison who had a contract to run coaches and lead coal, presumably using horses, and Kitchings also ran coaches. Of course, there had been many smaller such contracts for horse drawn traffic.

In 1834 John Graham took over the responsibility for day to day operation of the engines, as well as being Traffic Manager, and there is no doubt that Graham dealt directly with these contractors, not through Hackworth. Graham describes how, in August 1837, he visited R. & W. Hawthorn at their Forth Bank Works, near to Robert Stephensons, together with William Lister, in order to inspect a new locomotive offered for sale by Hawthorns for use on the "coaching" traffic. This was No.28 "Sunbeam", which was accepted and delivered to the railway about a week later.

The railway had, of course, also been developing their own works, which eventually became the Shildon Works Co., and as established in 1849 this was nominally an independent firm, with William Bouch, Oswald Gilkes, S. & D.R. secretary at the time, and his brother Edgar Gilkes, of whom more anon. But at the time we are now considering these works were under the charge of Timothy Hackworth until 1840. His son John Wesley Hackworth said that his father rented the premises from the railway. These works became the Shildon Wagon Works in due course, and were closed down by British Rail in fairly recent years.

John Graham says in his notebooks that the Railway decided in 1837 to take back all engines into their own hands, but this does not seem to have happened straight away and it is February 1839 before the Committee recommends that the Shildon Works be taken back, presumably from Hackworth, and it was April before the contract with Lister was cancelled. The Railway then carried out all these operations themselves until 1849, when the Shildon Works Co. began to operate contract work again. (Joseph Tomlinson, who served his time at Shildon from 1837, said in his Presidential Address to the Institution of Mechanical Engineers in 1890 that Timothy Hackworth was then [1837] Locomotive Superintendent and also one of the contractors for the working of the railway).

closed about twentyfive years ago. One of the family, another Nicholas, was a prominent member of the North Eastern Railway Association until early 1991, when he was killed in a tragic bus accident.

Thomas moved on in 1839 to go into partnership with George Fossick at the Stockton Engine Works on the Norton Road which eventually became Blairs, famous marine engine builders. Reading between the lines, I get the impression that Shildon was glad

Fig.2 Soho Works, New Shildon, in 1925. Part of the pattern shop, later used as a paint shop, and still existing as part of the Timothy Hackworth Museum, can be seen at the far left.

to see him go. I suspect that Timothy had a financial stake in the Stockton firm himself.

Hackworth had been buying property on the other side of the main road from the Shildon Railway Works from about 1830, and in 1832 started to build his own works there, in the northern part of New Shildon, and near to the main line. In 1833 he began serious work there. These works were given the name of the Soho Works, at, it is believed, the suggestion of Joseph Pease (Boulton & Watt's Birmingham works bore the same name in c.1760, and the first steam engine works in America, at Belleville, before the turn of the century, was also so named). He installed his brother Thomas there as manager, from about 1833, and the registered name of the firm was Hackworth & Downing, the Hackworth being Thomas, and Nicholas Downing his partner. Little seems to be known of Nicholas Downing; he was the founder (in both senses) of the Phoenix Iron Works in Shildon, in about 1832, which was taken over by Gilkes, Wilson, of the Tees Engine Works, in the mid 1860's, and he then remained there as manager for a few years, before opening another and very successful foundry at Stockton, which only

It would seem that Timothy Hackworth did most of the early design work for both firms, Shildon and Soho, and I have found it rather difficult sometimes to be sure at which of the two works any particular locomotive between 1833 and 1840 was built. The locomotives built later at Fossick & Hackworth also had a flavour of Timothy's design, for many years, and one can only conjecture about that.

It is clear that John Graham's notebooks, Craven's extracts from the Committee minutes and other sources such as the Whessoe official history, show that Timothy Hackworth, from 1833 until 1837, when the railway took back all the engines and rolling stock into their own control, was one, but certainly the main one, of the three contractors who actually operated and maintained the engines and coaches, while running and time-tables, etc., were controlled by John Graham working to the direct instructions of the Management Committee. I have mentioned the passenger coaches above. At the same time there were originally many small operators of both coaches and coal trains who supplied their own horses and drivers, and probably vehicles as well, paying the railway for the use of their track. The various collieries served also owned most of their own trucks. No wonder there was confusion! But Hackworth seems to have retained overall mechanical engineering control until 1840, when he resigned completely from the service of the S. & D.R.

He was succeeded by William Bouch, elder brother of Thomas Bouch, a civil engineer who was also closely connected with the Stockton & Darlington Railway and their associates, and who was responsible later for the design of the early and ill-fated Tay Bridge, a large part of which was in fact made and erected by associates of the railway, some of it in Middlesbrough. William Bouch was another idiosyncratic engineer, who very soon began to develop his own ideas, using Hackworth's as a basis at first, but the effect was to establish an individual Stockton & Darlington design flavour.

Timothy Hackworth died in July, 1850, after a very sudden illness, and although his son John Wesley Hackworth succeeded him the Soho Works were soon abandoned. They were bought by the railway in 1855 so that the Shildon Works could be developed as wagon works,and locomotive building continued at the Soho Works until well after the Darlington North Road Works opened in 1863, although at least some of these later engines were assembled there from parts made at and supplied by the North Road Works.

Many writers have decried Timothy Hackworth, saying that he was 'stuck in a groove', and that he added little to locomotive design after the "Royal George". Some have even said that he produced nothing better than a series of crude 0-6-0 mineral engines with return tube boilers, but this is manifestly untrue. He had the patience and pragmatism to realize that there was little point in change for its own sake, and that if a thing worked well and economically it was silly to elaborate on it, but there is no doubt of his design ability or inventiveness, where called for. He left a far greater legacy to future locomotive development than is generally admitted, and much greater acknowledgement of this is long overdue.

William Bouch was born in 1813, at Thursby near Carlisle, although the birth seems to have been registered at Whitehaven. I am told that his mother was actually in labour as they moved to the new home at Thursby. His father was a sea-captain. Bouch served his apprenticeship with Robert Stephenson & Co., in the company of many other young engineers who became well known later. It would appear that between completing his time, which would probably have been in 1834, and coming to Shildon, he served with, but not as a member of, the Imperial Russian Navy, where he held the rank of Chief Engineer for several years. This service involved two lengthy visits to Russia. But he was still only 27 when he joined the Stockton & Darlington Railway, at first as Foreman at the Shildon Works under Oswald Gilkes, and later as one of three contractors (the Shildon Works Co., 1849). After this contract ceased (date not known but about 1860) he became Locomotive Superintendent, and held this position both with the Stockton & Darlington Railway and then with the Darlington Committee of the North Eastern Railway until his death, making a total length of service with them of almost 36 years. During this time he also held several other positions and directorships with other firms, in addition to his main career with the S. & D.R., and he became the first President of the newly formed Cleveland Institution of Engineers in 1864, as recorded in a short obituary by the President at the time of his death, Thomas Wrightson, although officially I believe that John Dixon has been credited with this honour.

The other firms concerned included W. & T. Bouch, Locomotive Manufacturers, of the Woodbank Works, Carlisle, in 1840, and later, when this firm became Cowans, Sheldon, and moved from Woodbank, the brothers remained directors both of Cowans, Sheldon and of Darlington Forge, set up by Cowans, Sheldon because they did not have enough space at Carlisle.

The Craven/Hoole notes say that he invented, among other things, a multi-jet blast pipe with a steam reservoir to give a more even and continuous blast, but I have seen no mention of this in any of the descriptions of the engines. Ahrons mentions it, and I include his illustration, taken from the "Engineer" (Fig.3). Bouch's chimney jacket feed heater and various types of steam brakes will be referred to in the text.

He died at Weymouth in January, 1876, and his demise effectually ended the separate existence of this famous railway, which had continued to exert its independence even after the merger with the North Eastern Railway in July, 1863, and the Management Committee ceased to function soon after his death, its allotted time having lapsed. It lingered on until 1879, but with little effect.

Theodore West, who had been employed by the Darlington Section of the N.E.R. from 1865, held, I believe, the position of Chief Draughtsman (described as 'Engineer') at the North Road Works. He was a Quaker, born in Leeds in 1826, and during

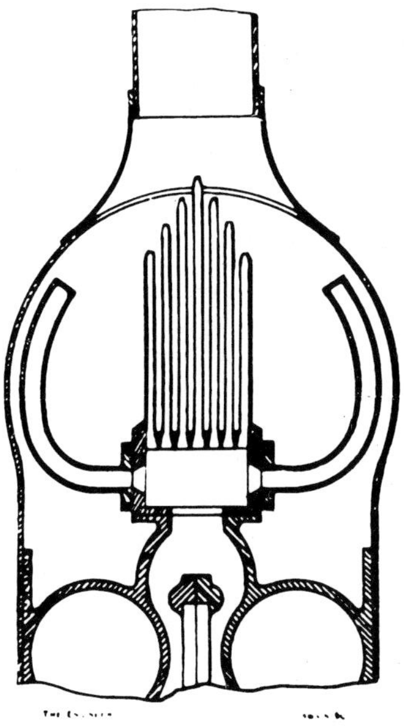

Fig.3 Bouch's multiple blast pipe.

his boyhood there had watched the Blenkinsopp-Murray rack locomotives working on the Middleton-Leeds colliery railway. He published in the 1880's a number of sheets of sketches of locomotives showing their development from the earliest days; two of these sheets showed locomotives of the Stockton & Darlington and the early North Eastern Railways. They are reproduced here as an appendix, but although very interesting, and in many cases the only illustrations extant of several of the engines, they should be used with great care. In many cases the name and place of business of the engine builder is incorrect, and the dimensions are not to be relied upon. If it is possible these facts should always be checked elsewhere.

However, as a result of this publication, he was asked to present a paper on the development of the locomotive engine to the Cleveland Institution of Engineers mentioned above, and in fact he presented two linked papers, the first one on general development, on March 1st., 1886, and the second, on Stockton & Darlington and N.E. Railway locomotives and also American locomotives, on April 5th., 1886. These papers were attended by several eminent railway personalities of the day, among them Mr. J. M. Oubridge, who had some interesting comments to make about Timothy Hackworth and others, and Mr. W. J. Cudworth, connected with the Kitching family and son of Mr. W. Cudworth of the S. & D.R.

These papers were read at the old Middlesbrough Literary and Philosophical Society Hall in Corporation Road, which was destroyed about thirty years ago, a loss to the whole area, but not as well used as it should have been during its later years. William Dresser & Sons, printers and stationers of Darlington, still existing in name there as booksellers and stationers, published a reduced size edition of these sheets of sketches, including some brief biographical notes on Theodore West, c.1900, but this book is quite scarce. My appendix copies were provided originally by Ken Hoole. Theodore West died in Darlington in 1898.

Another but later Victorian who produced a number of drawings of Stockton & Darlington Railway locomotives was Clement E. Stretton, C.E., who wrote a popular history of the steam locomotive called "The Locomotive Engine and its Development". This ran into at least six editions, the first being in the early 1890's and the last in the early 1900's. This gained him a certain reputation and some international kudos, and some of his work was undoubtedly good. Unfortunately, if he did not know or was uncertain about anything he just made it up and used it as if it was an established fact. I shall refer later to his drawing of No.6 "Experiment". He is almost certainly the originator of several sheets of locomotive drawings held by the Science Museum, although they may have been redrawn many years ago by a Museum draughtsman. One of these, reproduced in the appendix, is of ten of the Stockton & Darlington engines, beginning with No.1 "Locomotion", this a well known and widely used drawing, and finishing with No.165 "Keswick". The sheet of drawings includes a list of data. I am including it after West's sheets of sketches, by courtesy of the Science Museum, as it is of some interest.

I should also refer here to the Darlington Quakers generally, especially, of course, the Pease family and their relatives, friends and associates. It was their foresight, drive and capital that brought the Stockton & Darlington Railway into being, and their careful planning that extended it and kept it running for nearly fifty years.

General Note

It should be borne in mind that the Stockton & Darlington Railway did not number or name its locomotives for several years - the first one referred to by name was "Royal George" in Hackworth's notebook in December 1827, and the name of "Locomotion" was not mentioned officially until January 1832, when John Graham used it in one of his reports. Until then he had referred only to an engine's number (1-9). The first reference to an engine number is in the sub-committee minutes for October 5th. 1827 (No.2.) and Timothy Hackworth's notebook refers to No.1 locomotive on October 26th.

1827. Other references followed. In 1833 a list was supplied to the Comte (often referred to as the 'Chevalier') de Pambour for use in his book, "A Practical Treatise on Locomotive Engines", which was not published for another two years, possibly only in French at first, but certainly in English in 1840. This is the first known complete list of numbers and names, and it is obvious that these must all have been in use by 1833 at the latest. I reproduce this list below, as it gives some information regarding the engines about types of boiler and tubes, as well as commercial data. Unfortunately this boiler data conflicts in some respects with other information.

When numbers began to be used, they were applied to all engines in service (except at first for "Royal George", the only one to have a name then), until the passenger trade got under way. In February 1839 a decision was made that 'coaching' engines should carry names only, and the coal engines should have both names and numbers. This meant that No.26 "Swift", No.27 "Queen" and No.29 "Sunbeam" lost their numbers, which were not used again immediately. I assume that engines that were used for both passenger and coal duties kept their numbers. There were gaps in the list, and No.28 "Tees", No.30 "Beehive" and No.31 "Briton" were renumbered 14, 12 and 11 respectively to fill these gaps. This decision was reversed in May 1847, when the 15 'coaching' engines then in service received numbers from 40 - 54. I have not tried to show all these changes in Appendix IV, the list of locomotives in numerical order, except in so far as changes of number have been listed without the dates concerned, and each engine is shown against the number originally received, with its alternative number or numbers. Other changes of number took place for various reasons, and several of the numbers given are uncertain.

Instructions were given to Timothy Hackworth at the end of May 1831 for numbers (large figures, projecting from the chimney) to be fitted to the coal engines for identification purposes, but although John Graham obviously knew the numbers, whether in fact they were so fitted, and how quickly, is another matter. In the very beginning an engine would have been identified only by the regular driver's name, its builder, or some peculiarity about it, as in the name "Old Elbows" for "Experiment".

Samuel Smiles says that the original purpose of putting numbers on each engine was so that the public could identify an engine and thus also the driver when a nuisance was caused, such as excessive spark discharge from the chimney, a frequent cause of lineside fires, but the Management Committee Minute on the subject says that it is so that "the misconduct of the drivers can be more easily reported", and the most frequent items of misconduct recorded are inebriation, carrying illicit passengers (including girls!) on their engines, and excessive speed.

Again, when numbers began to be used, as engines were scrapped or sold (or lost by explosion!) new engines were given the vacant numbers as they

Statement of the work done by the locomotive engines on the Stockton and Darlington Railway, from July 1, to December 1, 1833.

Number of the engine.	Name of the engine.	Number of days that the engine was — In activity.	Number of days that the engine was — In repair.	Total number of miles travelled by the engine.	Tons of coals carried to one mile going down, by the engine.	Gross tons carried to one mile on a level, including the waggons and return.	Amount of the repairs made to the engine during that time. £.	s.	d.	Amount of the repairs per gross ton carried to 1 mile on a level.	Observations.
		days.	days.	miles.	tons.	tons.	£.	s.	d.	d.	
1	LOCOMOTION.	80	52	5,300	146,011	292,022	41	19	7	·035	Boiler with a flue and two returning tubes.
2	HOPE.	63	69	3,100	82,305	164,610	57	5	5	·084	—— with a single flue.
3	BLACK DIAMOND.	27	105	1,000	26,920	53,840	14	0	5	·062	—— with a single flue.
4	DILIGENCE.	2	130	80	1,906	3,812	13	18	3	·877	Engine taken to pieces.
5	ROYAL GEORGE.	11	121	700	23,733	47,466	161	7	8	·816	Boiler with a flue and one returning tube.
6	EXPERIMENT.	70	62	4,400	122,442	244,884	53	1	2	·052	ditto
7	ROCKET.	64	68	3,940	109,512	219,024	57	0	9	·062	ditto
8	VICTORY.	107	25	10,600	349,150	698,300	58	3	10	·020	ditto
9	GLOBE.	60	72	3,120	70,683	141,366	36	4	6	·061	Boiler with 120 returning tubes.
10	PLANET.	27	105	1,200	20,429	40,858	53	7	5	·314	—— with 88 ——
11	NORTH STAR.	55	77	2,400	47,546	95,092	32	5	10	·082	—— with 88 ——
12	MAJESTIC.	47	85	2,880	90,422	180,844	131	2	3	·175	—— with 104 ——
13	CORONATION.	52	80	2,940	97,687	195,374	46	16	2	·057	—— with 104 ——
14	WILLIAM IV.	55	77	4,060	134,440	268,880	78	19	8	·071	—— with 104 ——
15	NORTHUMBRIAN.	59	73	4,480	143,885	287,770	67	14	11	·056	—— with 104 ——
16	DIRECTOR.	91	41	5,860	202,492	404,984	107	19	11	·064	Boiler with tubes, Napier's patent.
17	LORD BROUGHAM.	62	70	4,780	155,729	311,458	62	5	10	·048	Boiler with 104 returning tubes.
18	SHILDON.	63	22	4,720	159,400	318,800	49	16	3	·038	—— with a flue and two returning tubes.
19	DARLINGTON.	88	44	6,180	200,110	400,220	45	0	6	·027	—— with a flue and two returning tubes.
20	ADELAIDE.	71	61	3,700	126,390	252,780	90	11	7	·086	—— with 104 returning tubes.
21	EARL GREY.	110	22	7,960	276,462	552,924	14	19	6	·007	—— with a flue and two returning tubes.
22	LORD DURHAM.	84	48	6,480	213,737	427,474	67	13	8	·039	—— with 104 returning tubes.
23	WILBERFORCE.	55	9	4,200	141,534	283,068	51	17	11	·044	—— with 104 returning tubes.
	Total	1403	1518	94,080	2,942,925	5,885,850	1393	13	0	·057	

occurred, and so numbers did not follow consecutively as the new engines were added to the list. This policy was changed in January 1855, when it was decided that new numbers should be consecutive, leaving blanks as older engines disappeared. This was reversed again in 1865, and blanks in the list were used, with chaotic results as far as numbering of classes of mineral engines are concerned.

Another interesting list of engines appears in John Graham's notebooks for October 1836, in which he lists 22 engines, by name only, so as to tabulate the results of a check on axle and wheel wear. The amount of wear, whether axial or radial, is almost incredible, and many engines had wheels not all of the same diameter! Three engines known to be there at the time are not listed; No.9 "Globe", probably out of action again and under repair, and Nos.10 "Planet" and 11 "North Star", both more or less condemned a year earlier, so not on the active list.

There has been for a long time a persistent rumour that No.1 was called "Active" at first; Longstaffe, in his 1854 "The History and Antiquities of the Parish of Darlington in the Bishopric", seems to have been the first to state this, and Samuel Smiles repeated it in his first (1857) and second editions of "The Life of George Stephenson", and also in the first edition of the combined book "The Lives of the Engineers, Vol.3, George and Robert Stephenson". He corrected it in later editions to "Locomotion". Robert Young also mentioned it in "Timothy Hackworth and the Locomotive". Others refer to it, and in general dismiss it as an error, but I am not so sure myself. It may be that someone at Darlington or Shildon gave it a name which did not get adopted officially. There was certainly an "Active", No.48, in the locomotive list later on, one of Kitchings' passenger engines (see list) and this could have been a revival of the name. Anyhow, it was never used officially, and No.1 "Locomotion" became the listed description in due course.

Stock valuation lists exist for the years 1837 - 1844, and possibly for other years. At first this was done by Jas. Kennedy, then for four years by John Dixon. In 1842 the valuation was done by Graham, Gilkes, and for the next two years by Graham, Bouch. In these lists, from 1839 on all the mineral engines are numbered as well as named, but not those described as 'coaching' or 'merchandize' engines. When an engine was moved from the list of mineral engines to one of the others, it ceased to be numbered.

These stock valuation lists will be referred to from time to time later. A further and very valuable list, dated 1849 but covering several more years, comes from John Dixon's manuscript book, giving some valuations and also many details of the origin of most of the engines, their designer, builder and date, and in many cases the date and place of rebuilding. All the engines in this list are both numbered and named.

I have also a printed List of Locomotives on the North Eastern Railway - Darlington Section, given to me by Ken Hoole, listing the engines from No.1 (blank) to No.220 (1868). It is interesting to see that at this late date, five years into N.E.R. days, most of the 'Tory' and 'Miner' classes were still listed. I have as well a typed list from Darlington Library, from No.1 "Locomotion" to No.196 "Roseberry" (wrongly named as "Whitby") which makes it about 1866, containing several errors but with a lot of useful handwritten notes. I have been unable to discover its provenance.

Other interesting lists were published in the "Engineer" magazine for October 31st. and December 26th. 1879, the first one, dated November 1840, giving dimensions and data, much of which differs from my other sources, and the second one, dated June 1839, giving mileages and commercial data. The first list was apparently contributed by George Graham, and is included here. The Craven/Hoole notes contain several lists of various sorts.

Other lists do exist, but I have not yet seen them myself. Craven refers to a list in B.R. archives, dated November 1840, which gives many details of the locomotives at the time. These are now in the P.R.O. at Kew, but I have no doubt that this 1840 list is the same one as that mentioned above as published in the "Engineer".

Timothy Hackworth's notebook, referred to earlier, covers a period of almost one year, from autumn 1827 until the following summer. It is not entirely homogeneous, since it contains entries in other handwriting, some of them of a much later period, but most of it does appear to be the work of Hackworth himself, although written in the third person as notes to himself. The entries are not in chronological order, the dates are often vague and much of it is in pencil and so difficult to decipher. It contains much about the Etherley winding engine, then undergoing modification, and many lists of items purchased, etc. The early engines, their drivers and the accidents to them are referred to frequently. There are also many dimensioned sketches of engine parts, ideas and suchlike, including one sketch of a malleable iron locomotive wheel centre, his notes and sketches for his spring loaded safety valve and a couple of locomotive arrangements. I shall refer to these notes in the following chapters.

E. A. Forward's extracts from the early Robert

Dimensions of Locomotive Engines upon the Stockton and Darlington Railway, November, 1840.

	Engines' names.	Builders' names.	Built in the year.	Cylinders. Diameter of cylinders.	Cylinders. Length of stroke. (ft. in.)	Position of cylinders.	Boilers. Length of boilers. (ft. in.)	Boilers. Diam. of boilers. (ft. in.)	Copper tubes. Length of C. tubes. (ft. in.)	Copper tubes. No. of C. tubes.	Copper tubes. Diam. of C. tubes. (in.)	Fire tube. Length of F. tube. (ft. in.)	Fire tube. Diam. of F. tube. (ft. in.)	Return tube. Length of R. tube. (ft. in.)	Return tube. Diam. of R. tube. (ft. in.)	Fire-box. Length of F. box. (ft. in.)	Fire-box. Breadth of F. box. (ft. in.)	Fire-box. Height above the bars. (ft. in.)	Driving wheels. No. of wheels.	Driving wheels. Diam. of wheels. (ft. in.)	Supporting wheels. No. of wheels.	Supporting wheels. Diam. of wheels. (ft. in.)	Area in inches of heating surface.
Common return tubes.	Locomotion	Stephenson	1825	9	2 0	Vertical	12 0	4 0	—	—	—	11 0	2 0	11 0	1 4	—	—	—	4	4 0	—	—	16,585
	Hope	Kitching	1835	11	1 8	Vertical	14 0	4 0	—	—	—	13 6	2 1	13 6	1 3	—	—	—	6	4 0	—	—	20,356
	Royal George	Hackworth	1827	11¼	1 8	Vertical	13 0	4 4	—	—	—	12 6	2 1	12 6	1 6	—	—	—	6	4 0	—	—	20,260
	Rocket	Stephenson																					
	Victory	Hackworth	1829	11½	1 8	Vertical	13 0	4 4	—	—	—	12 6	2 1	12 6	1 6	—	—	—	6	4 0	—	—	20,260
	Middlesbro'	Lister	1839	12¾	1 6	Oblique	14 0	4 4	—	—	—	13 9	2 2	13 9	1 8	—	—	—	6	4 0	—	—	23,844
	Auckland	Hackworth	1839	12⅝	1 6	Oblique	13 6	4 4	—	—	—	13 0	2 1	13 9	1 8	—	—	—	6	4 0	—	—	22,052
	Beehive	Hackworth	1838	11	1 8	Vertical	13 6	4 4	—	—	—	12 6	2 1	12 6	1 6	—	—	—	6	4 0	—	—	20,260
	Briton	Hackworth	1838	11	1 8	Vertical	13 6	4 4	—	—	—	12 6	2 1	12 6	1 6	—	—	—	6	4 0	—	—	20,260
	Tees	Kitching	1837	11	1 8	Vertical	13 6	4 4	—	—	—	13 0	2 1	13 0	1 3	—	—	—	6	4 0	—	—	19,601
	Tory	Hackworth	1838	12⅝	1 6	Oblique	13 6	4 4	—	—	—	13 0	2 1	13 0	1 8	—	—	—	6	4 0	—	—	22,052
	Whig	Kitching	1838	12¼	1 6	Oblique	14 0	4 4	—	—	—	13 9	2 2	13 9	1 8	—	—	—	6	4 0	—	—	23,844
	Etherley	Lister	1840	12¾	1 6	Oblique	14 0	4 4	—	—	—	13 9	2 2	13 9	1 8	—	—	—	6	4 0	—	—	23,844
	Enterprise	Kitching	1835	11	1 8	Vertical	13 6	4 4	—	—	—	13 0	2 1	13 0	1 3	—	—	—	6	4 0	—	—	19,601
	Ocean	Lister	1840	12¾	1 6	Oblique	14 0	4 4	—	—	—	13 9	2 2	13 9	1 8	—	—	—	6	4 0	—	—	23,844
Multitubular tubes.	Black Diamond	Lister	1838	10⅝	1 6	Oblique	10 8	4 0	4 2	103	1¾	6 6	2 4	—	—	—	—	—	4	4 6	—	—	35,538
	Diligence	Hackworth	1834	14½	1 4	Vertical	10 4	4 4	7 8	90	1¾	9 8	2 3	—	—	—	—	—	6	4 0	—	—	57,322
	Dispatch	Thos. Hackworth & Co.	1839	14½	1 6	Oblique	13 4	4 4	10 0	70	2	12 6	2 3	—	—	—	—	—	6	4 1	—	—	68,044
	Coronation	Hawthorn																					
	Director	Stephenson	1832	14½	1 4	Vertical	9 6	4 4	6 5	80	1¾	9 0	2 3	—	—	—	—	—	6	4 0	—	—	44,854
	Darlington	Hawthorn	1832	14¾	1 4	Vertical	9 6	4 4	6 6	88	1¾	9 0	2 3	—	—	—	—	—	6	4 0	—	—	48,724
	Adelaide	Stephenson	1832	14¾	1 4	Vertical	9 6	4 4	6 11	88	1¾	9 0	2 3	—	—	—	—	—	6	4 0	—	—	51,142
	Earl Grey	Hawthorn	1832	14¾	1 4	Vertical	9 6	4 4	7 0	88	1¾	9 0	2 3	—	—	—	—	—	6	4 0	—	—	51,626
	Lord Durham	Stephenson	1832	14¾	1 4	Vertical	9 6	4 4	6 6	80	1¾	9 0	2 3	—	—	—	—	—	6	4 0	—	—	45,293
	Wilberforce	Hawthorn	1832	14¾	1 4	Vertical	9 6	4 4	6 7	100	1¾	9 0	2 3	—	—	—	—	—	6	4 0	—	—	54,418
	Magnet	Hackworth	1835	15	1 4	Vertical	11 0	4 4	7 6	100	1¾	9 6	2 3	—	—	—	—	—	6	4 0	—	—	61,076
	Pilot	Kitching	1840	14½	1 6	Oblique	13 4	4 4	10 1	72	2	12 0	2 4	—	—	—	—	—	6	4 0	—	—	69,936
	Wilton Castle	Neasham and Welsh	1840	14½	1 6	Oblique	13 8	4 4	10 1	80	2	12 7	2 2	—	—	—	—	—	6	4 0	—	—	75,619
Coach engines.	London	Hague	1839	11	1 4	Horizontal	7 0	2 4	7 4	71	1¾	—	—	—	—	2 7¼	3 3	3 7	4	4 6	—	—	42,174
	Sunbeam	Hawthorn	1837	12¼	1 6	Horizontal	8 0	3 2	8 4	104	1¾	—	—	—	—	3 5¼	2 4	3 2	2	5 0	2	4 0	63,310
	Arrow	Hackworth	1837	21	0 9	Horizontal	7 9	3 4	8 0	133	1¾	—	—	—	—	2 7½	3 6	4 2	2	5 7	4	3 6	76,153
	Dart	Hackworth	1840	13	1 4	Horizontal	8 2	3 3	8 7	122	1¾	—	—	—	—	2 5	3 6	4 0	4	4 6	—	—	74,811
	Ruby Castle	Kitching	1839	12	1 6	Horizontal	8 0	3 3	8 3	102	1¾	—	—	—	—	2 7½	3 6	3 9	2	5 0	4	3 6	63,380
	Queen	Kitching	1838	14	1 6	Horizontal	8 6	3 8	8 10	86	2	—	—	—	—	2 7	3 6	4 5	4	4 6	2	3 0	66,315
	Manchester	Sharp, Roberts, & Co.	1840	12	1 6	Horizontal	8 0	3 3	8 4	89	2	—	—	—	—	2 8	3 7½	3 7	2	5 0	4	3 6	63,744
	Comet	Kirkly and Co.	1840	13	1 6	Horizontal	8 6	3 6	8 9	118	1¾	—	—	—	—	3 1	3 8	3 5	2	5 0	4	3 6	76,377

November 1840 List

Stephenson & Co. works ledger are very useful, giving an approximate date for each engine as well as the cost of all except No.1, but the dates given are just that, ledger entries, and not necessarily the actual dates of delivery. Works numbers are not given against each engine, but have to be lifted from other entries for parts, freight etc.

The reminiscences of George Graham, an early driver of No.1 engine and son of John Graham, the first Traffic Manager of the line, recorded by a Mr. Harold Oxtoby in about 1896, gives much useful information and detail about the day to day operation of the line in the early years. These memories are based in the main on his father's notebooks, referred to above, a set of four held at the Science Museum (together with a later resumé) which, having been written at the time are rather more reliable. Nevertheless, George Graham adds much from his own experiences in the way of later adventures and behaviour.

I refer from time to time to the Hoole/Craven notes, including relevant extracts from the minutes of the various committees, locomotive deterioration book etc. These came into my possession when I thought that the book was complete, and I have explained them further in the list of "Acknowledgements" at the end of the book.

History and Development

In such a brief and superficial survey over-simplification is inevitable. I do not intend to comment on the financial and economic history to any depth; others, including Dr. Maurice Kirby recently, have covered it very well.

The Stockton & Darlington Railway was built as a mineral railway, and for no other purpose. The philosophy was simple; to get coal from the South Durham coal-fields down to the River Tees for distribution and later for export at as low a cost as possible. A canal had been considered and rejected, and the advocates of a railway had their way. Since they were first in the field on any sort of scale, the railway did succeed in this purpose and continued to be profitable for a long time, but the first few years were touch and go.

The adoption of steam locomotive haulage was a bold step, and very nearly failed because of the unreliability of the early engines, but as it turned out to be cheaper than horse traction it was worthwhile to persevere with it.

As other railways were built and competition became fiercer, profits fell, and the discovery of the Cleveland ironstone deposits and the means of using that ironstone arrived just in time to save the line from commercial failure. So it still remained a mineral railway.

As a railway, it was not the first in any single aspect, but in combination and on a larger scale than heretofore, it was a major step forward. Edge rails and flanged wheels had been in use for a very long time. There were at least three earlier public railways. The Middleton-Leeds Railway had been using steam locomotives successfully for thirteen years when the Stockton & Darlington opened, but that was a colliery line, not public.

The passenger business was by way of being an accident. It had not been considered as a commercial proposition, and at first all passenger traffic and much of the coal traffic was contracted out to individuals, who ran their own horse drawn wagons and coaches on the line, and paid the Company for the privilege. It was a surprise to all concerned when this passenger service proved to be so popular, but it was not for this reason that the Company bought out most of the contractors later but because of the huge nuisance of the horse-drawn traffic under separate and largely independent control causing trouble and hindering any sort of regular steam hauled traffic. John Graham's notebooks and his son George Graham's reminiscences contain many examples of such incidents relating to horse drivers. I believe that the first official steam drawn passenger train was run at the end of 1830, at the opening of the new Middlesbrough line; they did have a suitable coach by then. If so, there were no more recorded until November 1833. It took many more years to get rid of all the horses. I am told that there was still a horse drawn passenger service on the Middlesbrough Branch in 1864, in N.E.R. days! In the meantime the railway began to run parcel trains, both by themselves and jointly with the passenger trains, and these were very successful. Later on, Parliament insisted that any new railways or major extensions should carry passengers.

Here again, they were not the first passenger railway. The Swansea & Oystermouth Railway (later known as the Mumbles Railway) had started up a public horse drawn passenger service in 1807, and there was also the Duke of Portland's Railway from Kilmarnock to Troon, built by William Jessop, which was carrying passengers in special vehicles in 1810. As to steam, Trevithick had run a temporary steam locomotive drawn passenger coach for public use in London in 1808, although it was more of a funfair ride than a serious railway.

Except for the first complimentary tickets issued for the inaugural trip in September, 1825, the Stockton & Darlington Railway did not use tickets for passengers for some years, probably not until 1833. The Liverpool & Manchester Railway, built as a passenger line, used them from the opening in 1830. Of course, the word 'ticket' may be subject to differing definitions, so some may disagree with me on this point.

The Stockton & Darlington Railway was the first to use wrought iron edge rails to any large extent, and as the advantages of these became obvious, most of the track was changed to wrought iron as soon as possible and convenient, and the weight of the rail increased to cope with the increasing traffic, although the few remaining sections of cast-iron rail continued to give trouble for years. We have George Stephenson to thank for this foresight; as Consultant Engineer he insisted on the use of wrought iron rail for at least half the original track, against his own immediate financial interest, since he was a partner and co-patentee with William Losh in a firm that made improved cast-iron rails! His reputation and trust in him gained from it, all the same. He had also an interest in the Bedlington Ironworks and their wrought iron rails, so he probably made up later what he lost at the time.

This is not the place to list all the meetings and discussion which took place before the line received formal approval, nor all the arguments and in-fighting; this has all been well documented by many people. However, to over simplify, Stockton had wanted a canal, and was over-ruled, but gave some money and business to the line in due course. Most unfortunately, as some of the more far-sighted Stockton business men had foreseen, the railway benefited Stockton for only a short while. It was soon necessary to move down river to deeper water for use by bigger ships, and I suspect that this had in fact been intended from the beginning. The opening of the Middlesbrough Branch together with the new coal staiths at Port Darlington was the beginning of the end of Stockton as a port, and the birth of Middlesbrough as a port and a town.

Thirty years later the Railway also created Saltburn as a seaside resort, but rather as a side issue (and a Parliamentary necessity) to the need to build an efficient branch to the Pease owned ironstone mines in the area.

At the opening of the line in September, 1825, four sections of the system were put into use, under the first authorising Act of 19th. April, 1821, extending from Witton Park Colliery at Etherley to Cottage Row at Stockton Wharf. The Brusselton Incline and the West Auckland Junction, and the connection to Darlington Depot, were also included. The original spur line to Yarm and the Black Boy Branch, covered in the same Act, followed very shortly.

The second Act, of 23rd. May, 1823, brought in the line from what Inness called Albert Hill Junction (this name did not exist then, and is not very accurate in any case), at Darlington North Road and east of the present North Road Station and Museum, to Croft, but it was more than three years before this was opened. After that there was a constant succession of new lines, frequently managed nominally by a separate company and leased to the Stockton & Darlington Railway, until February, 1863, when the system, as soon to be merged in July, 1863, with the North Eastern Railway, was virtually complete. Within the scope of this book, of course, and under the N.E.R., other additions and changes to the Darlington Section took place.

There is a minor puzzle at the Stockton end of the original line. Inness starts his list of openings with "Etherley, Witton Park Colliery, to Cottage Row, Stockton-on-Tees", and I have also mentioned this above. Cottage Row is shown on the 1855/7 25" Ordnance Survey map as a terrace of cottages between the lower east side of Castle Field and the coal wharves.

Overton's original survey of the route seems to have included this and on the maps usually published of his route and Stephenson's there is a double track shown at this point. However, the first (1821) Stockton and Darlington Railway Bill referred merely to "from the River Tees, at or near Stockton . . .", which leaves it fairly open. I understand that the Stockton party discovered that it was intended to terminate the line at St. John's Well, where Stephenson had cut the first sod. This would not have been much use to them in the matter of shipping coal, and they threatened to take legal action to force the Railway to extend the line further. Apparently they were successful, as the 1823 Act had the wording amended to include "One of the said Deviations in the said main Line of the said Railway or Tramroad, to be made from or near a certain Place called Cottage Row, situate in the Township of Stockton aforesaid, to join or communicate with the Line of the said Railway or Tramroad, as authorized to be made by the said recited Act, at or near the Western extremity of the Boundary Fence, between the Lands and Grounds belonging to the Lord Bishop of Durham, situate in the Township of Stockton aforesaid, and now in the occupation of Martha Howson, and the Lands and Grounds belonging to and in the occupation of Mr. John Jackson, and situate in the same Township:". This was tying it down indeed!

One can only ponder over the reasons for the Railway's attempt to deprive Stockton of an effective coal outlet; no doubt there were commercial considerations. Suspicions still linger! Eventually the line was extended further along the quayside to the end of Bishop Street, with various sidings, and no extra permission seems to have been needed for this.

Some of the lines which became part of the S. & D.R. system had a very confused beginning, one of the most complicated being the Consett connection. Originally this section, from Carr House through Consett to Stanhope Kilns and including the famous Nanny Mayer's Incline, was built and operated by the Stanhope & Tyne Railway, opened in mid 1834, which had a very short independent life. Robert Stephenson had had a very large stake in this Railway, due to taking payment in shares instead of cash, and was lucky to be able to sort matters out without them leading to his financial ruin. It had been built without Act of Parliament, and the cost of way-leaves escalated severely, one of the worst and greediest owners being the See of Durham, and the section included so many difficult inclines, both stationary engine operated and self operated, costly and complicated to run, that by 1842 the line was on the point of bankruptcy. They then sold this section of the line to the Wear & Derwent Railway and formed a new company called the Pontop & South Shields Railway to operate the remainder, thus surviving. Part of this latter railway was used later by George Hudson to spoil the Pease's plan to open a through line from York to Newcastle.

The Wear & Derwent Railway was owned by the Derwent Iron Co., later to become the Consett Iron Co., and I have little doubt that the S. & D.R. were behind the purchase. They leased it from the Derwent Iron Co. from the beginning in 1845 and purchased it soon after. The Derwent Iron Co. also put through the Weardale Extension Railway from Crook to Waskerley, but the S. & D.R. bought it from them before it was opened, and together with the Wear & Derwent Railway it then became the Wear & Derwent Junction Railway. There was no authorizing act for this latter; probably the way-leaves were obtained without difficulty, but in any case much of the territory concerned was Quaker owned land, some of it already belonging to the Pease family or their partners. Thus no Parliamentary sanction was needed. It was 1859 before the deviation line from Burnhill to Whitehall Junction made it possible for the Nanny Mayer's Incline and others to be abandoned.

The section from Crook to Tow Law also included the steep Sunniside Incline, stationary engine worked, and having gradients of up to c.1:13. Permission for a deviation line was obtained in 1862, but this was not built until April, 1867. In the meantime, trials were held in 1858, using No.125 "Gazelle" (mentioned in Chap. 7.) to see whether the latest locomotives could manage this incline unaided, in both dry and wet weather; these tests were successful, although it is doubtful whether much of a load could have been hauled. However, they were not followed up, as the decision was made to go ahead with the deviation line. The lower part of the incline was left to serve the many mines in the area, mostly Pease owned (see remarks re No.25 "Derwent" in Chap. 5.).

Most of the separate companies making up the system at that date were amalgamated into the parent line in September 1858, just after the Hownes Gill Viaduct was opened. The S.D. & L.U.R., the Eden Valley Ry. and the Frosterley & Stanhope Junction Ry. were amalgamated at the beginning of 1863. The Middlesbrough Dock system was incorporated in 1849. I list them below.

Wear & Derwent Railway.
Wear & Derwent Junction Railway.
Weardale Extension Railway.
Bishop Auckland & Weardale Railway.
Shildon Tunnel Co.
Wear Valley Railway.
Middlesbrough Dock and the connecting line to Old Town Junction (built by the Owners of the Middlesbrough Estate)
Middlesbrough & Redcar Railway.
Middlesbrough & Guisborough Railway.
Darlington & Barnard Castle Railway.
Eden Valley Railway.
South Durham & Lancashire Union Railway.
Frosterley & Stanhope Railway.

At the far western end of the system, the S. & D.R. directors also had a stake in the Cockermouth, Keswick & Penrith Railway, opened in 1863/4, giving them access to the west coast for mineral traffic, and a passenger connection.

The main line north through Darlington was formed by the Pease backed Great North of England Railway, as far as Parkgate Junction, and including the original but temporary Bank Top Station. From Parkgate Junction on, the Newcastle & Darlington Junction Railway, Hudson backed, continued north, but this is another story and has no place here. The line from York to Darlington was opened on 4th. January, 1841, in both directions, the trains being hauled by four Stockton & Darlington locomotives, two per train. This included a final short stretch bought from the S. & D.R. from Croft Junction to Parkgate Junction. It is known that other S. & D.R.

locomotives worked on the Great North of England railway, on loan or hire.

The Newcastle & Darlington Junction Railway opened on 15th. April, 1844, and the Pease hopes of a through line from York to Newcastle vanished.

At Stockton the Leeds Northern Railway, formerly the Leeds & Thirsk Railway, independent of both Hudson and the Pease group, opened their final stretch from Melmerby through Northallerton to Stockton, joining the Clarence Railway at Norton, on the 2nd. June, 1852, and including the predecessor to the present (so far) Stockton Station. The S. & D.R. then realigned part of their track from Eaglescliffe to Bowesfield Junction, to run adjacent and parallel to the L.N.R. line, from 25th. January, 1853.

In N.E.R. days and on the fringe of the system, the establishment of the Rosedale Branch and the Redcar to Whitby coastal route were by no means straightforward. The former had its beginnings in the North Yorkshire & Cleveland Railway Act of 2nd. July, 1855, which eventually resulted in the connection of Picton on the Stockton to Northallerton line to Grosmont on the Whitby & Pickering Railway, and the connection later of Nunthorpe junction to Battersby Junction. The section of line from Stokesley to Ingleby Greenhow was opened on 1st. February, 1858, and the Ingleby Mining Co. then opened a private narrow gauge line from Ingleby Junction to their nearby Rudd Scar mine, near the foot of the present incline bank, on 6th. April, 1858. This line was promptly bought by the N.Y. & C.R. and a further Act obtained for the line to Rosedale, including the Ingleby Incline, probably with N.E.R. backing. The N.E.R. took over the N.Y. & C.R. on 8th. August, 1859, and carried on with the construction of the remainder of the line to West Rosedale, opened 27th. March, 1861, and to East Rosedale by 18th. August, 1865, thus providing a final and chilly home for several of the S. & D.R. long boiler 0-6-0's at more than 1000 feet above sea level.

The Nunthorpe Junction to Battersby connection was not completed until 1st. June, 1864, and then carried very heavy mineral traffic for many years.

The coastal line was first proposed by the Whitby, Redcar & Middlesbrough Union Railway in their Act of 16th. July, 1866, but part of the necessary work had already been done by the N.E.R. with their extensions to the old Cleveland Railway from Guisborough to Brotton, which was reached by 23rd. February 1865, and then on to Skinningrove and its zigzag in the same year. The next section, to Loftus, where it was to join the W.R. & M.U.R., was opened 27th. June, 1867, after great difficulties.

In the meantime, the W.R. & M.U.R. was steadily running out of money, and by 1874 was in such a state that they asked the N.E.R. to take them over, which was eventually accomplished by the Act of 19th. July, 1875. The rest of this line was then completed with great difficulty and much rebuilding, and opened on 3rd. December, 1883, enabling ex S. & D.R. locomotives to reach Whitby either by the Esk Valley line or the coastal route. It was beautiful, but never profitable.

The connections from the Cleveland line to Saltburn from Brotton and from Priestcroft Junction near Boosbeck were opened in 1872 and 1873 respectively, completing the circular line to and from Middlesbrough which was used in both directions for many years.

It is interesting to look at the political, social and industrial history of the period covered by the life of the Stockton & Darlington Railway. In 1825 the country had been at peace for ten years, after more than twenty years of almost continuous war with France, and often also with France's allies and with America. The country was full of discharged service-men, many disabled. George IV. was still on the throne, to be followed by William IV. in 1830. Victoria succeeded in her turn in 1837.

The open field system had given way to enclosure during the previous 75 years or so. Factory towns had come into being, and rural and town life were separating more and more. Farming was becoming much more efficient and productive. During the long years of war the price of corn and other staples had risen to very high levels. When the war ended the price of corn fell suddenly, causing much hardship among the rural communities, but apparently the price of fodder did not follow so drastically. It had been mostly the high price of fodder, more than efficiency in bulk movement of goods, that caused mine owners and others to turn to the possibility of steam hauled transport, which continued to be cheaper, since the price of coal fell as the cost of transporting it lessened, so fuel continued to cost less than fodder.

Health was improving with advances in medicine and hygiene, lives were longer and the population was increasing, but at the same time unemployment, most especially among farm labourers, was widespread and rising, which kept wages down. There was industrial depression in the 1820's, which recovered slowly by 1850 but returned in the mid 1870's.

Literacy was beginning to be seen as desirable by the more intelligent workers, especially the artisan class (pace George Stephenson), and the better employers, led by the Quakers, encouraged this, and endowed and supported Mechanics' Institutes and such.

During this period both Reform Bills came into being, in 1832 and 1835 respectively, the Factory Acts of 1833 and 1847 were passed, and trade unions became legal in the year of the Railway's birth, 1825.

The present police system started under Sir Robert Peel in early 1830, and the present Post Office system with the penny post in 1840.

Gas lighting was already in existence; Watt's Cornish agent Murdoch had lit his house at Redruth with coal gas in 1792. Volta had produced a primitive electric cell for storage and production in 1800. In 1831 Faraday made and demonstrated the first dynamo.

In 1825 rolling mills for iron had been in use for some time, but not so wide-spread as they became in the 1840's, while the puddling furnace method of producing iron was still normal. Blast furnaces and the Nasmyth steam hammer were about to appear. Machine tools, with Henry Maudslay and Matthew Murray the leaders in the field, were being improved, but much of the work was still carried out by hand.

Afloat, William Symington's first steam driven boat had been tested on Dalswinton Loch in 1788, and his "Charlotte Dundas" had towed two sloops in windy weather on the Forth & Clyde Canal in March, 1803. Fulton's paddle steamers (with British engines) had been successful in America in 1807, and Henry Bell's "Comet" sailed on the Clyde in 1812. One of the Wylam locomotives, said to be "Wylam Dilly", had been turned temporarily into a paddle-driven keel-tugboat on the Tyne during the keelmens' strike in 1822.

Photography was being developed (pun) during the 1830's; it is believed that the first railway photograph was taken about 1845, and certainly railway photographs before 1850 are rare (see later notes). The electric telegraph came in 1837, was in use by the G.W.R. by 1843 and by 1848 was wide-spread, especially along the railways, which in most cases acted for many years as the telegraph agents, with the office at the local station. On the Stockton & Darlington system the Middlesbrough & Guisborough Railway was the first to have a telegraph line along its length, and Huby's account of life at Nunthorpe Station in the 1890's tells that not only was the station still the telegraph office but also the newspaper shop, and this latter was still the case until about 1930.

A full list of all the locomotives covered, with their principal original particulars, is given as an appendix following the text. Lists of details not all mentioned in the text, and details of changes where known, are given at the end of each chapter.

It should be noted that all statements made or dimensions given in this book are based on published data or information from other sources, but there can be no guarantee of the accuracy of such data and proof is very difficult to establish. Even now, records and lists may still surface which invalidate previous beliefs. It is known that the Inness articles contained some errors; he himself was aware of this soon after they were published, but never corrected them, and so I have probably repeated some of them.

I have not attempted to give all the details of springs, journals, plate thicknesses etc. included by R. H. Inness for many locomotives. Nor have I included reboilering dates, unless of interest. These details are all available in the original articles for anyone who is willing to search for them.

A list of lines opened with dates of opening and other details is given in Appendix V.

Track Development

John Birkinshaw of the Bedlington Ironworks first produced his patent improved wrought iron rail in 1820, and by the use of cams operating the rolls he was able to give it the fish-bellied form used in the cast iron rail first introduced by William Jessop. It could be produced in lengths of about fifteen feet, a huge improvement, and was, of course, far less fractile. The first type used at Bedlington was of 17lb./yard weight, but that used on the Stockton & Darlington Railway was heavier, according to Nicholas Wood, and of 28lb./yard, soon increased to 35lb. Charles E. Lee says that by 1837 they were using parallel rail of 50lb./yard, and Whishaw says that in 1842 64lb./yard rail was being used, secured by wedges in iron chairs of the same sort that were then being used on the Great North of England Railway. Tomlinson says clearly that in 1841 Bolckow & Vaughan's new Middlesbrough Iron Works was supplying the railway with rail of 73lb./yard weight.

Birkinshaw's was not the first wrought iron rail; square section rail had been used in several places for some years, and the Edinburgh engineer Robert Stevenson had proclaimed its virtues far and wide, but Birkinshaw's was far better in use.

Stone blocks were used as sleepers from Etherley to Darlington, from Etherley and Brusselton quarries. These had two holes each, with oak pins hammered in for securing the cast iron chairs. The chairs were supplied by Robert Stephenson & Co., and Hackworth's early notebook is full of comments about the errors made with these, supplied to the wrong pattern and with the wrong number of holes. The rails were secured to the chairs with iron pins through them; wedges were not used until later. In 1835 the size of these blocks was increased. Many of these blocks survive in various places; in the sea wall at Saltburn, behind the old depot house at Yarm, and many others. Oak sleepers from old ships' timbers were used for the remainder of the track bed, possibly because of cost of transport.

Liveries

Very little is known with any certainty of the colours used or when a definitive livery came into use. There seems to be a general body of opinion that green was used for locomotives, possibly back as far as No.23 "Wilberforce". The Locomotive Publishing Co. produced an early postcard of this engine, based on the photograph used for Fig.47, with the engine coloured green and the smokebox and chimney black.

Ernest F. Carter, in his "Britain's Railway Liveries", published in 1952, included just over three pages on the Stockton & Darlington Railway. but gives very little of source to back up what he says. Some of it was certainly taken from the "Locomotive" magazine's special Centenary Supplement of 1925.

In recent years, the well known photograph (Fig.196) of No.160 "Brougham" at Alpha Place, Saltburn, in c.1861, has been used as the basis of a coloured postcard in conjunction with Saltburn's first Victorian weekend celebrations, commissioned by Langbaurgh Council from a local artist. He has used a dark green with black and white lining, and yellow (brass) splasher rims, etc., dark red frames and black smokebox and chimney, which all corresponds with Carter's account. (See also Chap.4, No.9 "Globe".)

Timothy Hackworth's 1827-28 notebook contains a list of materials purchased for each month in 1827 from Rt. Kitching. This list contains a lot of white lead, probably used as a jointing compound, 1 stone of Spanish white, $11^1/_2$ stones of white paint, 1 lb. of yellow paint and 3 lb. lamp black. This latter could have been mixed with white paint to produce black easily enough. No green paint is mentioned. Kitchings' locomotive cost books mention paint but not the colour. The boiler cleading must have been treated with some preservative, possibly paint but more likely linseed oil, which could be rubbed in with a cloth.

The drawing of the engine No.29 "Queen", built in 1836, shows some lining to the splashers. This drawing was published in the "Locomotive" magazine many years later, and its source is unknown, but in style it resembles other known Kitching sketches. Apart from this, the earliest known photograph of an engine with boiler bands and lining is that of No.50 "Meteor", built in 1843, but this photograph almost certainly shows it at Saltburn, so cannot have been earlier than 1861. The photograph of No.38 "Rokeby", built in 1847, possibly shows lining applied, but is of the engine as rebuilt later.

Kitching built passenger engines from 1848 to 1851 still had the boilers cleaded with wood, and sheet metal cladding does not begin to appear until a few years later. No.98 "Pierremont" and No.101 "Marske" of 1855 had such metal boiler cladding as built. The early photographs of passenger engines built in 1856 show lined out panels on footplate sides and tender, with painted on numbers. 0-6-0 mineral engines seem to have been treated in the same way, but there are so few photographs of them in original condition, and these of engines built from 1856 onwards, that what little evidence there is is inconclusive.

The Stockton & Darlington Railway never put the Company's name or initials on either engine or tender. The numbers seem to have been painted on at first, sometimes in a decorative framing, and later a large oval plate with beaded edge came into use, on both engine and tender. During the second half of the 1860's N.E.R. initials began to appear on tender sides, as the engines were renumbered.

Some tentative but interesting notes on this subject by the late Laurie Ward were published in "North Eastern Express", Vol.15, No.61 of November 1975.

Locomotives at the 1875 Jubilee

Ken Hoole, in his "North Road Locomotive Works", has pointed out that there appears to be no complete record of the locomotives on show at the Jubilee celebrations. There was no procession as such.

The 1925 and 1975 celebrations and processions have been very well documented, and there is no need to discuss them here (see note in bibliography).

It is interesting, however, to speculate on the 1875 exhibition; several clues exist, and many photographs were taken, a lot of which found their way into the collection of John Kitching, grandson of Alfred and a consulting railway engineer himself. It is known that he designed some locomotives for the N.E.R. in 1875, and also carried out some railway specification and contract work elsewhere, so he was, no doubt, in touch with many other railway engineers and able to acquire such photographs easily. His main collection has disappeared, but parts of it still exist here and there. Besides this, an album of photographs was produced at the time as a presentation volume for important official guests at the celebratory dinner, as well as the specially produced volume of J. S. Jeans' "Jubilee Memorial - The Railway System", sub-titled "A History of the Stockton & Darlington Railway and a Record of its Results". According to Theodore West, who was there, the first attempt at the photographs for the album was a failure due to poor conditions, but it is obvious that another and more successful attempt was made in time. The Goldsmiths' Library of the University of London has a good copy of this album.

Jeans listed those engines which were intended to be shown, but it is known that many substitutions were made, some of which I have referred to in the text. Ken Hoole mentioned other engines known to be present.

Jeans listed the following (# indicates a known photo.):
No.10 "Auckland", 0-6-0.#*
No.1033 "Shildon", 0-6-0, said to have a Napier boiler.#*
No.1041 "Dart", 0-4-0.#
No.1050 "Meteor", 2-2-2.#
No.1058 "Woodlands", 2-4-0.#
No.1062 "Southend", 0-6-0.
No.1066 "Priam", 2-4-0.#
No.1068, 2-4-0, of the 'Gamecock' class.
No.1080 "Duke", presumably the 0-6-0 ex-banker.
No.1082 "Hawthorn", 0-6-0.
No.1084 "Albert", 0-6-0.
No.1160 "Brougham", 4-4-0.
No.1221, 0-6-0.
No.1270, 4-4-0, of the "Ginx's Babies".
No.1291, 0-6-0, long wheel base.

* In fact, as pointed out in the text (Chapter 5), the first was No.26 "Pilot" in disguise. The second, No.1033, ex No.33 "Shildon", may have been fitted with a Smith's boiler, since that shown in the photograph is certainly not the original, nor are the tenders shown in their correct position for the engine as rebuilt, so there was a little bit of misrepresentation here. The Napier boiler would have been a Hackworth multitubular return tube type, as fitted originally.

The following engines were also there and photographed at the time:
No.1175 "Contractor".#
No.1032, 0-6-0, built that year.#
No.1093 "Uranus".#
No.1035 "Commerce", 0-6-0.#
No.1089 "Huddersfield", Bury 0-4-0.#

No.1 "Locomotion" must have been moved from its North Road Station plinth to the works, as Henry Pease declared the exhibition open from it. It was raised off the rails so that the wheels could be revolved by steam from an auxiliary boiler in the works. No.25 "Derwent" had not yet been returned from Crook, so was not present.

Other locomotives present, according to Ken Hoole and as verified by photographs, were the following, all from other sections of the N.E.R. or other railways:
N.E.R. No.273, 0-4-0WT Crampton.#
N.E.R. No.910, 2-4-0 Fletcher.
N.E.R. No.174, 0-6-0.
N.E.R. No.1431, 0-4-4WT.
L.N.W.R. No.2187 "Penrith Beacon", 2-4-0.
L.B.S.C.R. No.151 "Grosvenor", 2-2-2.
G.N.R. No.47, 4-2-2.
G.S.W.R. No.106, 4-4-0.
M.R. No.1160, 0-6-0.

The photographs referred to above were those taken at the time. Other photographs exist of the other engines present, of course.

The celebrations were far from being entirely a Stockton & Darlington occasion, although the Darlington Committee had not yet been dissolved and William Bouch was still alive. The North Eastern Railway and the Pease family made it a propaganda event for themselves, and sincerity was somewhat absent. Theodore West, in a long and self illustrated letter to a lady friend, was surprisingly bitter and cynical for a member of the Society of Friends; he accused the organizing committee of rank hypocrisy and favouritism, and of breaking promises to the men who had served them so well by failing to look after them and include them fully in the celebrations as promised.

Chapter 2. Principles and Valve Gear.

Basic Mechanics

For those of my readers who are not familiar with the way that the simple steam engine works, I shall try to give a simple explanation. The principle of the cylinder and piston is familiar to most people, if only in the form of a bicycle pump or garden syringe, but the way in which steam is used to drive the piston backwards and forwards may not be so clear. I am ignoring here the very early low pressure steam engines which used a combination of vacuum and atmospheric pressure for their operation.

The Stockton & Darlington Railway never got to the point of using more than two cylinders (except in one very early and short-lived engine), compound engines or superheaters. So here we have only to consider the simple double-acting steam engine.

A double-acting steam engine is one in which steam is used at both ends of the cylinder, so that power is applied in each direction alternately. This is essential in order to produce even rotary motion at the crankshaft, even with two cylinders, unless a large flywheel is used, such as Richard Trevithick fitted to his early single cylinder steam engines. So we have two cylinders, each fitted with a piston and piston rod, and each piston rod having a connecting rod hinged to it and driving at the other end a crank on a wheel or crankshaft. The cranks are arranged at 90° to each other, so that there is a power stroke at each quarter turn of the crankshaft or wheel axle, one stroke in each direction for each driven wheel.

Simple steam power works by using the expansion of the steam which is produced by boiling water; steam as such is not visible, and the white stuff that you can see is really water vapour; this visibility means that the steam is condensing back into water. However, everyone calls it all steam. Here we are dealing with what is called dry saturated steam, which nevertheless contains quite a lot of water in suspension. Water is not compressible, and can cause mechanical damage to the working parts, so it is desirable to eliminate as much of it as possible both before it enters the cylinders and later. So, in order for this steam to be as dry as possible it is taken from as high a point in the boiler as can be arranged, usually from inside a steam dome, and then passed through a simple deflector or sieve type dryer. Hence also the clouds of steam often seen near the cylinders as an engine is starting off and the driver has opened the cylinder drains to get rid of what water may have carried over into the cylinders or condensed there because they were cold.

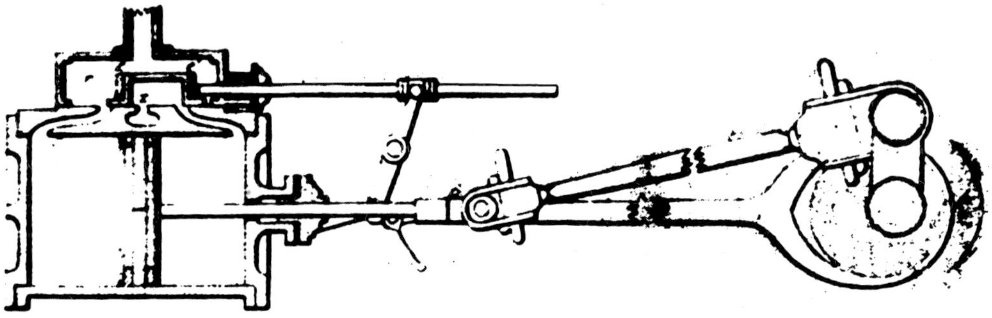

Fig.4 Diagrammatic illustration of a simple early double acting engine, showing cylinder, valve and valve chest, connecting rod and crank, eccentric and rod, operating the valve through a typical early lever device, of c.1835.

In the early days of steam engines steam was admitted to the cylinder under pressure during the whole working stroke, and then allowed to escape, which is very wasteful. Later on means were developed of allowing only as much steam to be admitted as would drive the piston along the cylinder against the work load by expansion alone. This is called expansive working.

The admission and exhaust of the steam to and from the cylinder is controlled by a valve for each cylinder (see Fig.4) This valve may be in the form of a flat hollow plate containing steam passages, or of a smaller piston shaped like a dumb-bell with a space between the ends. The plate type is called a slide-valve, and moves against a flat surface next to the cylinder, with steam passages in it at each end through to the cylinder. The other type is called a piston valve and moves in a smaller cylinder arranged in a similar manner to the slide valve chest described above. Both types of valve work in the same way; they are moved by means of mechanism linked to the rotation of the wheels, and roughly at 90° to the crank, so that the valve is half way along when the piston is at either end of its stroke. The valve allows steam to enter the cylinder when the piston is at that end of the cylinder and then allows it to escape when the piston reaches the other end of the cylinder, at the same time allowing steam to enter at the other end as soon as the exhausted steam has escaped.

The point at which the valve stops the steam from entering the cylinder is called the 'cut-off', and at first this point was fixed when the engine was assembled. Other terms that you may meet are 'lap', the amount by which the valve moves past the inlet and exhaust ports, and 'lead', which is the angle of relation of the valve position to the piston position, in other words, how far in advance of the piston in rotational position it is set. I do not think that it is necessary for me to discuss these terms here.

The next stage in the development of valve gear was to make it possible for the 'cut-off' of the entry of steam to the cylinder to be adjusted by the driver as necessary for the working conditions. When starting a train away from rest, the power needed to overcome the huge inertia of the train requires all the steam possible to provide enough energy, and so steam is allowed to continue to enter the cylinders for as long as the setting of the valve will allow: this is called 100% cut-off. As the train gathers speed and momentum less steam is required to keep it moving, and the cut-off point is brought back to the smallest percentage possible to keep the train moving without losing speed. With the old type of Stephenson link gear this could be reduced to about 25% when conditions allowed, which meant a great saving both in fuel and water. The gradient of the line has also to be taken into account, of course. This adjustment is done by altering the ratio of leverage of one of the parts of the valve operating mechanism. The control is so arranged that reversing can be carried out by the same mechanism, and the system used is called 'link motion', because it uses a sliding link between the forward and reverse valve operating rods as the adjustable lever.

The other main control of the engine is the throttle valve, used to regulate the flow of steam from the boiler to the cylinders. This valve is usually situated inside the boiler, and is operated by a long handle which traditionally takes the form of a lever and is called the 'regulator'.

Gab Gear

The valve gear used in most early steam locomotives was of what came to be called 'gab' or 'fork' type, although Trevithick and Hedley used simple pegs as 'strikers' to move the valves up and down, copied by George Stephenson in his very earliest engines. There were other types. I shall try to explain the gab gear, as much in order to clarify my own mind on the subject as to impart information to my readers. It was used at first in conjunction with one or two eccentrics of 'loose' type, loose because they were not firmly secured to the axle or spindle, but caused to revolve by a peg in a slot, and free to turn for the length of this slot. When the engine was moving forwards the peg would be at one end of the slot, and when moving backwards the peg would be at the other end of the slot, thus altering the valve operating point in relation to the wheel rotation. In order to change direction it was necessary to detach the driving rods to the valve gear by lifting the gabs, operate the valves manually to get the engine moving in the other direction, and replace the gabs when the peg had reached the other end of the slot and was rotating the eccentric in the other direction.

Here I shall be a little repetitive. My own experience during my apprenticeship was almost entirely of Stephenson link motion on both vertical and diagonal engines, but it has left me with an understanding of the way in which the sliding link allows each valve to be operated by one or the other eccentric rod as required, and with some residual memory of lap, lead and cut-off. From there I can, as it were, work backwards to double eccentric gab gear, and say, "Ah, yes, some silly so-and-so has disconnected the eccentric rods from the sliding link, discarded that, and fitted hooks to hang each rod on to a peg on the valve rod in turn". Paradoxically I find it more difficult to visualize single eccentric gab gear, although the principle is simple enough, in theory. Stop the engine, lift both gabs, operate the valves manually by the handles provided until the engine is moving in the other direction (this takes skill), then the eccentric and its attached rod stay still until the eccentric fetches up against the other stop and begins to revolve again in the other direction, reconnect gabs. However, as many writers have said, it needed a highly skilled man, preferably with four hands and the ability to see in the dark, to do it smoothly and without problem. These gabs were usually fitted with catches to keep them in place.

Early gab gear had no guiding forks fitted, which must have made relocation in the dark very hazardous. The conjugated single eccentric drive valve gear fitted to No.1 (Figs.6 & 7) seemed to have so

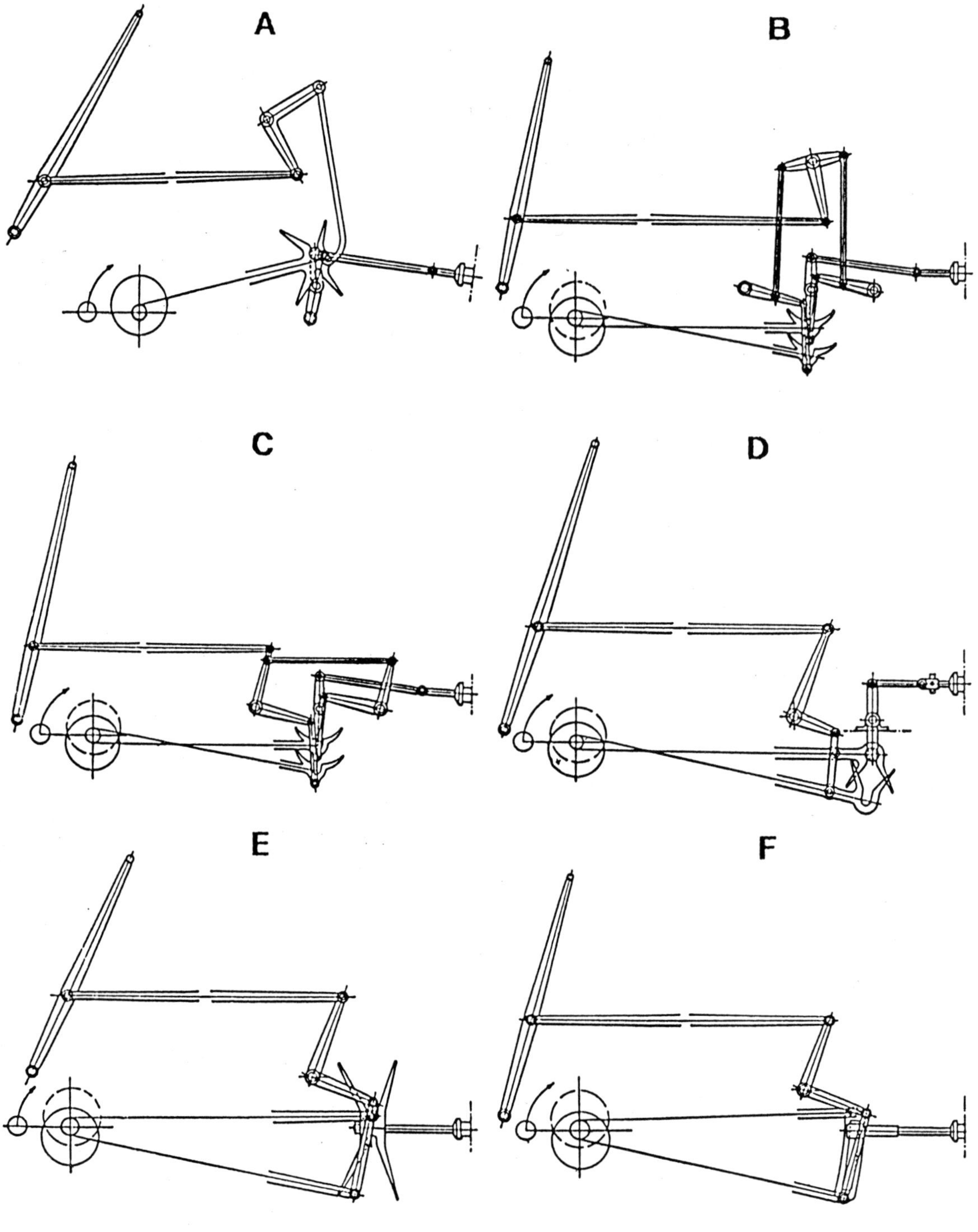

Fig.5. Evolution of the Link Motion

The Gear is shown for one side of the engine, i.e. for one cylinder, in each case.

A. Two fixed eccentrics, Carmichael's gear as first applied by Robert Stephenson & Co., 1835.

B. Four fixed eccentrics, with forked gabs and single reversing lever, as applied by Stephenson.

C. „ „ „ as before, but with reversing shafts below boiler.

D. „ „ „ with forks opposed, and single reversing shaft, as applied by Buddicom.

E. „ „ „ with gabs combined on valve spindle, as applied by Stephenson, 1841.

F. „ „ „ united with slotted link, by Williams and developed by Howe, 1842.

many handles that one man, even with four hands, would have had difficulty, and the gab pockets were very small, just big enough to fit over the pegs, but obviously it is possible - Mike Satow * said so! Some drivers were very good at it; Jem Stephenson, George's brother, seems to have been one such, although liable to get himself into hot water in other directions. Certainly he drove many of the early S. & D. locomotives, from No.1 as far as No.7 "Rocket", but he may also have been Stephensons' resident representative, with the duty of testing and demonstrating each new engine from the Forth Street Works, as well as being paid by the S. & D.R. as a driver. Another such expert driver was the William Chicken mentioned by George Graham in his reminiscences.

C. E. Stretton has two very interesting sections in his "The Locomotive and its Development", in which he first describes gab gear as fitted to the 'Planet' class, with a 'treadle' on the footplate which disengaged the eccentric rods and a lever which moved them from one stud or stop to the other, in addition to the gab handles, but this was comparatively sophisticated. The famous Rainhill Trials "Rocket" had something similar. Later he describes the development of the link motion, apparently first evolved by William Howe, one of Stephensons' fitters or pattern-makers, from an idea by William Williams, an apprentice draughtsman, in 1841, but it is known so universally as 'Stephenson's' that there is little point in insisting on its origin. Whether it was actually patented by Stephensons seems to be doubtful, but certainly they challenged its use by other firms; nevertheless, its advantages were quickly seen, and many other railway engineers copied it, with some modification to get round the original design. The Shildon Works Co. produced a version in 1847 (Fig.9), and Bouch redesigned it later (Fig.10). The first version did not provide any adjustment of cut-off and was really just an improved gab gear.

Dunlop, in his series in the "Engineer" referred to later this chapter, illustrates another intermediate version of the sliding link gear, which was patented by Timothy Hackworth in 1849, but as far as I know this was not used on any S. & D.R. engines.

The early engineers really had no option than to copy from each other; news of improvements soon spread, and it was more a case of "Why didn't I see that first? I can do it better!" than of stealing others' ideas.

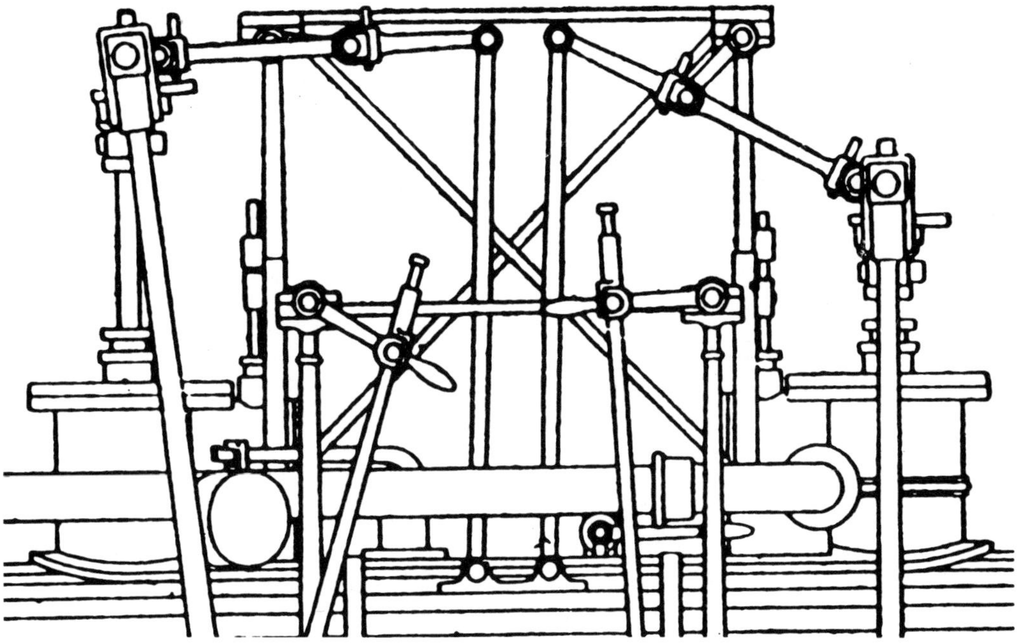

Fig.6. Arrangement of valve gear and gabs as now existing on No.1 engine, from a drawing by J. Dunlop ("Engineer")

Both Hackworth and Stephenson evolved a four fixed eccentric single lever reversing control, very much at the same time, but both were probably derived from someone else's. J. B. Snell in "Railways - Mechanical Engineering", attributes this to George Forrester of Liverpool in 1834, which is a little late; J. G. H. Warren mentions Forrester and also R. & W. Hawthorn, among others. Warren's chapter on valve gear development in his "A Century of Locomotive Building by Robert Stephenson & Co., 1823/1923" is excellent, and I could not approach its scope and clarity. However. I give some illustrations to help the descriptions above (Evolution of the Link Motion, Fig.5). I would recommend to anyone interested that they obtain and study some of the articles published in the "Engineer" magazine in the first half of this century, especially J. Dunlop's in 1920 on "The Development of Locomotive Valve Gear", including that chosen for No.1, and E. A. Forward's articles in the series "Links in the History of the Locomotive" in 1940, 1941 and 1943.

It is worthwhile pointing out that where a rotating eccentric shaft is driven from parallel motion gear, as it was in "Royal George" at first, there are problems of geometry beyond the simple one mentioned by Dunlop in his articles referred to above. He says only that since the foot of the eccentric drive rod, which is attached to a fixed point on the grasshopper beam, moves in a slight arc, not in a straight line, there must be some allowance made in bearing clearance on each side. However, a greater problem is caused by the fact that when the grasshopper beam is at mid position the crank on that end of the eccentric shaft will not be at 90 degrees to the vertical but slightly below it, which means that rotation of the shaft is not even and that there is some conflict between each side of the drive. No doubt this accounts for the modification to this valve gear which soon took place.

There were so many variations in the form of the gab gear that one becomes very easily confused; I have, I think, chosen the simplest for my illustrations, as used by Warren (Fig.5). It is interesting to see that the final form of the Stephenson gab gear before the use of the sliding link had the gabs fixed to the valve spindles, not the eccentric rods, which were joined by a solid link carrying the pegs. This was a reversal of the earlier forms, and a very short step to the final sliding link gear.

It would seem that the word 'gab' is in fact a dialect form of 'gap', referring to the notch or gap in the end of the rod which fits over the peg, whichever way round they are arranged. I have also been told that it refers to the 'mouth' formed by the gap, as in "the gift of the gab", but this seems to me to be much the same.

*Michael Graham Satow, O.B.E., C.Eng., F.I.Mech.E., who had a brilliant career with I.C.I., finishing as Managing Director, I.C.I.(India). Magnificent machinist, perfectionist, lover and supporter of steam railways everywhere and originator of the schemes to build replicas of No.1 "Locomotion", the Rainhill Trials "Rocket", Hackworth's Rainhill Trials "Sans Pareil" and the other competitor, "Novelty". Drove No.1 at Rail 150 in 1975. Died in November 1993.

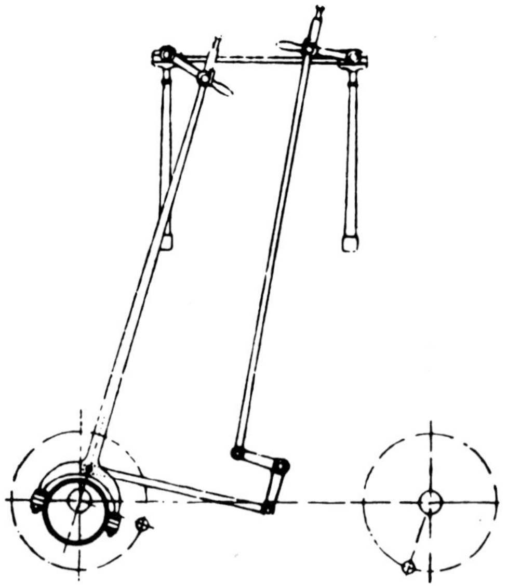

Elements of valve gear drive for two cylinders from one eccentric. Possible original connection of rear valve drive from foot of main eccentric rod; drive in engine today is from the back of the eccentric strap.

Fig.7. Dunlop's sketch of proposed valve gear for No.1, by courtesy of the "Engineer"

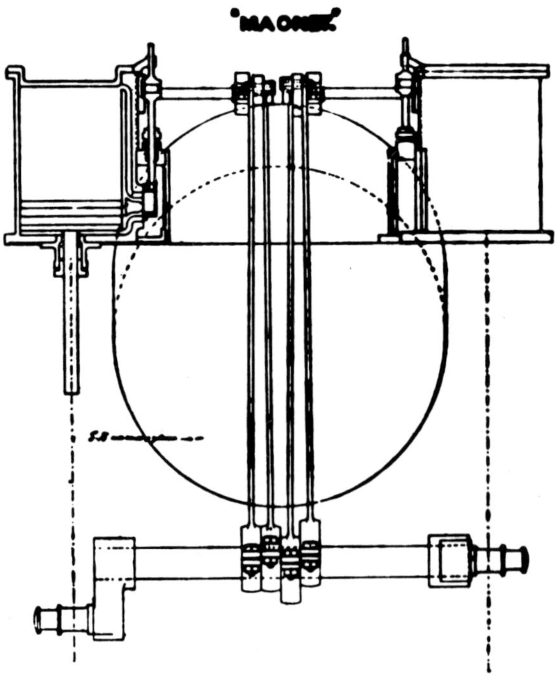

Fig.8. Countershaft and valve gear fitted to No.24, "Magnet"

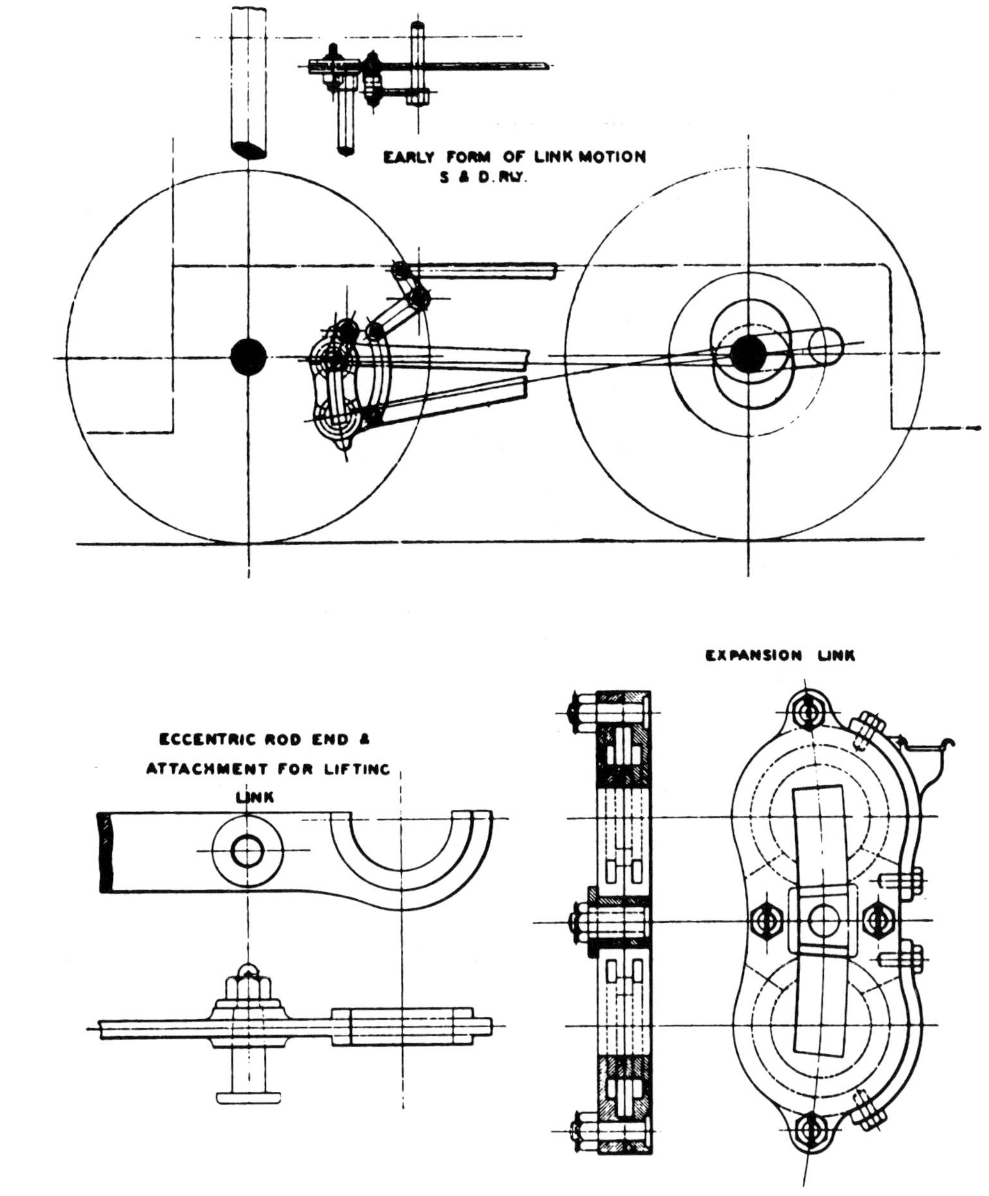

Fig.9 Early (1847) S. & D.R. link gear of hook type, without cut-off adjustment.

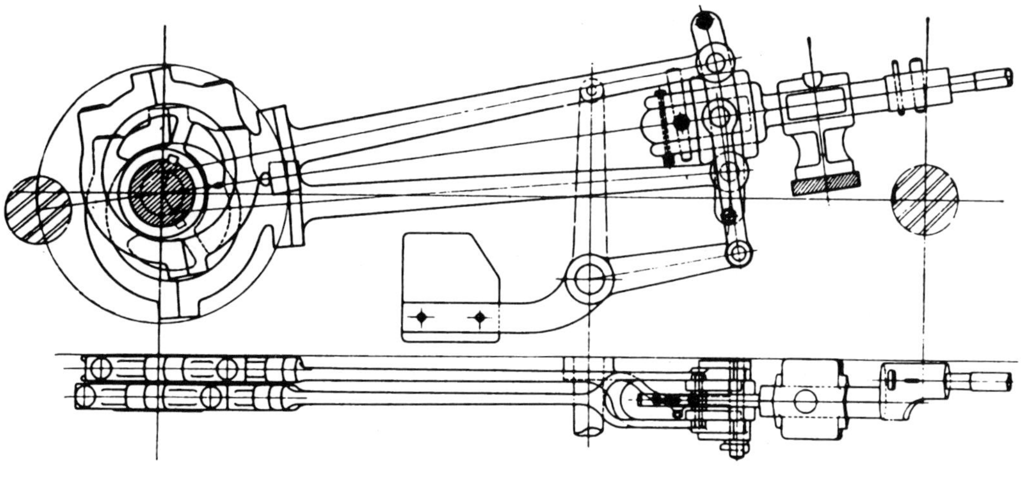

Fig.10 W. Bouch's adaptation of Stephenson link motion, marine type.

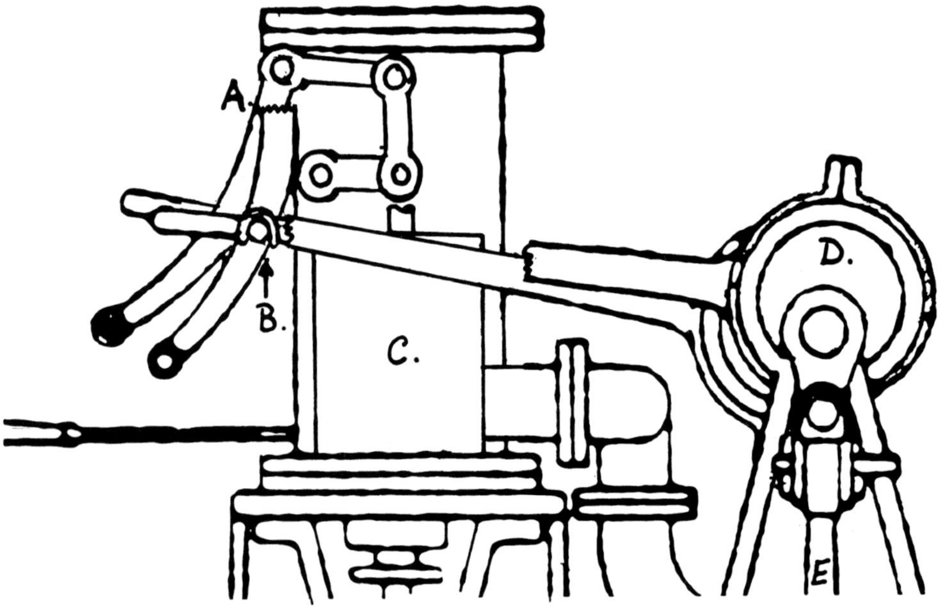

Fig.11 Probable arrangement of valve operating gear as fitted originally to No.5 "Royal George". It may be assumed that the valves were operated by arms on the lower cross shafts as shown in the photograph of the "Sans Pareil" replica above. This part is similar to the Rastrick sketch and to Alan Prior's drawing (Figs.26 & 30), neither of which include the curved rocking levers and handles clearly shown on the Science Museum drawing (Fig.27) and on the line drawing on which I have based my reconstruction. These lower cross shafts would be moved by a pair of parallel arms and links as shown, from the curved rocking levers, which also acted as manual valve operating handles when the gabs were lifted. These were pivoted higher up, probably on a higher pair of cross shafts or a bracket on the centre line. Rastrick's sketch was drawn from the original engine, and it seems strange that he did not show these rocking levers. Maybe they were a later addition, or guessed at by the later artist (possibly Stretton), in which case there seems no reason why more simple and direct motion could not have been used.

A. Valve operating levers | **B. Gabs on eccentric rods**
C. Slide valve chest | **D. Loose eccentrics** | **E. Eccentric shaft drive rod.**

Fig.12. Valve operating gear as fitted to "Sans Pareil" replica; Timothy Hackworth Museum, Soho Cottages, Shildon. No.7 "Rocket" and No.8 "Victory" would have been similar.

Fig.13. Four fixed eccentric gab gear as fitted to No.25 "Derwent", as now preserved at the Darlington Railway Museum, North Road Station, Darlington, arranged for operation from a single lever on the footplate.

Chapter 3. The Early Engines: 1825-6

Design

The design philosophy for the very early engines could not have been more simple. Firstly, they had to work, in both senses of the word. They had to be able to haul sufficient loaded wagons, coal chaldrons then, of course, to pay for themselves and have a bit over. They had to be reliable, a vain hope at first, and they had to be easily repaired and maintained, as manpower was expensive and skill was scarce. Fuel economy was not a priority, and for much of the early life of the Stockton & Darlington Railway fuel was the driver's responsibility; he was paid an allowance out of which he had in turn to pay his fireman or men and to supply fuel, oil and firebars. This, of course, led to many abuses, but in general, it worked.

Insulation was not seriously considered; conservation of energy was a strange and new concept, although the first engines did have their cylinders partly inside the boiler to retain heat. This was probably more due to Trevithick's lead than to deliberate design. Wooden slats were soon fitted to the boiler sides and top, but more to prevent serious injury to the enginemen than to conserve heat.

The main aim was simplicity, both for the sake of constructional ease and for reliability. Most of the machine tools available were crude and restricted in the size of material that they could handle (but see below). Rolling mills had been in use for some time for bars and it was now possible to roll small plates and sections, but it is likely that most plates were still beaten out by hand hammer until the mid 1830's, and would have been even smaller than rolled ones. The only welding known or practicable was by means of the blacksmith's hammer: the parts to be welded had to be brought up to white heat and sparking, which restricted the size of welded parts. Brazing had long been used, of course, but I have no knowledge of its application to early locomotive construction. Riveting was well known, but had to be improved and developed further than had previously been necessary, and the same applied to caulking, the only way of making boilers steam tight (except by the use of oatmeal in the water!).

There has been a lot of nonsense written about the problems associated with screw threads, even Rolt saying that each nut was individual to its bolt. This is just not true. Screwcutting machines had been in use since the Wyatt brothers' 1760 patent. Lathes, drilling machines and planer/facing machines were available, with steam power to drive them. Maudslay had produced his screwcutting lathe soon after the turn of the century, and an American called David Wilkinson had done the same at much the same time. Die plates for small screws had been in use for quite a while, and stocks with split dies came in very soon after. Taps for cutting internal screw threads had been in use since at least the late 1600's. The days of the blacksmith's swage for large bolts were nearly over, and chasing tools were in general use for finishing threads in the lathe if necessary. This is not to say that every small workshop was equipped with all these tools, but most of the bigger ones were, and used them. Certainly works such as Forth Street would be fully equipped, and the rest soon followed. Kitchings were using two sizes of lathe and other machine tools in the early 1830's, and there is a record of such machine tools, offered for sale by Thomas Hackworth at the Soho Works, being inspected by one of Kitchings' engineers.

The relationship, almost a symbiosis, between progress in design and manufacture on the one side, and development of machine tools and metallurgy on the other, had always existed, but the speed of development was beginning to increase. Neither can exist without the other; this is still true. Newcomen, one of the very few true inventors, had struggled for years before he was able to make his engines work reliably, in spite of the brilliance of his mechanical insight. There was a known need for improved materials and methods, and Abraham Darby I, another Quaker, was the first to supply and satisfy this need at Coalbrookdale in the early 1700's: soon iron castings were available of a better quality and greater size than ever before, which made it not only possible to make engine parts but also the machine tools to finish them with. And so the progress curve steepened, each development allowed further progress in design and manufacture which in turn improved the technology of supply.

Trevithick could not have built his engines and boilers without Coalbrookdale, and later, Harveys of Hayle and others. Blenkinsopp could not have produced his successful locomotives without Matthew Murray's design work and improved machining methods.

The Wylam and Killingworth engines were crude, and limited in application, but served their restricted

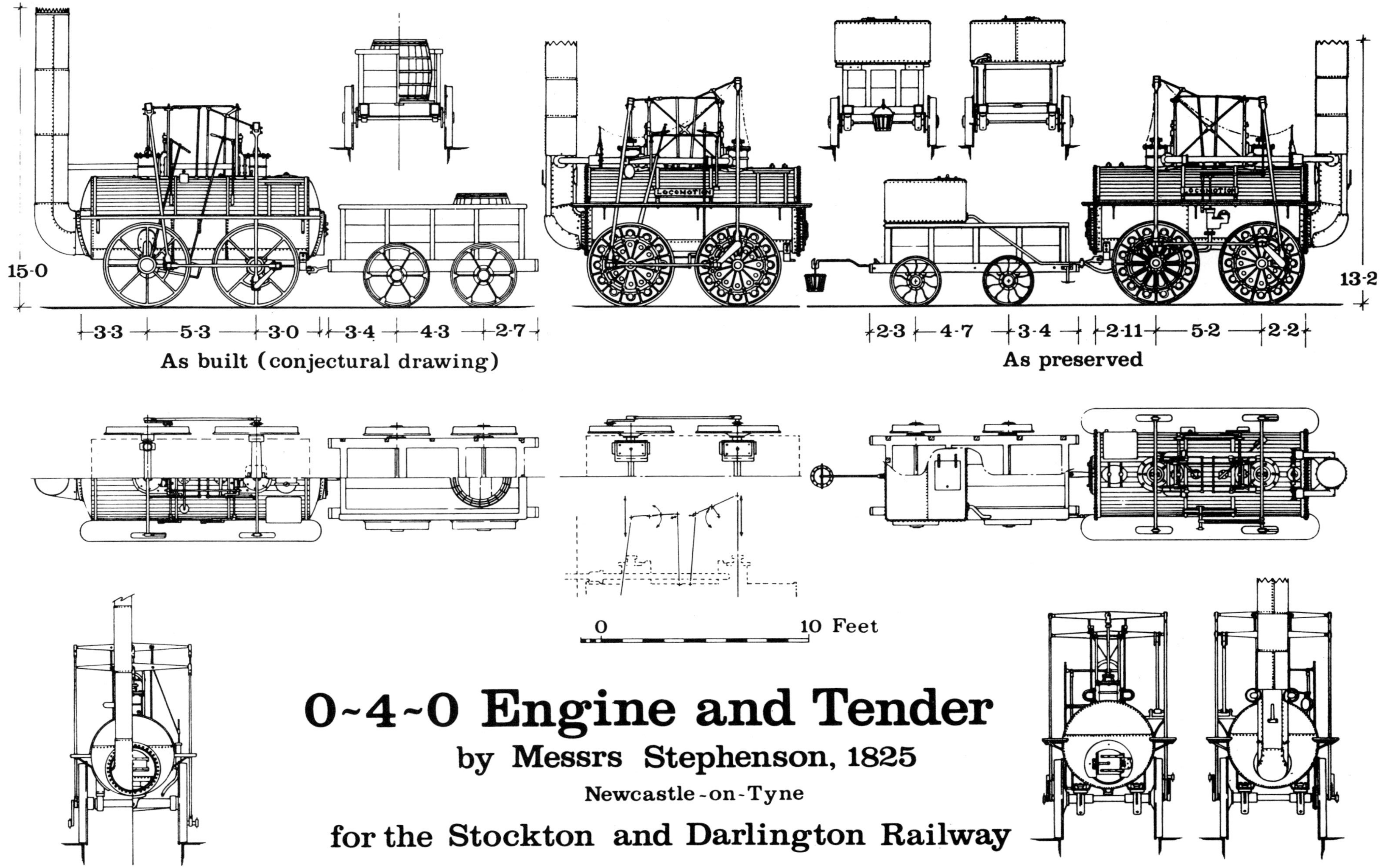

Fig.14 Conjectural Drawing of No.1 "Locomotion" as delivered, and drawings of the same engine as preserved.

purpose well enough. From them came the first Stockton & Darlington locomotives and the application of Stephenson's and Hackworth's experience, enriched by the benefit of the other minds now entering the field from all over Europe, and soon from all over the world. The speed of development between 1825 and 1845 was incredible; I have no doubt that there has never been any such rapid improvement in mechanical design and understanding at any other time. These days each new development gets thoroughly examined and checked before it goes into service, but the early railway engineers had no time for such niceties. There were many failures, of course, but also a great number of successful improvements.

The first main problem was steam production, which had also held back Newcomen. A boiler needs two main features, apart from the purely mechanical ones of strength and steam-tightness. There must be enough heating surface to produce steam more quickly than it can be used, and there must be sufficient surface area of the water to allow the steam to be released as it is produced - this is called evaporation area. All the first boilers were deficient in heating area, until Trevithick's return flue, and this was the first design problem. There was no separate firebox, but just a crude grate built into the entrance to the flue tube. The first major development, then, was to increase this area by doubling back the flue tube so that the heated combustion gases passed through the boiler twice. This, however, meant that the grate and the chimney were at the same end, which was not convenient for the controls, and so the double tender arrangement came into being, with the fuel and the fireman in one and the water and the driver in the other.

It took George Stephenson a long time to accept this necessity for increased heating surface; all his locomotives until late 1827, after the S. & D.R. had been in operation for some time, had single flue tube boilers when delivered - a step back, in fact. Even this later engine relied mainly on a single flue tube.

In the "Royal George", Timothy Hackworth made an improvement to the blast pipe. Trevithick had arranged his steam exhaust pipe to discharge into the chimney, probably to reduce noise, and it was natural to turn the end of the exhaust pipe upwards. He noticed that at each exhaust puff the fire burned more brightly, and recorded this, but did nothing further about it. Neither did Hedley, but Stephenson did try the idea in one of his early Killingworth engines, then rejected it because of noise and discharge of hot cinders with consequent excessive fuel consumption, obviously because of the single straight through flue tube. Hackworth took the idea a stage further, and fitted a nozzle or restriction which accelerated the exhaust steam by means of the venturi principle, and thereby increased the draught to the fire, producing greater heat but at the same time increasing the speed of the combustion gases through the flue tube. This, of course, produced more steam more quickly, but also burned much more fuel, much of which was wasted and discharged from the chimney as hot ashes, as Stephenson had discovered, but since there was more resistance to gas flow with his return flue boiler it did work. The return tube boilers a little later were able to use the blast with great success.

So the next requirement was both to slow down the combustion gases, and to retain them in the boiler for longer, and at the same time to increase the heating surface further. The obvious answer was to fit more and smaller tubes instead of the large flue tube.

Many engineers were working on this problem at the same time; Seguin in France, Napier, Robert Stephenson and Hackworth in England, and probably others. Hackworth tried out several designs before settling on a return tube type similar in many ways to the later marine Scotch boiler, and very similar to the Napier development. Tubes were now available of improved quality, first of copper and soon of iron, but water quality was still poor and a major problem, so very small tubes were out of the question because of rapid furring up on the outside, and choking with combustion products inside. This design still called for double tenders, since the tubes were only on the return side, from a large flue tube below. Robert Stephenson, on the other hand, after many abortive experiments and following advice from Henry Booth, it is said, settled on a true single pass multitubular boiler, probably derived from Seguin's, which eventually became the standard locomotive boiler, and this required only one tender to carry both fuel and water. However, the need to retain the combustion gases in the boiler for as long as possible still concerned the engineers, and for heavy mineral locomotives both Hackworth and Stephenson evolved forms of the long boiler locomotive. Stephenson first applied it to the 'Patentee' 2-2-2 class in the mid 1830's, and patented the principle. It was left to William Bouch to apply the Stephenson type to 0-6-0 mineral engines on the Stockton & Darlington Railway with the 'Commerce' class in 1847, since the earlier 'Miner' class of 1845, although of long boiler appearance, was still using the Hackworth return tube boiler. Both Bouch and Stephenson continued to build this type of engine for many years, and the last Stockton & Darlington ones were the class built by Dübs, Avonside and the Darlington North Road Works between May 1874 and March 1875. I have never been able to find out whether the S. & D.R. paid Stephensons a licence fee for the use of the design, which the latter had patented, or whether there was some agreement between them on it to waive this (see later note in Chapter 7.).

The long boiler locomotive had its purposes and served them, but for general railway use at higher speeds it was really not suitable, since the overhanging ends caused great instability, and poor distribution of the axle weight. There was also the inherent design problem that if the exhaust gases are retained for too long in the boiler they can cool down to the dew point and deposit acid products of combustion which cause tube corrosion. No doubt the short wheelbase had its uses on colliery line curves.

Returning briefly to the water supply, this, both from the quality and quantities available where most required on the railway, continued to cause severe problems for many years. On the Redcar line it was not unknown for sea-water to be used. Not until the Middlesbrough Owners built a water-works and reservoir at Middlesbrough in late 1843 was there much improvement.

Regarding fuel; many early engines used coke, probably because it generated much less smoke, (a requirement of the Railway Act) but it was also more than four times as expensive as coal, as pointed out by de Pambour in his comparisons between Stockton & Darlington Railway practice and that of the Liverpool & Manchester Railway. Coke was specified in the rules for the Rainhill Trials and continued to be used afterwards. The S. & D.R. had access to a steady supply of cheap coal, and the smoke was no problem to them as long as there were no complaints. Here we have the difference between a mineral line and a passenger line. This is oversimplification, of course; calorific values and grate design come into it as well, but in general it covers the facts. However, it was not until much later (Matthew Kirtley; the brick arch, 1860's) that firebox design was improved sufficiently for coal to be used generally. Some of the 'Tory' class were designed to burn coke, and a policy decision was taken at about that time to build only coke burning engines; there was some mention of Fossick's patent coke burning tube. This policy did not last very long.

The Locomotives

When the Management Committee of the Stockton & Darlington Railway had made the decision to use steam driven locomotives and had put in hand the necessary measures in the second enabling Act of Parliament for this purpose, replacing the original Act, they ordered from Robert Stephenson & Co. in September, 1824, the first two engines, which were delivered in September and November, 1825. A second order was placed, at about the time of delivery of the first engine and the official opening of the line, for a further two engines, but it would appear that in fact three more were built, delivered and paid for; the third of these, given Stephensons' works No.7, never appeared on the S. & D.R. books, and there is no obvious record of it running on the line in its original form. It seems not to have been used for some months after delivery. According to Tomlinson (and confirmed by the minutes) the directors had also applied to Fenton & Murray of the Round Foundry, Leeds, for a quotation for the supply of locomotives, knowing of the success of the Blenkinsopp/Murray engines on the Leeds-Middleton Railway, but this application was not accepted, thereby losing this firm a great opportunity.

Nos.3 & 4 were delivered in April and May, 1826, respectively. The other engine, according to Forward, was delivered in March 1827, and in my opinion for a few months carried the No.5. However, the first recorded reference to No.5 is in November 1827, when Hackworth "got No.5 Locomotive taken of the Carriage & set upon the Line in 2 Hours & 3/4s." Tomlinson, Warren, Forward & Dendy-Marshall all disagreed about this, but in general believed that it was another engine. I believe that this was the same engine as that booked out in March, but where had it been and what had been happening to it in the meantime? Forward says that the one delivered in March was billed later that month, and paid for on September 19th., at £550-0-0d., £50 less than the previous two, but the Stephenson works ledger of the time merely states it as entered on March 19th., 1827. The delivery date is not given. Nevertheless, it must have been there when the Prussian engineers made their inspection, or they could not have seen six engines. Neither could it have been "Experiment", as has been suggested, since they had also seen this engine under construction at Forth Street. I shall return to this.

No.1, later to be named "Locomotion", has been so well described elsewhere that there should be little to be said further about it. I refer later, in "Attributions", to Brian Reed's monograph, which covers the subject very comprehensively. However, a few things may be added here.

I have said earlier that Timothy Hackworth may have been concerned with the design and construction of No.1 engine, and nearly every major railway historian from the mid-1920's on has believed that George Stephenson himself was not directly involved with the design of No.1, except insofar as it followed generally the Killingworth type, although the means of linking the wheels together was changed to simple outside coupling rods, and the crosshead guide arrangement and valve gear was much modified. Robert Stephenson is known to have worked on its construction, before going to South America, but it is doubtful if his involvement was any more than this. However, two drawings exist which indicate that George Stephenson may have had some thoughts about it earlier, probably in late 1822. The first of these was published in a University of Newcastle Archive Teaching Unit folder, No.3, "Railways in the Making", in 1969, and I am obliged both to the University and to the Tyne & Wear Museum service for its use here (Fig.15). It was presented to the Museum service by a Mr. J. D. Andrew of Grainger Street, Newcastle-upon-Tyne, in July 1934, with other material. The other and later drawing (Fig.16), showing different arrangements of drive and valve gear, has been used by both Warren and Reed, and is described as a works drawing. It is obviously based on and extends the earlier one referred to above, which shows one possibility using a complicated drive arrangement of rocking levers, the purpose of which is to connect the two pairs of wheels together in a cumbersome way, which I shall return to a little later. Copious notes in what appears to be George Stephenson's own handwriting cover much of the area. It is of some interest that Thomas Tredgold, a well known writer on the steam engine, used a somewhat similar rocking lever arrangement on a design for a road locomotive some four years later, but this seems never to have got beyond the design stage.

The second drawing shows a number of alternatives, including both sides of the lay-out of similar rocking levers to the first drawing, also outside coupling rods and the possibility of a cog-wheel connection between the axles. Both drawings show parallel motion of Freemantle type for the crosshead guides instead of slide bars as used previously. It shows also alternative arrangements using straight connecting rods from each end of the crosshead bars such as were actually used, and the left hand sketch of the pair of side elevations seems to have been the basis for the final design for the arrangement of the valve gear. The question regarding the reason for the change to parallel motion instead of the slide bars used on the Killingworth and Hetton engines, and fitted later to the Mount Moor/Springwell engines being built at Forth Street at the same time as No.1, still stands unanswered, but Stephenson's own comments on the sketch clearly mention it, saying that it should be as fitted to the Etherley (stationary) engine. Dendy-Marshall suggested that it was due to the possible problems of lubrication and retention of the oil on vertical guides on an engine intended for a comparatively long continuous run, as against the fairly short distances worked by a colliery engine, which may well have had something to do with it.

It has also been suggested that James Kennedy, Stephensons' works foreman, who later joined Bury at Liverpool, may have been involved, but it was Timothy Hackworth who had previous experience of parallel motion at Wylam. Nevertheless, he was still at Walbottle Colliery, I believe, when these drawings were made, and when the Etherley engine referred to in the notes was built. Joseph Locke is another distinct possibility. The 'works' drawing is on paper watermarked 1822.

The text of the first drawing reads: "this scetch will shew you my Ideas in the way I would combin the tow engines to gether." (It was quite usual to refer to each cylinder and crank set as an engine, a practice used in marine engineering until quite recent years.)

"I have placed tow Beams upon one pin - on one side of the engine each has a long & a short end so as to alow the Crank pins to work clear of each other. the black line shew the work on one side and the doted one on the opposide side of the engine. the long end of the Beams will work a longer stroke than the short one, and the tow wheels on one side must be made to sute as shown in the scetch. the Beams will work back to back like a pare of shears and they should project a little fourther out than the wheels so that the connecting rod may not grind against the wheels you will precieve that a black line and a doted one work together one on the fore(?) wheels and the other on the hind ones and on contrary sides the Beams are a little lent (bent?) to bring them to a square with the exels of the wheels. you can put on the paralile motion like the Eatherley engines." (Supplied in October 1823 for the Etherley Incline, and probably under construction at the time of this drawing.) "I have given a Scetch of it but not the proper lengths.
the excels for the Beams ought to be a bout 2 Inchs diameter and the Beams about the same strength as the present ones that the Pestons Heads are attached to. the tow levers on the other side of the Engine, that will be requred for the paralile motion may be very slight as they will only have to carry the p - - s (pistons?) Rods. I think it will be better to put on ball and socket joints on to all the beams so that they will

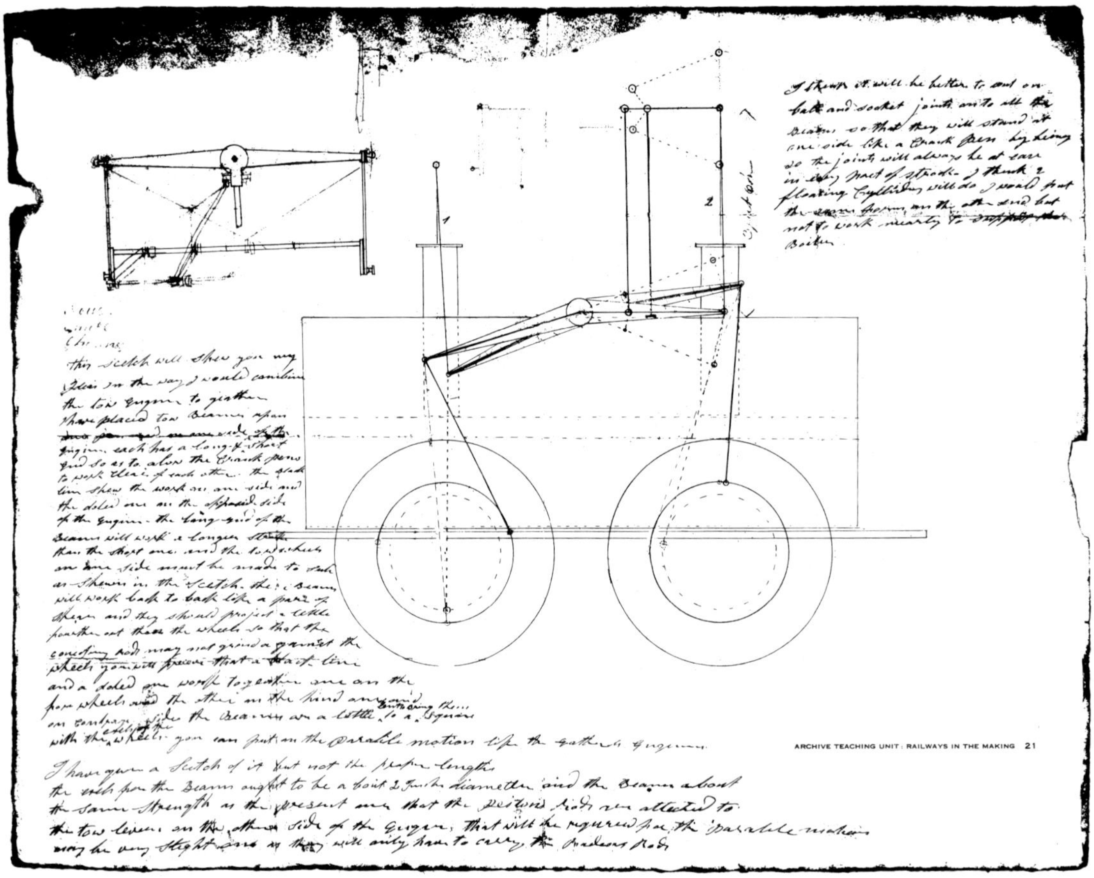

Fig.15 George Stephenson's early design sketch for Stockton & Darlington engine.

stand out on one side like a Crank Pin by being so the joints will always be at ease in every part of stroke. I think 2 floating cylinders will do. I would put the same forms(?) on the other end but not to work mearly to support the boiler."

Note here that he refers to 'floating cylinders', by which almost certainly he means the 'steam springs' used on the later Killingworth engines, but which were not in fact used for No.1.

The method of operation seems to be as follows: although he says clearly that the two beams are on one side of the engine, and his dotted lines on the other side are not complete, it is obvious that there must have been levers of similar length but in two independent halves on the other side. The drawing shows a side elevation, with no handing. On the side shown, the crosshead bar from each cylinder drives through a short connecting rod on to the short end of the rocking lever. The long end of each rocking lever drives through another short connecting rod on to a crank pin on the wheel at the other end of the engine. On the other side of the engine, a similar short connecting rod from each crosshead beam drives on to the wheel at the same end and is guided by a lever of similar length to the same (short) end of the opposite rocking lever. This necessitates a smaller radius crank throw, since the rocking levers are not equally balanced and the lever at that end is shorter. In this way, each cylinder is driving on to one wheel at each end but on opposite sides, and the two cylinders are connected by the levers and the linkage of the short connecting rods.

The unequal lengths of the rocking beam ends were for the purpose of allowing the end bearings to pass each other without taking up too much space. There need only have been one pair of equal length levers on the other side to balance the drive. It is just as well that this idea never came to fruition, or there would probably have been even more amputated limbs than did happen, but it does show the fertility of George Stephenson's mind. The later 'works drawing' also shows simple outside coupling rods, but it is possible that he himself ignored them as not worth consideration, and for once someone else knew better.

Certainly the cylinders, valves and driving gear fitted follow the Killingworth and Hetton engines closely, although they are more compact. The use of outside coupling rods instead of a chain or gearing to keep the wheels revolving together is the first known example; the coupling rod and cranks had been thought of earlier, and included in the

Dodd/Stephenson joint patent of 1815, working inside and using cranked axles, an unreliable scheme at the time. The Mount Moor/Springwell engines were also fitted with such outside coupling rods, possibly because of the success of those fitted to Nos. 1 and 2, and the practice from then on became almost universal. Of course, the fact that there were two connecting rods from the crosshead beams to each pair of wheels meant that the coupling rods could only work by means of a flycrank on one wheel on each side, i.e. one of each pair, another ingenious addition which cannot be attributed to any single person, but which may have been George's own idea.

According to Reed, since the front axle carried the single eccentric it must have been mounted otherwise, probably in the same sort of simple plummer block bearings that are fitted now. However, the Forth Street drawing (Fig.16) does show this axle, complete with eccentric on the inside of the wheel as if part of it, running in just such a pivoted tube. It seems more likely to me that such pivoted tubes on both axles were dependent on the fitting of the steam springs or floating cylinders, and since these were not fitted it was necessary to secure one axle in fixed bearings, the other being left pivoted to accommodate itself to track level etc. A drawing by Nicholas Wood exists, showing one of the Killingworth engines modified by fitting it with straight plate springs, but it is possible that the cart type leaf springs of the time were not of good enough quality to carry the greater weight of a locomotive engine, although it was not very many more years before they came into general use for this purpose.

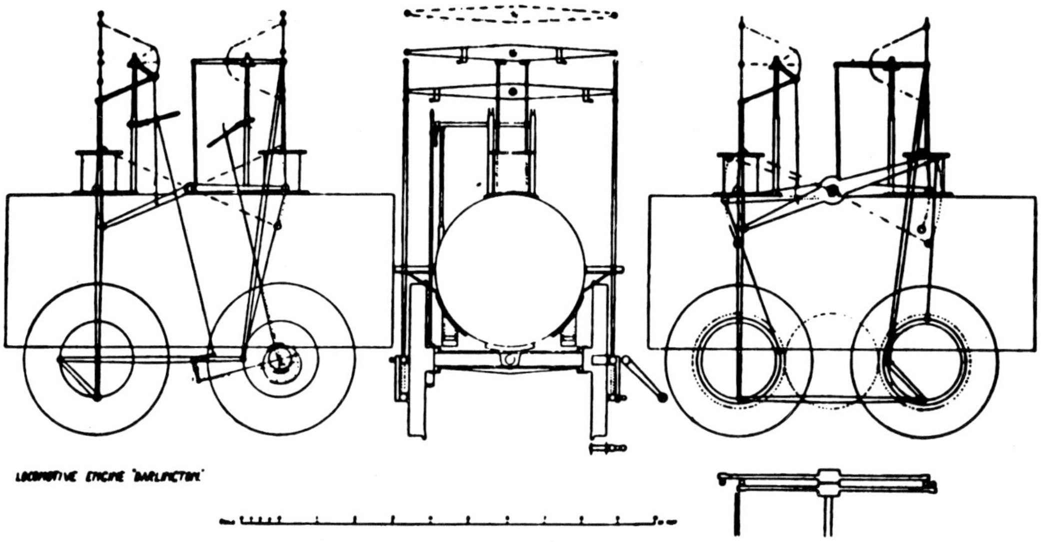

Fig.16 Forth Street Works drawing developed from Fig.15.

These drawings seem to me to leave no doubt that George Stephenson did take sufficient interest in the engines intended for the Stockton & Darlington Railway to have done a little design work on them, in spite of his numerous and massive other involvements at the time, although the detail and completion was left to others.

No.1, when delivered, probably had a boiler with dished ends, and plain spoked wheels of cast iron, similar to the Killingworth engines. Alan Prior's drawing (Fig.14) shows what it may have looked like. The flue tube was single and straight, without either separate firebox or smokebox, the grate being built into the rear end. The water level was just above the flue tube, with a pair of test cocks to check it, leaving plenty of steam space and a large evaporation area, which would be necessary with such a crude boiler and very limited heating area. The pistons were packed with hemp, a practice which ceased very shortly after this.

Originally, again, one pair of wheels was probably mounted in a tubular axle box, pivoted in the centre.

The chimney was also longer at first, and a single exhaust steam pipe led to it on one side. The valves had lap and lead, although, due to the unbalanced single drive to the two valves, this was unequally distributed between the two cylinders. The eccentric was, of course, loose, and operated two rods, one upwards to the front cylinder and the other horizontally and through a bell crank to the rear cylinder. From time to time since the engine was first preserved this bell crank has been fitted the wrong way round, whether deliberately or accidentally is not known (see Figs.17 & 18).

In 1889 John Wesley Hackworth was in correspondence with a Mr. Thos. Greener of London for the purpose of establishing some point regarding the first engines, and had taken or was about to take 'testimonies' from John Greathead, an early apprentice to Timothy Hackworth, and George Jennison, one of the first engine drivers on the line, regarding these engines. Both of them say or are led to say that No.1 engine had flat ends to its boiler, and that the wheels were of cast iron without tyres, among other known facts. Unfortunately the dates when each of them commenced employment are not filled in, and it looks as if these 'testimonies' were drafts only, not yet read by the two men, so it is impossible to say that they are correct. Jennison gives details of the very different valve drive on the succeeding engines which agree with Forward's description, which I shall come to later, but it is likely that Forward had access to this record, inserted in Timothy Hackworth's 1827/8 diary by his son. There is also other evidence.

As preserved, the changes, although obvious, are really superficial (see Henry Pease's remarks, given on page 39). The chimney was shortened and fitted with a portable plate at the bottom for cleaning purposes, a double exhaust steam pipe fitted to improve the blast (Dunlop says with upturned nozzles), and the boiler rebuilt with flat ends or replaced, as it probably was more than once. The most obvious feature is the replacement of the spoked wheels by the two part cast iron wheels with separate wrought iron tyres, probably originated by Robert Wilson and quickly taken up and developed by Hackworth as more serviceable and easy to maintain. The axle bearings are all fixed. One of the present wheels is odd, having an open spoked centre instead of the usual type with holes in the web only, but this is not of any importance. The usual wooden slats were fitted to the top half of the boiler as protection and lagging.

No.1 is believed to have cost £500, as did No.2. No.1, however, lasted for a long time. With modifications, it was used on the S. & D.R. as a locomotive until 1846, when it hauled the train at the opening of the new Redcar section, although it has been suggested to me that in fact the engine on this occasion was one of the later engines disguised as No.1. This would have been very difficult, for several reasons which will become apparent. It was probably little used after 1840 until this event. Later in 1846 it was being used at Howden Station as a pumping engine, "fitted up with pumps from the engine at Middlesbrough". It was sold or loaned in 1850 to the Peases' West Durham Colliery as a pumping engine, and was put back into more or less the original condition in late 1856, at an estimated cost of £50. It was preserved in 1857 on the well illustrated plinth at North Road Station, with many journeys to various

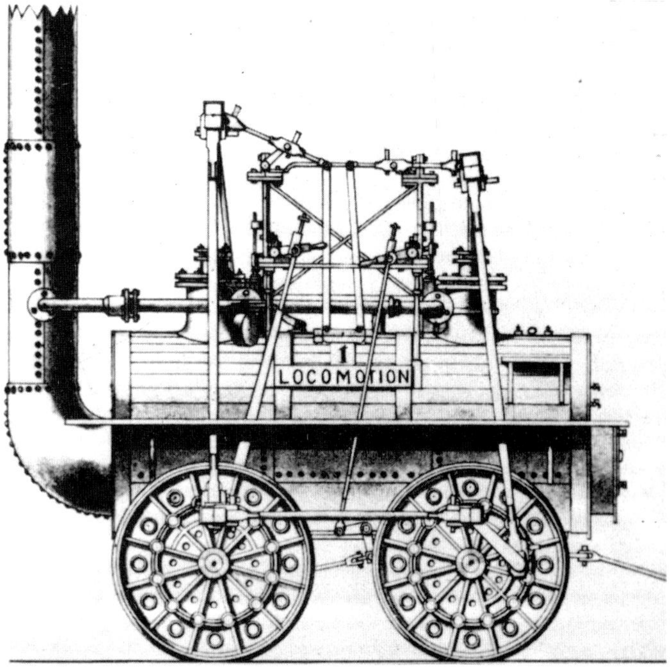

Fig.17 No.1 "Locomotion with valve rod to the rear cylinder correctly connected.

celebrations and centenaries, as far afield as Chicago and Philadelphia (for the U.S.A. Centenial in 1876) and Paris in 1889.

In 1892 the engine was moved to a plinth at Bank Top Station, joined there a little later by No.25 "Derwent", and stayed there until 1975, when it was borrowed by Stockton in conjunction with the Rail 150 celebrations, Darlington having not moved quickly enough. Although Mike Satow's replica actually ran in the Shildon procession, No.1 itself was hauled up and down a length of track at Preston Park by two Fowler ploughing engines, watched by the Duke of Edinburgh. Neither engine returned to Bank Top; Nos.1 and 25 are now housed in the North Road Station Railway Museum, and the replica No.1 is at Beamish Open Air Museum.

One of the original wheels broke very soon after the engine was put into service. The single flue was at one time replaced by a flue with two smaller return tubes, then by a double (return) tube, and finally back to the original type of straight-through flue. The flue tube burst on at least one recorded occasion from excessive steam pressure, while taking water at Aycliffe Lane on July 1st. 1828, killing the driver, John Cree, and maiming the water pumper, Edward Turnbull. The engine was derailed at least twice, at Aycliffe Lane in November 1837 (damage £22-16-10 d.) and at Middlesbrough in October 1839, when it fell on its side and was badly damaged. An engine referred to as "Cree's engine", and Cree was the regular driver of No.1 for a long time, fell over at

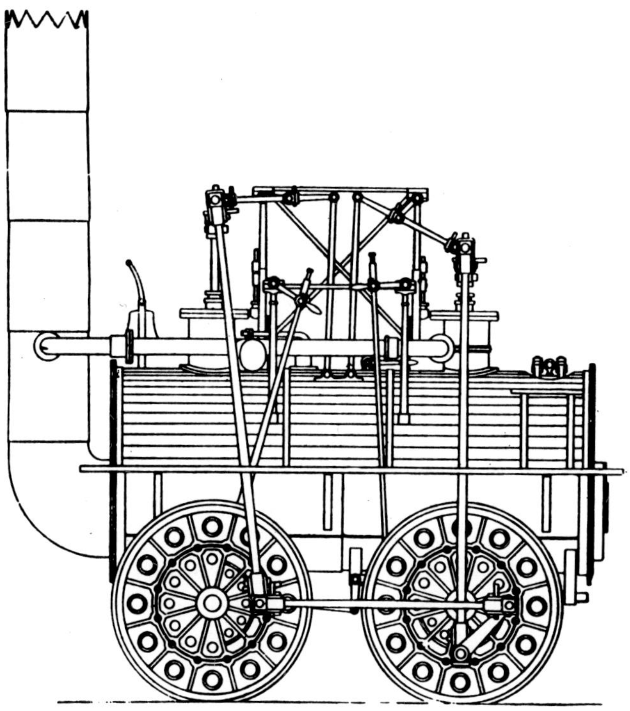

Fig.18 No.1 "Locomotion" with the valve rod to the rear cylinder wrongly connected.

New Shildon and needed repair on one other occasion. There was also an incident near Goosepool in 1833, when the engine hit a donkey, and the fireman lost a foot.

The straight fly-cranks may have been replaced for a while by curved ones in 1827, when the Prussian engineers saw it.

Francis Whishaw, in his early survey, "Railways of Great Britain", 1840, said that he had seen this engine mounted on six wheels, but although Forward accepted this, its accuracy is doubtful. The Hoole/Craven notes agree, and say rightly that the 1840 list shows it then as having four wheels only. Whishaw's account of the locomotives contains other errors. The stock valuation lists referred to earlier show No.1 consistently as having a much lower value than the other three survivors, except for No.3 "Black Diamond", before it was rebuilt. This indicates to me that no very extensive rebuilding of No.1 took place, although John Graham says in December 1834 that three engines, not spring mounted, had been rebuilt, one at Shildon Works and two at Kitchings - more about this later. The preserved engine still has the same type of valve gear as was originally fitted, indeed, some of it may actually be the original gear. The cylinders, although increased in bore from 9" to 10" diameter, and a few other such pieces, may also be original, but it is doubtful that much else can be. I would suggest that it was this valve gear, quite different to the later engines, that made it not worthwhile to rebuild the engine to any great extent, and also why it is unlikely that it was any other engine that opened the Redcar line; the valve gear could not easily be disguised.

Longstaffe and Samuel Smiles, as I have said above, called No.1 the "Active" at first, but by 1833 it was certainly called "Locomotion", and it was never referred to by any other name thereafter.

The second engine was slightly different in several ways. Firstly, and most obvious, the valve gear was much modified. Both axles were now fitted in tubular axle boxes, one pivoted, and Forward, as I said earlier, thought that this made it difficult to fit an eccentric on the front axle. I think it more likely that this was part of a general improvement in design. Instead of the single eccentric and conjugated drive, the valve drive was taken by means of a rod from the middle of each coupling rod upwards to a crank on each end of an eccentric shaft mounted across the top of the boiler. This was shown in an engraving by Swain of the "Engineer" magazine in October 1941, taken from a woodcut illustration published in 1829 in "Brewster's Parochial History of Stockton-on-Tees", the only known contemporary illustration of Nos. 2, 3 & 4 (Fig.19). E. A. Forward's 1941 conjectural drawing (Fig.20) shows this in detail. This method provided a more balanced and direct operation of the valves. Apparently this type of valve operating gear was patented by Phineas Crowther of Newcastle about 1800, and was widely used on pithead winding engines and other stationary engines; it would appear to be the source of the eccentric cross shaft mounted on top of the boiler as used on several S. & D.R. locomotives in the early days, from Robert Wilson's engine and thence to No.5 "Royal George", until it was found better to fit the eccentrics on the rear axle, the driving axle of the coupled pair. This Crowther gear is also described in George Jennison's 'testimony' in June 1889, mentioned earlier.

It is also known that the pistons were fitted with metallic packing, probably brass rings with internal springs, instead of hemp as used on No.1.

No.2, eventually named "Hope", had a chequered career. It took a week's work to make the engine operational when delivered to the railway. It is known to have run away once at Stockton on the quayside on 1st. October 1827, and was badly damaged. The press report, which did not identify the engine, stated "exceedingly injured and other serious damage done. . ." (see later). I find it surprising that this incident was not recorded by Timothy Hackworth in his notebook at the time. The fireman was discharged on the 5th. October, which may well

Fig.19 Woodcut of No.3 "Black Diamond" from Brewster's Parochial History of Stockton-on-Tees, 1829, showing valve gear and wooden wheels.

have some connection with the incident. Both engine and tender were again badly damaged in collision with waggons in November 1835. It or its successor was rebuilt in 1835 by Kitchings into an 0-6-0 with increased heating surface, and with vertical outside cylinders like "Royal George" and "Victory". Some of this is confirmed by John Dixon's 1836 list, which shows both "Hope" and "Diligence" as then having six wheels. Such a rebuilding would in effect have produced an engine which would include very little of the old one, but it would stay on the books as the same engine. It is not known exactly when the engine was scrapped; Inness says that Nos.3 & 4 were disposed of in the early 1840's, but the valuation lists show No.2 still there in 1844, although put on the reserve list the previous year. There is a puzzling report that the number 2 was allocated in 1831 to an engine named "Bedlington", which almost immediately became No.12 "Majestic". The earlier No.2 was certainly in existence then, and should also have held a number, although, as stated earlier, the name "Hope" may not have been listed yet officially. If, in fact, as mentioned later, it had been intended to call the first No.2 "Bedlington", and it became "Hope" because of the accident to the original No.2, things would be clearer, but this is almost certainly not the case.

During the 1939-45 war, No.1 "Locomotion" together with the second No.25, "Derwent", was taken to Stanhope in 1941 for safe keeping for several years.

"Chittapratt"

Late in 1825 an engineer named Robert Wilson, who at that time had a workshop in Forth Street, Newcastle-upon-Tyne, which cannot have been far from the Stephenson works, offered to the Stockton & Darlington Railway a locomotive of his own design and construction which he claimed to be a great improvement on anything so far produced. Certainly it was different in many ways.

The railway accepted it for trial for one month, the price asked being £380, and operated it for some weeks, possibly longer. During that period, at the end of 1825, the great early French engineer Marc Seguin, accompanied by his brother Paul and by his cousin, M. de Montgolfier of balloon fame, visited this country to survey progress in railways, in which Seguin had a commercial interest, and amongst others visited the Stockton & Darlington Railway at Darlington and/or Shildon. There he saw this locomotive and made brief notes and a rough sketch, which is the only known contemporary depiction of this very interesting engine. Very soon afterwards it was decided that it was not suitable, probably because it was not powerful enough, and it was bought from Robert Wilson at the agreed sum for the use of its parts, taken to pieces and put aside for the time being. (Brian Reed says that the actual purchase was not until April 1827.)

Marc Seguin, known in France as Seguin l'Aîné (the elder) since he was the eldest of five brothers, returned shortly thereafter to France and his notes disappeared from sight for almost a century.

Apart from the generally agreed fact that the boiler of this engine was used as the basis for that of "Royal George", with possibly the wheels and some of the valve gear, there seems to have been very little recorded about this engine, and many erroneous accounts gained circulation. Theodore West attributed it to a Leeds works, confusing the builder with another Wilson, and Francis Whishaw confused Wilson with the boilersmith John Wight, of Lumley Forge, Durham, who later rebuilt the boiler. Others correctly identified Robert Wilson, but placed him in Gateshead and other locations.

The text accompanying a list in the "Engineer" of 31st. October 1879, and referring to the year 1840, believed to have been supplied by George Graham, says that "No.$1^1/_2$", obviously "Chittapratt" (why not $2^1/_2$?), was built at Leeds (Theodore West's source?) and had worked only a few months "when it was broken by a collision near Stockton". The only correct statements were that (a) it had four cylinders, (b) it had four wheels, and (c) that it was driven more or less directly on to the crankpins which were set at 90 degrees to each other. Zerah Colburn, a highly accurate engineer in general, stated that it was

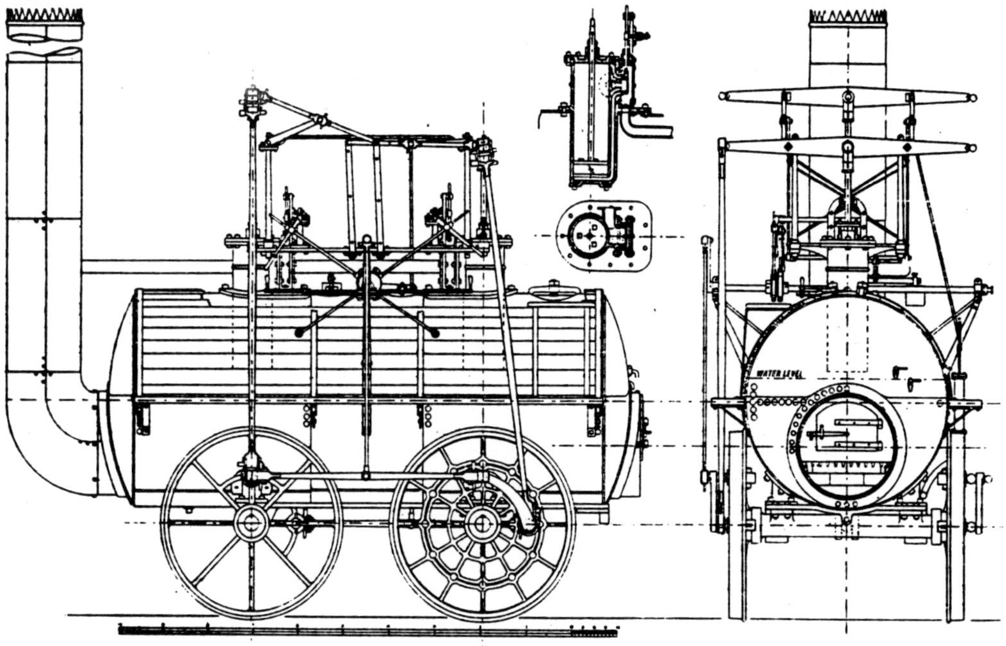

Fig.20 Conjectural drawing of Nos.2, 3 & 4, by E. A. Forward.

arranged with one cylinder to each wheel, or rather two cylinders to each pair of wheels. C. E. Stretton also said quite clearly that the engine builders considered outside coupling rods to be an objection, and repeated Colburn's error, to the extent of publishing a very detailed drawing showing it as No.5 "Stockton" (see later and App.III.) James Dunlop, in an article in the "Engineer", said that the boiler exploded, inferring that this was why the engine was laid aside, and this was also mentioned in George Graham's reminiscences, giving the date of the occurrence as 1826, but since the boiler was used again it is more likely that the 'explosion' was a flue tube rupture, quite serious in itself but not uncommon. J. U. Rastrick said in his diaries that the engine was taken off after a month because it was too heavy for the rails, but he was not there at the time, and wrote from hearsay.

The Prussian engineers von Oeynhausen and von Dechen also visited the Stockton & Darlington Railway at the beginning of 1827, and reported on the locomotives there; it would seem that they saw the boiler of Wilson's engine. They gave a description and dimensions which I shall return to.

This engine does not seem to have been given a number, but it was certainly the third locomotive delivered to the line. The men called it "Chittapratt" from the noise it made; I have used Inness's spelling, but most other writers omitted the final 't'. There is no record of whether this name arose from its

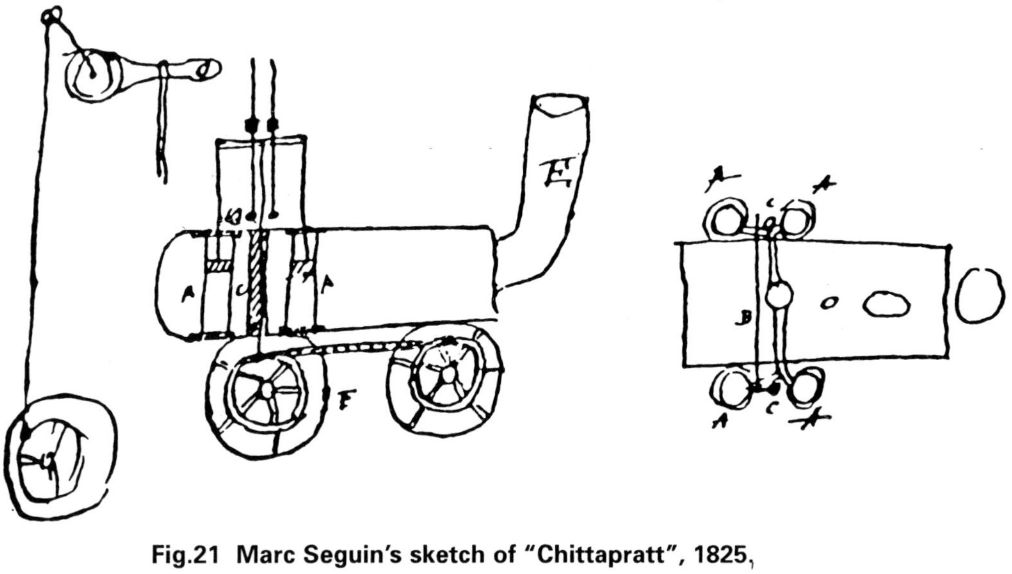

Fig.21 Marc Seguin's sketch of "Chittapratt", 1825,

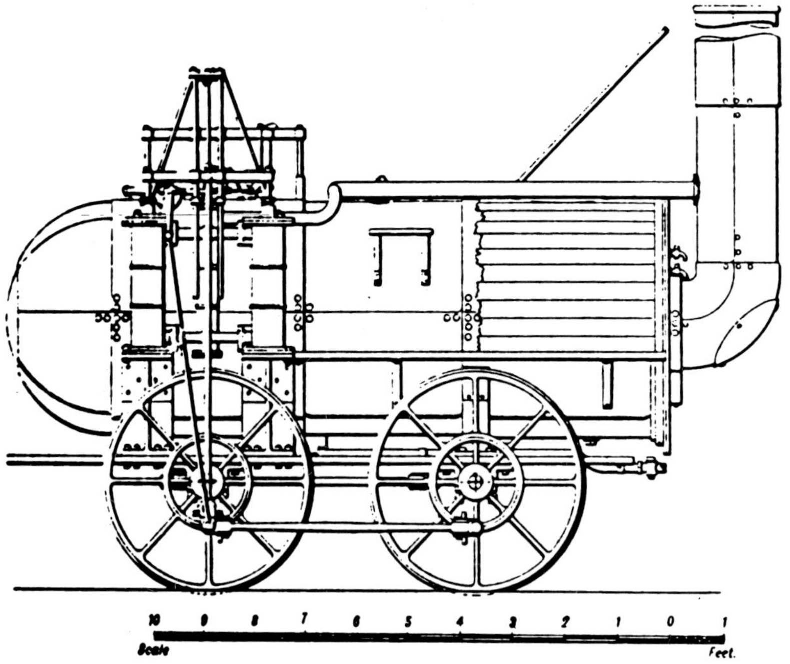

Fig.22 Conjectural drawing of "Chittapratt", by E. A. Forward.

mechanical noise or from the exhaust beat, but the mechanical explanation seems the most likely.

In 1927, Laurent Seguin, grandson of Marc, came to London with Ferdinand Achard, who had discovered copies of Marc Seguin's notebooks and had been studying them for some time. The Science Museum has in its files a letter from M. Achard to E. A. Forward in October 1923. referring to the matter.

On 22nd. February, 1927, they read a paper to the Newcomen Society at the Institution of Mechanical Engineers in London, entitled "British Railways of 1825 as Seen by Marc Seguin", part of which had to do with "Chittapratt", including the rough sketch, which I reproduce here (Fig.21) and also a translation of Seguin's notes. C. F. Dendy-Marshall very shortly commented on these in his 1929 book "Two Essays in Early Locomotive History", and amended and enlarged the account later in his 1953 book "A History of Railway Locomotives down to the Year 1831". E. A. Forward also contributed articles to the "Engineer" magazine in August 1940, which are very detailed and include a conjectural drawing

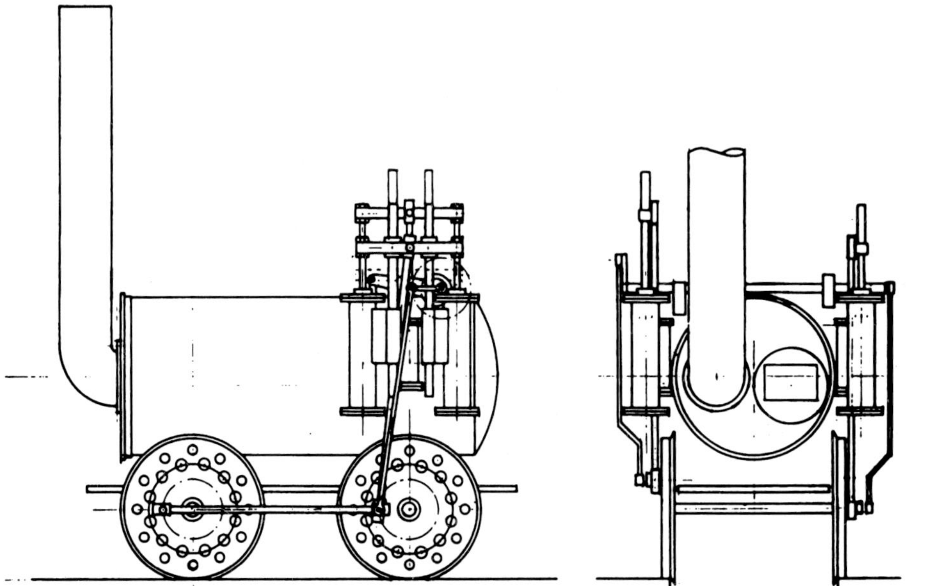

Fig.23 Author's conjectural drawing of "Chittapratt"

based not only on Seguin's notes and sketch but using mechanical design formulae and calculated weights of parts to substantiate his ideas. Before reading this I myself produced a conjectural but much less detailed drawing from Dendy-Marshall's remarks and the published paper from the transactions of the Newcomen Society, and I was pleased to find that they coincide in most major particulars, although my first attempt had some serious errors, especially in the the flue tube arrangement. I give both for comparison (Figs.22 & 23).

Unfortunately the original document, Seguin's notebook No.13, seems to have been lost again many years ago, probably stolen and sold, and Achard was working from copies, which also seem to have disappeared now, and so far I have been unable to study the original notes in French as I should have wished. I have to say that I am uneasy about the translation, which does not ring true to me. However, I give it here as published both by the Newcomen Society and Dendy-Marshall. A few words in French were shown under the sketch in the transactions, and that is all.

Notes explaining sketch:-

A, cylinder 6in. diameter.

B, a movable rod receiving a movement from a pin on the wheel. It actuates in a small cylinder a piston C, intended to distribute steam, as in the system of Martineau.

C, the above piston, including the steam passages opened and closed by gearing, whose detailed and enlarged drawing is in D.

(D is a rough sketch to the left of the main sketch of the engine)
E, chimney having a diameter of at least 18in. and a height of about 10ft., limited by that of the bridges it has to pass under.

F, the wheels. They are made of two concentric circles and the power is applied on the line of their junction.

Notes in text:-

"In the latest of these engines there are four small cylinders of 6in. diameter and - inches stroke*, into which steam is delivered under two atmospheres pressure, which steam is probably allowed to expand a little in the cylinders. This arrangement, of two cylinders on each side, with a connecting rod between them, has been adopted without doubt to facilitate the communication of the motion of the piston to the crank below the cylinders. The piston-rods, instead of being guided by rollers, are guided by straight rods, which are fixed to the middle of the yoke or crosshead, and pass through holes similar to stuffing boxes."(*apparently 20")

It would seem, from Forward's comments, that Seguin had just visited Taylor & Martineau in London, had ordered from them an engine and boiler for a steam boat, size not known, and had seen their latest developments, which included the use of piston valves with some amount of lap, and roller guides.

In the transactions referred to above there are also illustrations of two other locomotives which are captioned as intended for the St. Ettienne-Lyon Railway, on which Seguin was working a little later; one of them, attributed to Stephenson, is very similar in some ways to Hawthorns' "Swift" a few years later, but without the intermediate drive shaft. The other one, Plate No.XIII, (fig.24) although very different in layout and driving gear, is fitted with vertical cylinders, piston valves and roller guides. Forward has assumed that there is some link here with Robert Wilson's work, and that these cylinders and valves are of the same type as fitted to "Chittapratt". In any case, they seem to be of about the right size, and fit into the layout quite well. I may say that M. Pierre Seguin, great grandson of Marc, has told me that this plate is in some way connected to "Chittapratt": this information may have come from Ferdinand Achard, who was certainly assisting Forward with his research at the time. Certainly the hemispherical boiler ends of this engine are very similar to those drawn by Forward in his reconstruction drawing. The grasshopper drive beams on this engine seem to be fixed at the front end, which would cause great difficulty in the guide boxes, as no swing links are shown.

Robert Wilson is believed to have been the person who is on record as having seen the Trevithick 'Gateshead' engine in 1805, and he is probably the same Robert Wilson, described as Whitesmith and Bellhanger (in other words, a non-ferrous founder) whose premises were in Back Lane, Newcastle, in 1811. After 1847 his name seems to have disappeared from the records and directories.

He told Hackworth that when working on "Chittapratt", which he described as "my unfortunate one", he had had several ideas on possible improvements, and in 1827 he offered to construct another locomotive for them, but this offer was not accepted. Is it possible that the locomotive illustrated in Plate No.XIII was from his hand? Or did Seguin himself use some of Wilson's ideas?

Nos. 3, 4 & 5.

This is a good place to review the visit in early 1827 by the two Prussian engineers. They said that they saw six locomotives on the S. & D.R. at that time, five in operation and one in reserve, and that four of them had cylinders of 9" diameter x 24" stroke, which conforms with the original dimensions of the five engines delivered by the Stephensons' works. They said also that the boiler diameter of four of them was $4^1/_2$ feet with a length of $11^1/_2$ feet, which is probably not precise, since those locomotives did vary a little in such respects. It is believed that Nos.1, 2 & 3 had boilers of 4 feet diameter and that No.4 had a boiler of 4'-4" diameter. They went on to say that another engine boiler (serving for a similar locomotive engine) had a length of 10'-10", diameter 4'-4", closed at one end by a hemisphere and at the other by a cast iron plate, and containing a return tube flue. It is assumed that this engine was Wilson's, the "Chittapratt". They then go into great details regarding suspension, wheels and driving gear, but it is not clear whether they mean this to refer to all the engines seen, which is what I believe they were saying.

So, they have said that they saw six engines, but only described five. What was the other engine? I suggest that there is some lack of precision in translation, arising from the fact that 'everyone knew' that there were only four Stephenson locomotives, and that what they meant is that there were five engines actually in working order, and that the one in reserve was complete and on standby; the sixth engine was at that time in pieces, and was not the one that they referred to as in reserve.

They go on to describe the three designs which Hackworth had in mind for evaluation, one of which came to realisation very soon as "Royal George", and then another engine under construction at the Stephenson Works which was almost certainly "Experiment". As regards Hackworth's designs, they said that "at the beginning of 1827, the engineers were occupied with the construction of some other locomotive engines", and were then shown the designs for these by Hackworth, which could be interpreted to mean that not only "Royal George" was being built. There is no other indication of this, and they may not have understood what they were being told, but it could explain the apparent second No.5. The first design described was of a four wheeled engine, similar otherwise to "Royal George", and something like the Rainhill Trials "Sans Pareil", and the third design sounds vaguely like a basis for the later 'Majestic' class.

No.3, "Black Diamond", was delivered in April 1826, and No.4, "Diligence", in May. The Stephenson works ledger, which came to light in 1928, show that one further engine of this type was booked out to Darlington in March 1827, and paid for in

Fig.24 Locomotive supplied to Marc Seguin for the St. Ettienne-Lyon Railway in 1828: Plate XIII in the Newcomen Society's Transactions, 1927, covering the paper by Ferdinand Achard and Laurent Seguin.

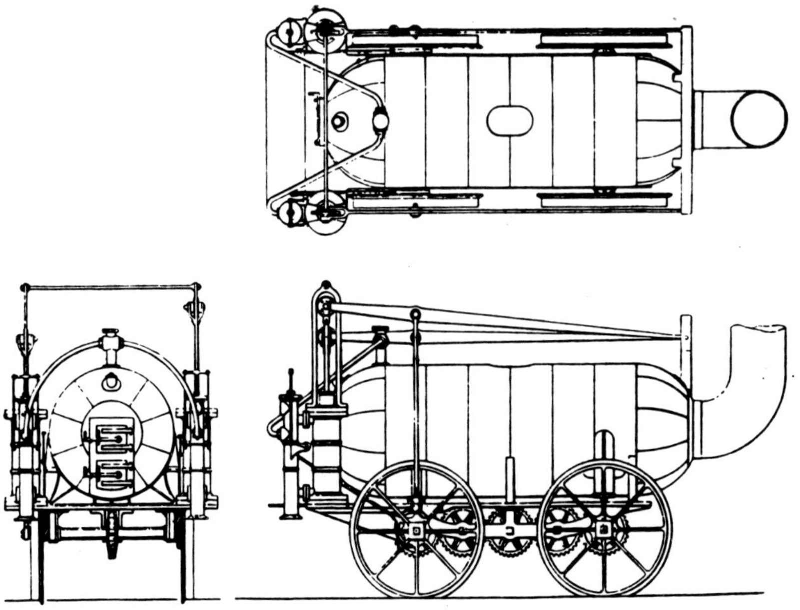

LOCOMOTIVE SUPPLIED TO SEGUIN FOR THE ST. ETIENNE-LYON RAILWAY, 1828.
(From an original drawing).

September, according to Forward, but except for Hackworth's notes there is no record of it running on the line (see later and list at end of chapter). This does agree with the Prussians' statement that "a sixth serves as reserve", and Forward and Dendy-Marshall both accept this. It seems strange that they did not relate it to the others or describe it more fully. Forward and Dendy-Marshall believed that this engine was laid aside for the time being, and eventually rebuilt by Hackworth into No.8 "Victory". I think this unlikely, for reasons which I shall go into when I get to that engine. However, I have been offered another theory, based on a report by William Bouch dated 23rd. November 1858, which contains a statement that between 1825 and 30th. June 1858, 16 engines had been sold and nine broken up, two of the latter having been lost by explosion. One of these two was certainly No.9 "Globe". It is believed, as stated earlier, that No.2 ran away on the quayside at Stockton in October 1827 and was badly damaged, but there is some contradiction between the press report and Tomlinson's account in his "North Eastern Railway". He says that it had carelessly been left standing with the safety valve fastened down, a common practice at the time to maintain steam pressure while running. The press report, in Thomas Richmond's "Local Records of Stockton and Neighbourhood", does not identify the engine and says that "it was wantonly set off" and that a reward is offered for the conviction of the offenders, implying outside interference. The only mention in the Hoole/Craven notes of the sub-committee minutes with any relevance to this incident is the report of the dismissal of the fireman mentioned above, which did refer to No.2. It is not known how badly it was damaged (see above); the word "grievous" was used by Tomlinson, and there could have been a boiler explosion which was not recorded. On the face of it, and since the press report might have been expected to include such a serious happening, this seems unlikely, especially as this engine is supposed to have been rebuilt and mounted on six wheels by Kitchings in 1835, but the engine rebuilt by Kitchings may not have been the same one, although bearing the same number.

This theory postulates, however, that the engine was destroyed in 1827, and replaced by the next in line, so that the 1835 No.2 "Hope", was in fact another engine. In this case the original No.2 was to have been named "Bedlington". I do not believe this.

There is also another but even more unlikely possibility, firmly believed in by a young Darlington friend of mine who has made a very detailed study of these early engines.

What few contemporary illustrations there are of the opening and of No.1 engine show an apparent Killingworth or Hetton type locomotive, usually explained by the obvious fact that very few people then knew anything at all about locomotive engines or what they should look for when drawing them, and artists and others tended to copy what existing drawings there were. There is only one known contemporary published description of the engine, in W. Newton's "London Journal of Arts & Sciences" of October 1825, and this is inconclusive; it reads like a press release rather than an eyewitness account. It refers to 'sweep rods', which could be either connecting or coupling rods, although I favour the latter, as the connecting rods were usually called 'side rods'. It refers to "these engines", when at the time there was only the one. Longstaffe's 1854 History of Darlington is more definite, saying that the engine's wheels were coupled together by a chain and sprockets and that the engine was called "Active", but this was 29 years after the event and we do not know if he is referring to a witnessed account or what. The only firm evidence is that of Henry Pease at the 1875 Jubilee celebrations, when speaking from the engine itself, he said that "The engine . . . is to all intents and purposes, as an ordinary observer would have seen it on this day fifty years ago", which is not particularly precise.

Most early drawings of Stephenson's engines show the chain and sprocket coupling covered by the 1815 and 1816 patents. A corner vignette on Otley's early 1830 plan of Middlesbrough shows such an engine with widely spaced cylinders hauling a train of chaldron waggons along the river bank approaching Port Darlington; the visual difference, even in a hazy drawing and at a distance, between the two types of valve guide gear, is recognizable immediately.

Suppose that the intended No.1 engine had not in fact been ready for the official opening date, and the railway insistent that an engine be provided. They had plenty of pressure, since Edward Pease had the largest financial share in the South Street Works. Works Nos.1 & 2 were intended for Mount Moor Colliery (later joined by Springwell Colliery and probably sharing a common railway system) and it has long been believed that their construction was delayed because of the urgency for the Darlington engines. They were eventually delivered in April 1826, just before the second pair of Darlington engines. Although when delivered they were fitted with outside coupling rods, this need not have been the original intention, and they were fitted with guide bars (slides) for the crossheads, similar to the earlier Stephenson engines, and also as shown in contemporary illustrations.

Suppose, then, that one of these two engines was almost complete at the crucial time. Spare axles, sprockets and chains would have been available. What easier than to let the S. & D.R. have it in order to open on time? Who would have known the difference?

It is known and well documented that No.1 engine "blew up" at Aycliffe Lane while taking water on 1st. July, 1828, as mentioned earlier. The extent of the explosion is not known, and it has been assumed that this was only the usual flue tube rupture and thus repairable. Forward said that a spare flue tube was received on August 13th., and one had certainly been ordered in May (see below). It *may* be that No.1 was *not* repairable, and was the other engine lost by explosion as listed by William Bouch. In this case Stephensons' works Nos.3, 4, 5 & 6 would then have assumed the intended railway numbers 1, 2, 3 & 4 when numbering began officially.

Now, Timothy Hackworth's notebook contains, besides the note of off-loading at Darlington in November, more than one reference to an accident on 19th. March, 1828, when No.5 engine burst its tube, due to the negligence of the driver, James Stephenson, in leaving the safety valve locked down when the engine was stopped. This incident led to a system of 10/- fines for such carelessness, at the next Committee meeting. Both men were compensated financially, but John Gillespy had been blown 24 yards and badly scalded, and Edward Corner had been blown 16 yards and his thigh broken. Gillespy died a few days later.

There is an earlier reference, for 7th. February 1828, to "No.7 viz the Quadrant Engine", which engine soon became No.6 "Experiment". There are several references to the "Royal George" engine, from December 1827 on, but without an engine number. As I have already said, it is my belief that the No.5 engine mentioned was in fact the last of the first five Forth Street engines, and that at that time "Royal George" was No.6, but since it had a name it was unnecessary to use the number. If, then, the No.5 engine referred to proved to be unrepairable, as it may well have been, all is explained; "Royal George" became No.5 and "Experiment" became No.6. The Hoole/Craven notes state that this No.5 was returned to the makers about June 1828. Hackworth's notebook states that a new flue tube was ordered (from Forth Street) for "the loco. that burst", which could have then only have been No.5, but obviously it was No.1 that was repaired and survived. Again, the Hoole/Craven notes support this. In fact, three spare flue tubes, two of them complete with flanges, were supplied by Stephensons at this time, one at the end of May, one in mid August, and one at the beginning of October. Since these flue tubes needed replacement quite often it does not tell us much.

John Wesley Hackworth stated both in public lectures and in the "testimonies" referred to above that No.4 "Diligence" blew up at the foot of Simpasture on the 19th. March 1828 (which would have been before it was named), killing John Gillespy, a new fireman on his first trip, and maiming Edward Corner, the old fireman, who was on his last trip. The driver, James Stephenson, escaped unhurt. It seems to me that J.W.H. had his facts confused, and that he had either forgotten, or did not know, that there were originally five of these engines. The name "Diligence" is meaningless here; several years too early.

Apart from the possible differences in boiler dimensions mentioned above, Nos.3 & 4 were very similar to No.2. "Hope", which, when rebuilt, gained another 26 sq. ft. of heating surface, vertical outside cylinders as "Royal George" and "Victory" and six-coupled wheels; Kitchings listed it as a new engine. Francis Whishaw saw all four of these engines in 1839, and described them in his book. No.1, he said, was then on six wheels, but there is no other confirmation of this at all, and as I have already said, the valuations of this engine do not show at any time the increase that might have been expected by such a rebuild. No.2 was as described above. No.3 was on four solid wheels and was fitted with sloping cylinders at 45 degrees and a multitubular boiler. John Graham says that it was rebuilt by William Lister in August 1838. The Hoole/Craven notes support this and add that the cost was £280 and that it was back in service by August, fitted with 4'-6" diameter wheels. Graham also mentions that 3 engines were rebuilt in December 1834, one at Shildon Works and two at Kitchings, not mounted on springs so must have been early engines. A pity that he gives no other details. No.4 had been rebuilt by Hackworth, according to Whishaw, and had outside vertical cylinders and six-coupled wheels. Craven quotes a minute of 11th. July 1833 which instructs Timothy Hackworth to rebuild it as (i.e. similar to) the "Wilberforce". No.1 still had the cylinders sunk into the top of the boiler. All of them are said to have had new boilers, Nos.1 & 2 being fitted with return tube flues, and No.4 with a return multitubular boiler, but I have some notes from Ken Hoole's files which say that Hackworth did in fact use the old boiler shell of No.4, and valued it at £15. This engine was considered for 'remodelling' in 1840, ie. fitting with springs on all six wheels, which would have meant a new cylinder arrangement, but instead it was offered for sale. These notes say that No.4 was fitted with new wheels and steam dome in the year ending 30-6-1838, and in those 12 months ran 17,209 miles, of which 56 were on coaching trains. In the following 12 months it ran 19,950 miles. No.4 "Diligence", was sold for £650 to Mr. Potter of the South Hetton Railway (Coal Co.?) in February 1841. It is shown in the 1840 list as having a 'circular' tender, whatever that may mean - a wooden barrel, perhaps? These lists show the other three engines on the list of mineral engines until 1842, when No.3 "Black Diamond" increases sharply in value and is listed as a 'coaching' engine, probably just rebuilt again. By 1843 Nos.1 & 2 were on the 'duplicate' list, where they appeared again in 1844. As I said earlier, Inness says that Nos.3 & 4 were withdrawn from traffic and replaced by heavier engines in the early 1840's. This applied to No.4, sold as above and replaced by the second No.4, "Stockton", of the 'Collier/Tory' class, but there were not another No.2 & 3 until Kitching built the 0-6-0 mineral engines "Graham" and "Times" of 1849.

No.3, "Black Diamond", was tested on Brusselton West Bank hauling St. Helens coal waggons with some success in May 1841, and was hired to a Frank Trewet for hauling earth waggons in May 1842 at 30/- a day. It was sold in early 1846 for £400 to John Anderson, contractor on the Redcar line.

There are another couple of points of interest that may be noted; the Brewster woodcut (Fig.19) showed wooden wheels fitted. It seems that Hackworth was asked to try out some wooden wheels with malleable iron tyres and cast iron naves made at the Bedlington Ironworks and supplied through Stephensons; these were fitted to No.3 on 1st. August, 1828, for trial, but only lasted for ten weeks, according to Rastrick. The price for the four wheels and two axles was £91-10-0d., much higher than the current cost of cast iron wheels. I can only think that much longer wear was expected from them because of the greater resilience of wooden wheels. Certainly this was not the only trial, and Stephensons used them on many engines, as well as supplying them to other engine builders, such as Fenton & Murray.

There are other mentions of wooden wheels; Hackworth used them for No.9 "Globe", the two 'Planet' type engines had them at first and some of the 'Majestic' class engines may have been fitted with them. The wheels shown in Forward's drawing are lighter cast iron wheels, of solid construction and with eight spokes for the front pair, similar to the Killingworth engines, and these were presumably fitted all round to all the first four (or five) when delivered. He shows on the rear axle a pair of light two part cast iron wheels, similar to those attributed to Robert Wilson, and it is likely that this type was very soon fitted to all of the engines, because of the much greater ease in changing the rims, a fairly frequent maintenance requirement, and the reduction in cost both of the casting replaced and the labour involved. The wheels were first secured in place on the axles with soft iron wedges, and then rust was induced in the joints by the application of a mixture of iron filings, sal ammoniac and urine to make them more solid, so it was a major job to remove them completely, if not impossible without breakage.

These two part wheels were later evolved into the ten and twelve spoked varieties with webs between the spokes which became standard usage for mineral engines for many years on this line, and were also used by other builders. I show Inness's drawing of them (Fig.25) but Colburn had illustrated them much earlier. No separate tyres were fitted at first, and the cast iron rim had a very short life, of months only. The development of a wrought iron tyre, long used on wooden wheels, is attributed by Bailey to Nicholas Wood, and was put into use by Stephensons in September 1826. This was a major improvement. When first used they were simple hoops with a flat section, shrunk on against the flange, but in 1829 the flange was made part of the tyre, and the slight taper of the tyre began to be used.

Hackworth's notebook also contains design sketches for malleable iron wheel centres in two sizes, and apparently some were made, but there is no indication of on what they were tried out.

The Prussian engineers said that the fly cranks were curved at the time that they saw the engines, and there is some evidence in the correspondence between Hackworth and Stephenson that these cranks gave trouble and had to be changed. This would seem to be a fairly minor matter compared to other problems, but everything added up, and there is no doubt that Hackworth and others had to work very hard to keep the engines running, not always with success. At times there were only 5 out of 12 or 13 engines available for use in 1832, and out of far more on 1834. The problem was not only reliability, but gross misuse. Frequent collision damage had to be dealt with, due to the macho attitude of the enginemen and the horse drivers, irresponsible behaviour and plain carelessness. Time and time again John Graham reports that "the engine was broken; the tender wheels were broken; some of the rods

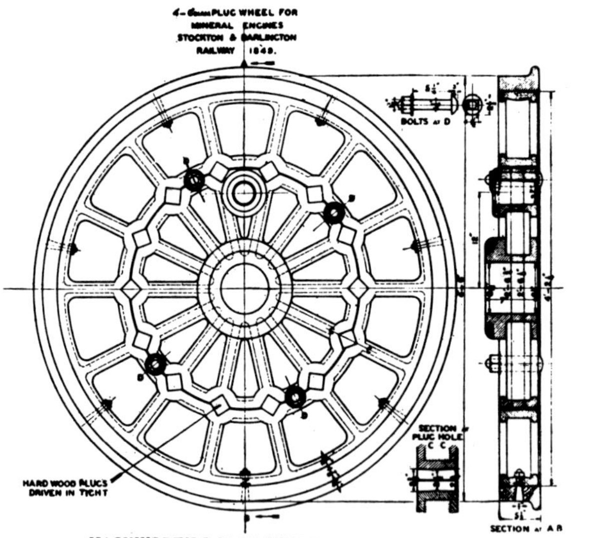

HACKWORTH'S PLUG WHEEL FOR LOCOMOTIVES 1849.

Fig.25 Inness's drawing of the Hackworth Two-part Cast Iron Plug Wheel, as later developed.

were damaged". Large number of waggons were derailed or collided with, let run 'amain' down inclines, sometimes deliberately. Racing between engine drivers was not unknown, siding against main line, and neither would give way when they came to the points!

Steaming capacity must have been a major problem with the single flues fitted originally, and it is not surprising that return flues were fitted very soon. What is surprising is that No.1 was in the end fitted again with a single flue before finally leaving railway service; maybe it just happened to be the only spare boiler available. Steaming capacity would probably not have been so critical on the pithead pumping duties over the next few years.

John Graham refers from time to time to the "square" engines, and it would seem that by this he meant the first four before they were rebuilt on more than four wheels.

Stockton & Darlington Railway Locomotives booked out from Robert Stephenson & Co. Ltd. between Sept. 1825 & April 1831

Works No.	R'y. No.	Name.	Ledger Date	Delivery Date (if known)	Price £.	Engine weight T. Dry Full	Comments
3.	1.	Locomotion	30-9-1825	16-9-1825	600	6½ 7¾	not named at first.
4.	2.	Hope	31-12-1825	1-11-1825	600	c. same	Do.
5.	3.	Black Diamond	17-4-1826	c. same	600	c. same	Do.
6.	4.	Diligence	18-5-1825	c. same	600		Do.
7.	5.	not named	19-3-1827 17-11-1827	c. same	550		Hackworth notes.
8.	7/6	Experiment	4-2-1828	c. same	550	6¾ (St.) 7⅜ c.8½	not named at first. (Hackworth)
	7/6	new frame	19-7-1828	c. same	108		6 wheels & springs.
16.	7.	Rocket	25-9-1829	c. same	550		6 wheels & springs. tender not mentioned. plus spares.
27.	9.	Globe	27-11-1830		515		S. & D.R. supplied wheels, axles, and boiler plates, and then completed engine.
28.	10.	Planet	23-11-1830		780		Second of Planet class.
36.	11.	North Star	9-3-1831		780		Planet type.

Data approximate. Listed in order of building.
n.k. = not known. n. = nominal.

No. Name	W.P. psi	Boiler Length Dia.	Surface Area F^2	Cylinders Bore Stroke	Wheels Dia.	Wheel -base	Weight Tons Cwt	Rebuilt by & Date
1. Locomotion	25	10'-0" 4'-0" +2x3" (dished	50/52 ends)	9" 24"	n.4'-0" a.3'-11"	5'-1"	6½ dry 7¾ full	– –
2. Bedlington/Hope	"	" "	"	" "	"	"	"	Kitching 1835 as rebuilt
3. Black Diamond	"	" "	"	" "	4'-6" 4'-0"	"	"	Lister 1838 (also 1844?)
4. Diligence	"	" 4'-4"	58	" "	"	"	Heavier; Tender not included	Hackworth, 1834
5. similar								
Flue tube c.18"	dia. x	10'-6" long						
– Chittapratt	"	10'-10" 4'-4"	n.k.	6" n.k.	4'-0"	n.k.	n.k.	

Details of No.2 "Hope" as rebuilt much as 'Enterprise' class, Chapter 5.

No.3 "Black Diamond" as rebuilt by Lister in January 1838, cost £280, 4 x 4'-6" dia. wheels, oblique cylinders 10⅜" x 18", boiler 10'-8" x 4'-0", firetube 6'-6" x 2'-4", tubes 103 x 4'-2".

No.4 "Diligence" as rebuilt by Hackworth in 1834, 6 x 4'-0" dia. wheels, vertical cylinders 14½" x 16", boiler 10'-4" x 4'-4", firetube 9'-8" x 2'-3", return tubes 90 x 1¾" x 7'-8", possibly similar to "Wilberforce".

Later Changes as Recorded:

De Pambour - 1833

1. Locomotion - fitted with 1 flue & 2 smaller return tubes

2. Hope - single flue tube

3. Black Diamond - " " "

4. Diligence - in pieces

Francis Whishaw - 1839

* Conflicts with other accounts and valuations.

1. Locomotion - 6 solid C.I. coupled wheels*, case hardened tyres. 24" dia. return tube flue. Vertical cylinders in boiler, 10" bore x 24" stroke.

2. Hope - 6 coupled wheels as above. 26" dia. return tube flue. Vertical outside cylinders, 11" bore x 20" stroke.

3. Black Diamond - 4 coupled wheels 4'-0" dia. Straight through boiler, 2'-3" dia. firebox. 107 x 1½" copper tubes. Sloping outside cylinders @ 45°, 10¼" bore x 18" stroke. Net weight 10 tons.

4. Diligence - 6 coupled wheels as above. 106 tubes (no other details). Vertical outside cylinders, 16" stroke (bore not given).

George Graham (in "Engineer) as at 1840

	Boiler L.	D.	Flue/tubes	H. surface Sq. inches	Cylinders Bore	Stroke	Wheels
No.1	12'-0"	4'-0"	Return flue 11'-0" x 2'-0" d. 11'-0" x 1'-4" d.	16,585	9"	2'-0" vertical	4 x 4'-0" d.
No.2	14'-0"	4'-0"	Return flue 13'-6" x 2'-1" d. 13'-6" x 1'-3" d.	20,356	11"	1'-8" vertical	6 x 4'-0" d.
No.3	10'-8"	4'-0"	Return tube (?) 6'-6" x 24" d. 103 x 1¾" x 4'-2"	35,538	10⅜"	1'-6" oblique	4 x 4'-6" d.
No.4	10'-4"	4'-4"	Return tube 9'-8" x 2'-3" d. 90 x 1¾" x 7'-8"	57,322	14½" x	1'-4" vertical	6 x 4'-0" d.

Les. Charlton (Industrial Locomotives of Durham)

No.1 R. Stephenson Works No.3, 1825, 0-4-0, vertical cylinders;
Sold to Peases' Lucy Pit, Crook, 1850, as pumping engine.

No.4 R. Stephenson Works No.6, 1826, 0-6-0, outside cylinders (as rebuilt);
Sold to South Hetton Coal Co., February 1841.

Stock Valuation Lists (£)

	No.1	No.2	No.3	No.4
1834*				
1838	320	650	150 (old material)	725
1839	280	750	680	760
1840	300 (with 1 tender)	720	580	680
1841	100 (& tenders)	750	600	-
1842	100	700	500 (coaching)	-
1843 (Dup.list)	100	650	-	-
1844 "	100	500	-	-
1849 (J. Dixon)	100	-	-	-

* No.4 Entirely broken up except for boiler shell.

No.1 as preserved : Boiler 10'-3" inside ends x 4'-0" dia.
Flue tube 22" dia., 58 Sq.Ft. surface.
4 x 4'-0" nominal dia. wheels.

Chapter 4. "Royal George" and Successors: 1827 - 1831

Timothy Hackworth knew very well that some improvement was necessary in the locomotives to enable the line to continue by steam traction; maintenance was excessive and the overall cost was higher than could be justified. He was aware that Stephensons were working on the same problem, but he had his own ideas about the solution, and when the Prussian engineers made their visit in early 1827 he had already prepared three possible designs for consideration. The Management Committee gave him permission to go ahead with his plans, and the result was "Royal George", which became No.5 (Fig.27), a locomotive that without doubt made sure that steam traction would continue and develop. It is generally agreed that he used the boiler of Wilson's "Chittapratt"; it was sent to John Wight at Lumley Forge, Durham, for enlargement. The length was increased from the original 10′-10″ to 13′-0″; Robert Young also stated that the diameter was increased at the same time, but this seems unlikely, as it would have been a much more difficult and expensive job. The Prussians said that the diameter of the other boiler they saw was 4′-4″, and this agrees with the known diameter of the boiler for "Royal George". Again, the Prussians said that it had a return flue,

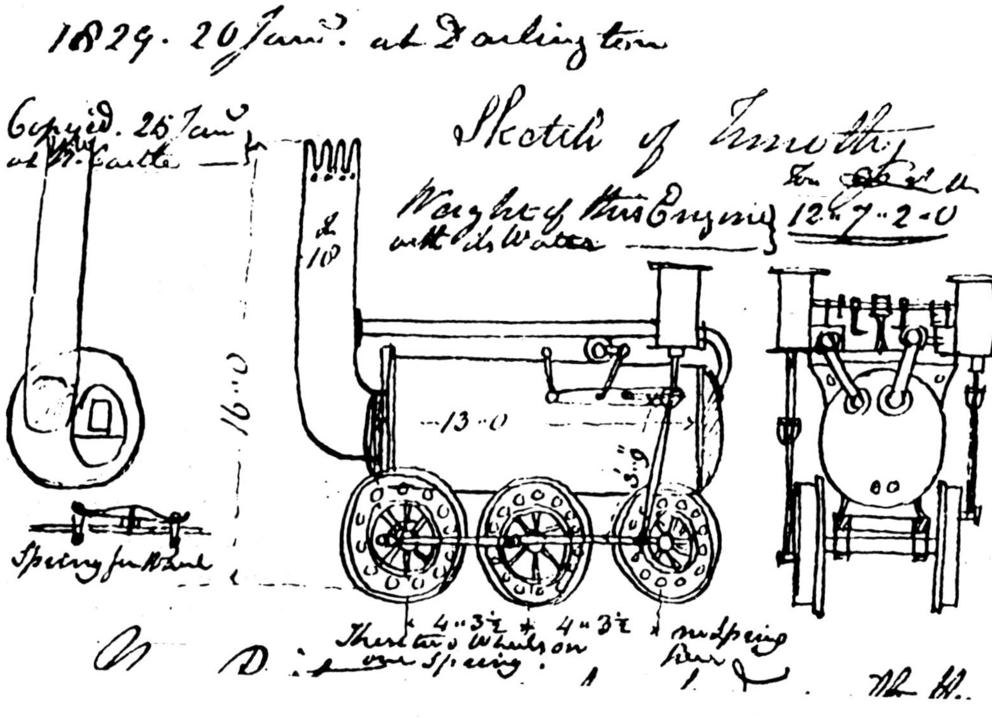

Fig.26 Sketch of No.5 "Royal George"; J. U. Rastrick, from his 1829 notebooks.

and it was unlikely that there had been any change of flue at that time, so although Robert Young said that it had originally had a straight through single flue tube, which was changed to a return flue when building "Royal George", it is probable that he was in error. Cost would have been a priority, and the more of the earlier engine Hackworth could use without major alteration the better. It would have been easy enough to lengthen the existing tube to suit the longer boiler.

I said in the last chapter that the original No.5 was an earlier locomotive, and at the time when "Royal George" was first put into use it had no number, since it was the only engine then to have a name. When the original No.5 exploded in March 1828, and was not repaired, "Royal George" became No.5, and retained the number for the remainder of its time with the railway.

There is no early drawing of "Royal George"; when seen and sketched by Rastrick in January 1829 (Fig.26), it had already been modified to some extent, and by then was certainly fitted with the heavier type of the two part cast iron plug wheels. Forward suggests that it may at first have been fitted with four wheels only when built in late 1827; he got this idea from Tomlinson, but does not offer any substantial proof. Certainly one of Hackworth's alternative designs was for a four wheeled engine, but it was much shorter, and could have been the basis for the later Rainhill "Sans Pareil". George Graham, who may have been there, says that it had six wheels when first set to work on October 27th. 1827 (29th. according to Hackworth's notebook). However, it could well have been fitted with the four wheels from "Chittapratt" to start with, plus two more similar; they would not have been difficult to copy. If we assume that Forward was right regarding the valve gear on "Chittapratt" there was a model for Hackworth to copy, or possibly even the original eccentric cross-shaft to use again, with a different drive (see Chapter 3 comments). The vertical cylinders chosen were easy of access for maintenance, and the faults inherent in them had still to be experienced; in any case, at the speeds usual for their purposes at the time the reciprocating effect and hammer-blow was minimal and could be ignored. Later on, with higher speeds and heavier engines it became a problem, and eventually the cylinder layout had to be modified, but at the birth of "Royal George" it did not matter very much. There is some indication that Hackworth had given the matter some thought, since the third alternative design had cylinders at the front of the boiler with connecting rods that "go out upwards", but it is probable that overall cost influenced the decision and simplicity is always cheapest.

Many small original features of this engine are in doubt. Robert Young claimed that a primitive feed water heating tank was fitted, through which part of the exhaust steam was passed, not unlikely, but none

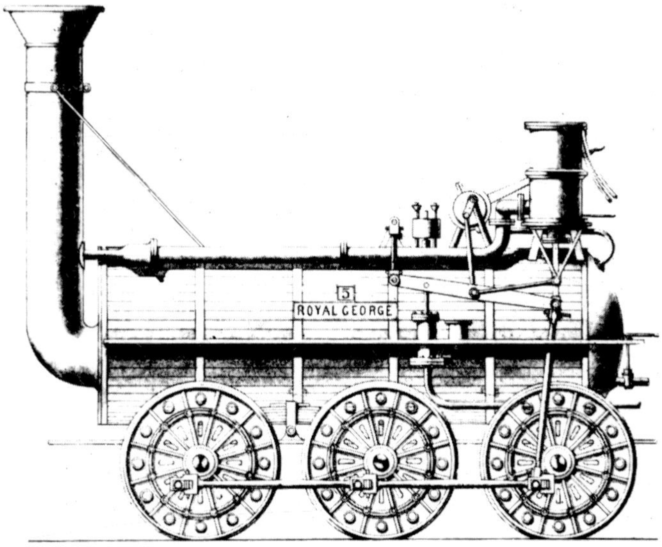

Fig.27 No.5 "Royal George", as running when seen by Rastrick.

of the known drawings show this. Alan Prior's drawing (Fig.30) shows details of the compensating springs fitted to the two front axles; it has been suggested by more than one writer that these were added later, possibly copied from No.6 "Experiment", but it is one of the points mentioned by the Prussian engineers and they saw the design for the engine in early 1827. Certainly it was there when Rastrick saw the engine, in 1829. Alan Prior believes also that the first two axles were carried in centrally pivoted tube bearings, which is certainly possible. However, the inconvenience and difficulty of removing a wheel-set from such a tubular bearing must have led very soon to the adoption of horns and conventional split bearing brasses.

One of the improvements was the use of Timothy Hackworth's own design of a spring loaded safety valve (Figs.28 & 29) in addition to the deadweight lever valves used generally up till then. The spring loaded valve could not so easily be adjusted or tied down by the driver, which greatly improved safety. It was not an original invention; it had been developed by James Fenton in 1812, with a spiral spring, and similar valves were in use on the Blenkinsopp/Murray rack locomotives already in successful service. As I said earlier, this was probably fitted in 1828.

His notes, although difficult to read, are roughly as follows: a note in the margin says "springs instead of weights", then "Weigh upon the safety valve leaver 22 lbs when the Boiler is charged at 50 lbs upon the square inch add 9 lbs will make 20·4 lbs more for a safety valve to prevent the Boiler to explode best done by 12 springs this view got in March 1828 it is safe." There is no way of telling whether he knew of Fenton's valve.

Certainly it was the first conventionally coupled six-coupled locomotive, and the direct ancestor of the long line of British 0-6-0 goods engines. (At least two other earlier six wheeled locomotives are recorded, one with chain and sprocket coupling between the wheels, and the other with cog-wheel coupling.) It had a blast nozzle in the chimney to improve the draught; again the principle was not original, as Trevithick had used the idea and recorded his observations of the results, but an improvement in that Hackworth used a venturi or nozzle to accelerate the exhaust steam and increase combustion. It was the first locomotive engine to have direct drive from the pistons to the wheel crankpins, and is also claimed to have been the first engine to have direct single lever control of reversing, without manual operation of the gabs. However, this feature cannot have been fitted at first, as it would not have been possible with

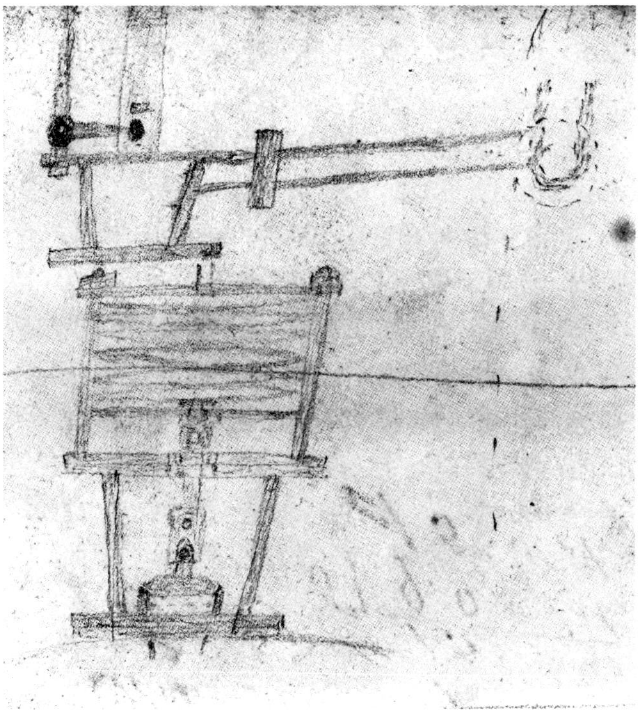

Fig.28 Hackworth's leaf spring loaded safety valve, from his 1827-8 notebook.

the original valve gear, and by the time that it was converted to four eccentric operation such single lever control was already being fitted to other engines.

This modification, with the eccentrics being fitted in a convenient position on the rear axle and rods led up to the gabs, replacing the drive from the parallel motion beams, was probably carried out after delivery of No.7 "Rocket", which had these features from the start, and at the same time the parallel motion piston rod guides were replaced by crosshead slide bars, also fitted to No.8 "Victory".

"Royal George" was a successful engine, and led the way in many respects. The increase in adhesion and hauling power afforded by six coupled wheels

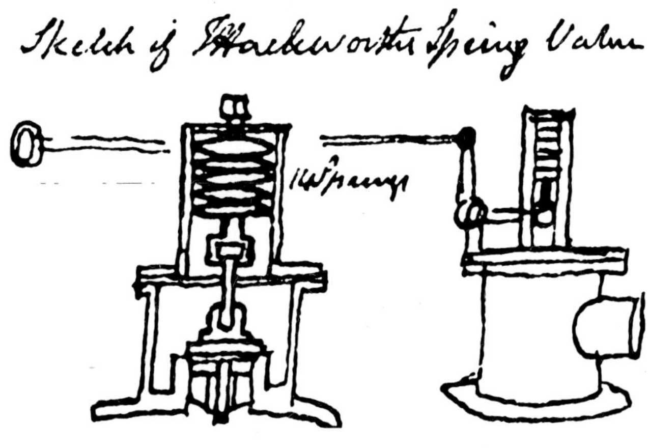

Fig.29 Hackworth's leaf spring loaded safety valve, as sketched by Rastrick in his 1829 notebooks.

and even the rudimentary partially compensated suspension fitted achieved Hackworth's object. According to George Graham, in one year (1838) it hauled 22,443 tons of coal over a 20 mile run, but he doesn't say at how much per trip. Unfortunately de Pambour's list was compiled at a time when this engine was undergoing major repair or rebuilding (second half of 1833, which possibly points to the date when all the main modifications were carried out) so the engine was out of action for most of it. The engine worked on the Stockton & Darlington Railway until the end of 1840, when it was sold to Wingate Colliery for £550, £125 more than its original cost. It continued to work there for several years, and was eventually sold to the Earl of Durham. A pity that more of its history is not recorded.

It seems to have suffered fewer accidents during its working life than most of the others, or maybe got them over earlier. Timothy Hackworth's notebook records that on December 13th. 1827, when quite new, the ring round the fire-tube fractured, and the whole front end plate was probably replaced later. The engine was derailed and badly damaged at Middridge Kilns in May 1837, but little else is recorded.

Joseph Tomlinson said that it ran uphill (Middlesbrough to Shildon) water-tender and driving end first. This may well have applied to most of the early engines, to save turning them at each end of the line. The firebox door was on the right, facing the chimney end, while the later return flue tube engines had it on the left.

Lists have been published showing another earlier No.5, named "Stockton", and Kidner referred to it in 1946, saying that it was Wilson's engine. I have found no substantiation of this at all. It is possible that confusion has arisen with the second No.4, "Stockton", of the Collier/Tory class, 1841 (q.v.). James Lowe, in his "British Steam Locomotive Builders", suggests that this was in fact the official name and number of "Chittapratt", but this is very unlikely, since there were then no official names or numbers, and "Chittapratt" had no Stockton connection whatsoever. C. E. Stretton, as mentioned earlier, did publish a very detailed but erroneous drawing of "Chittapratt" as No.5 "Stockton", which may account for the story.

"Experiment"

The Prussian engineers described in some detail an engine that they saw early in 1827 at the Forth Street Works, being completed for the Stockton & Darlington Railway, and according to them tested in the workshop. Most of their description coincides

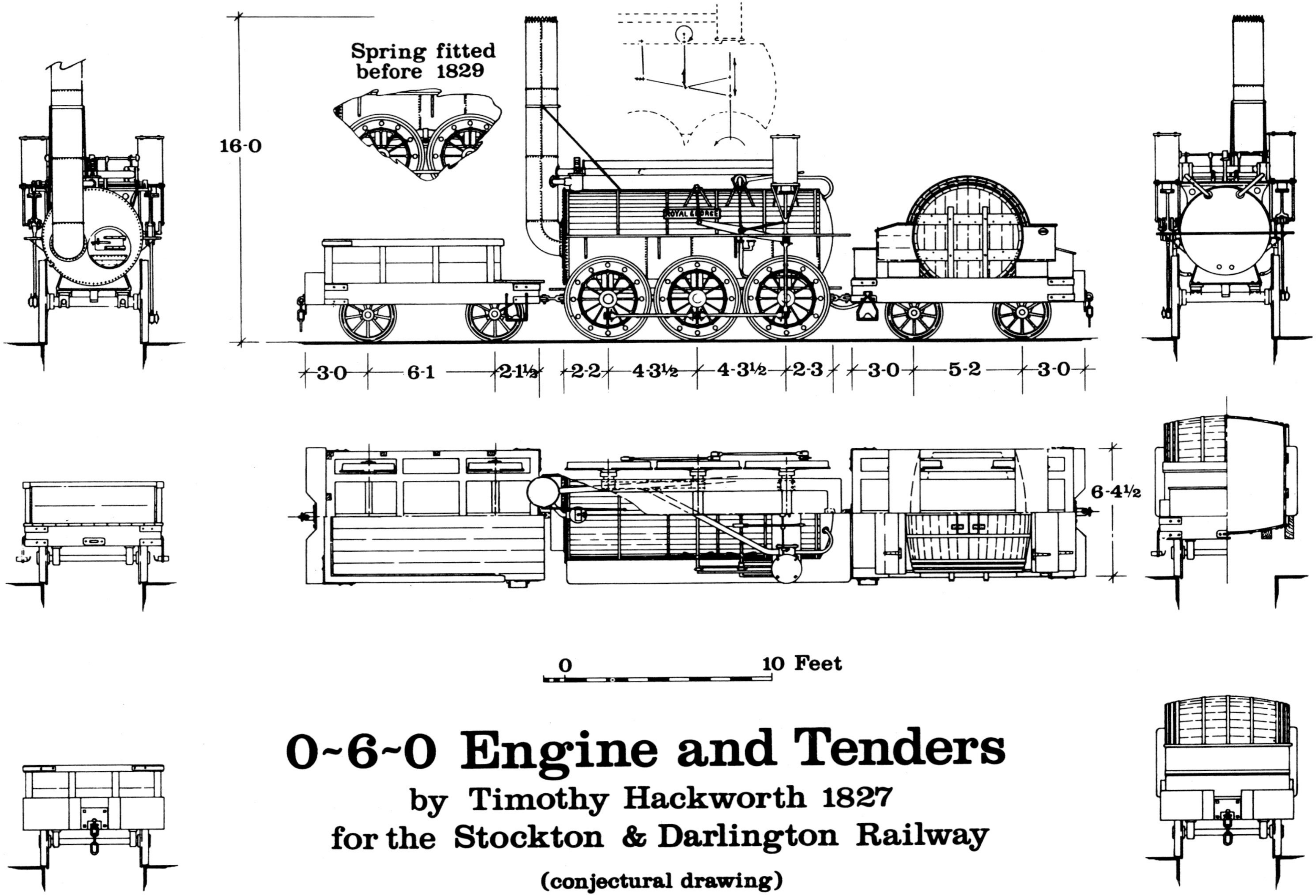

Fig.30 Drawings of No.5 "Royal George", Alan Prior.

very well with what is known about this engine. They did not actually say how many wheels it had, but inferred that it was four-coupled as the previous four engines; this again is in accordance with other accounts. It is not known clearly whose brainchild it was, since Robert Stephenson was still overseas in South America, and George had to some extent distanced himself by now from locomotive design, being heavily in demand for wider and more responsible duties and contracts, although it is known that he kept in touch, and probably was quite unable to let other people carry out all the design work. Joseph Locke may well have had a hand in it.

However, Hunter Davies, in his biography of George Stephenson, quotes from a letter to Robert dated 23rd. February 1827; "I think I told you about my new plan of locomotive it will be a huge job the Cylinder is intirely within the Bolier and neather Cranks nor cham (chain?) will be wanted. I have no fire door and I will not use more than the coals that has heather to been used. You will think that I have mistaken some ideas about this but I think not.

You may depend on it that if you do not get home soon everything will be at prefection and then there will be nothing for you to do or invent -". W. O. Skeat also includes this letter in his book, "George Stephenson - the Engineer and his Letters". This so obviously refers to "Experiment" that we must agree that George at least had a hand in the project.

The engine was delivered to Darlington a year later, Stephensons' works No.8, booked out on the 4th. February 1828 as S. & D.R. No.7. This was such a long time between apparent completion and delivery that the question must be raised as to whether in fact the engine that they saw may not have been the one booked out in March 1827, and believed to be similar to the first four. George Stephenson's letter, referred to above, would fit this idea quite well. The prices entered are both the same, £550 in each case. There is more confusion about the weights to muddy the waters even further. Forward more or less ignores this long completion period, but then goes on to claim that the No.5 engine unloaded at Darlington in November 1827 must have been "Experiment". If so why was it not booked out until five months later? I do not believe it. It was clearly numbered 7 at first, and "Royal George" would have been No.6 then had it had a number. Why would Hackworth call the November engine No.5, not just once but three times?

There was correspondence with Forth Street in late 1827 about the new engine proposed to be supplied by Stephensons; Hackworth was instructed to write asking for a guarantee that the weight of this engine would be less than seven tons dry, and a reply was received stating that the dry weight would be not more than 6 tons 15 cwt., and therefore the engine was on its way to Darlington for trial (Minutes 16th. Nov. 1827). The fact that it was on trial may account for the later ledger entry with the price.

Hackworth weighed this engine carefully at the end of January 1828, and found it to weigh c.8 tons 8 cwt. dry, this including an additional allowance of 4 cwt. for a grate door, frame, bearing bars and bridge, apparently not originally supplied. He assumed at least 1 ton for water, probably well on the light side. In view of this discrepancy, Warren believed that

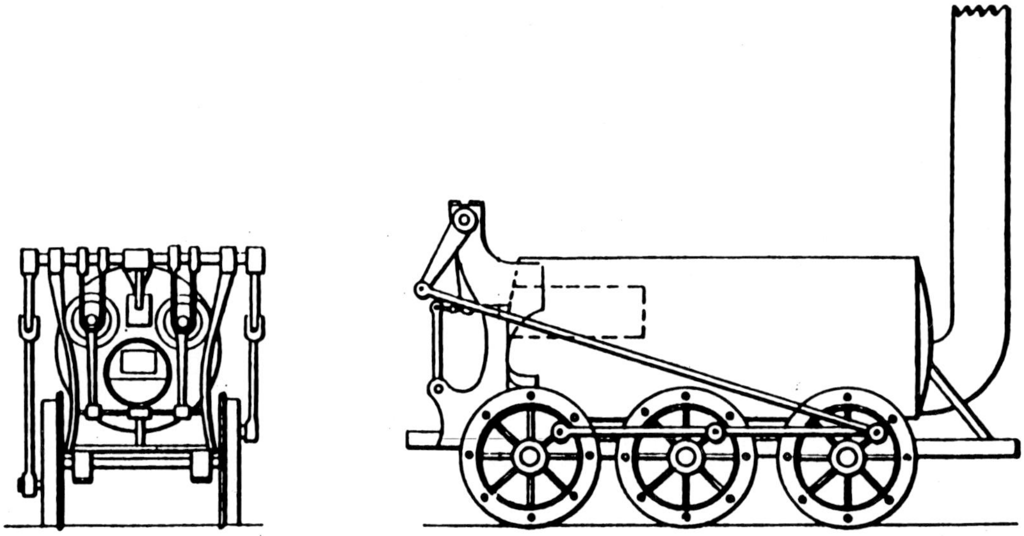

Fig.31 Drawing of No.6 "Experiment", as seen by J. U. Rastrick in 1829, prepared from his notebook sketch by E. A. Forward.

there had in fact been two different engines, the earlier one the lighter, both with the water filled firebars which were a feature of "Experiment", and that the first one was the No.5 that Hackworth reported exploded in March 1828. He then assumed that this engine was returned to Stephensons as unwanted, and this belief in a "Trial Engine" has persisted. Forward rejected this idea completely, claiming that Stephensons must have left parts off the engine when weighing it in order to present an acceptable weight to the railway. If Warren was right, this engine was never entered in their ledger. Of course, such a theory would explain the long delay between sighting by the Prussian engineers and completion, but it seems unlikely.

Parts for No.8 (the works number) Travelling Engine were entered in the works ledger on five occasions between June 30th. 1826 and December 31st. 1827. The last entry could have been for parts used after delivery. Forward mentions a toll charge of 11/- in November 1827, but I have not seen this. My belief is that the engine was not so far advanced as represented to the Prussians at the time, that the engine was not satisfactory at first and had to be largely rebuilt before delivery, which must be assumed to have taken place in mid November 1827.

The description given by the Prussian engineers includes the eight water tube firebars, each 4'-0" long x 1⅞" i.d., opening at the firedoor end (they called it the 'front' end) into a cast iron box connected with the boiler, this box containing access to the tubes for cleaning purposes, and at the other end into another 20" dia. tube running through the flue tube as far as the chimney and with connection at the top at the chimney end to the boiler. They also describe a crude feed heater waterbox under the chimney, heated by exhaust steam. Since they say that the wheels are arranged in the same way as the engines previously described it is assumed that there were then four only, and this is supported by later events, of course.

Obviously it was intended to increase efficiency, by improved heating surface and better combustion, and there was a clear attempt to improve appearance and to make the engine more compact. However, the concept of very indirect driving gear and the valve gear going through so many changes of direction and angles did not make for greater mechanical efficiency. E. A. Forward produced a very good pair of articles on it in the "Engineer" of August 1943, and I reproduce his conjectural drawing of how it may have looked on delivery (Fig.32). He assumed that the rear axle was carried in a pivoted tube like that used in the early engines.

It was soon found that the wheel loading was much too heavy for the track. Instructions were given in March 1828 that it was to be 'laid off' until altered, and it was taken out of action while Stephensons made a new six wheeled underframe with springs for it. Hackworth was at Newcastle discussing this in May 1828, when he also specified metallic pistons. At the same time it is evident that the watertube firebars must have been removed and a normal grate fitted. This new frame was sent to Shildon to be fitted, and took some considerable time to do; obviously it was later than May before the new parts were received. He recorded in mid May that the engine was in use, probably in emergency, also "Waggon men drinking & Quad. eng. cylinder bottom fails". By October that year it was back in regular service, and a committee minute at the end of October records that it was a great success, calling it "the engine on six wheels", and Hackworth was instructed to see that all the engines were so

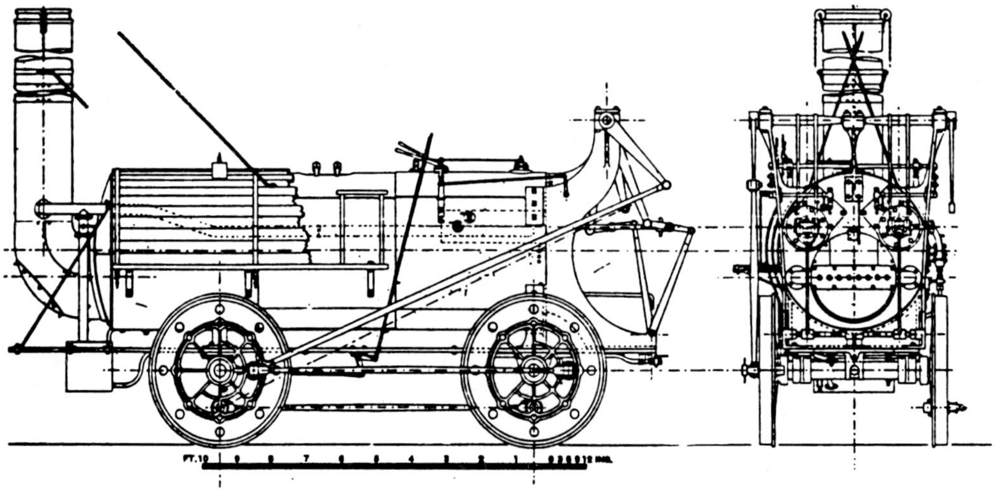

Fig.32 Conjectural drawing of No.6 "Experiment" as delivered; E. A. Forward.

modified as soon as possible, presumably with springs as well.

J. U. Rastrick saw this engine at Stockton in January 1829, and made a sketch of it mounted on six wheels, but with the valve gear unchanged, showing most of the working parts in great detail (Fig.31); he also made some notes which, together with the Prussians' description gave a fairly complete picture of it in its early days.

It seems possible that Hackworth suspected that the change to a conventional grate would increase the weight at that end, since he took the trouble to weigh the engine very carefully twice before this was fitted, the second time at each end. This may have been a factor in the decision to put it on six wheels. We have no information concerning the reason to remove the original water filled firebars; it has been suggested that one exploded, as might have been expected, but there is no clear evidence of this.

Dendy-Marshall says that it was given the number 7 at first, and Hackworth's notebook confirms this ("No.7 loco, viz. the Quadrant Engine"), when it was weighed at the end of January 1828, and that John Nicholson from the builders was altering it early in February. Forward says that "Experiment" was actually sent to Darlington for trial before "Royal George" was put into regular work, which seems reasonable. When the name "Experiment" was given to it is not clear, but must have been by 1833.

No.6 "Experiment" was just that, and certainly not a particularly successful one at first, although it obviously gave quite a lot of good service after its modifications. There are many mentions of it in the Graham notes. John Graham says that it was one of the two engines involved in a race at Oak Tree Batteries in September 1831, No.9 being the other, and that No.6 actually overtook No.9, but not by enough to avoid a collision! The same month it was damaged in a collision with a waggon at Simpasture, and had many arguments with horse drivers. In

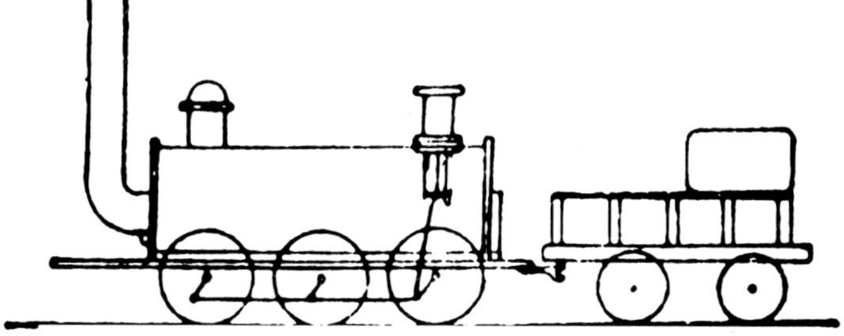

Fig.33 Theodore West's sketch of No.6 "Experiment" as rebuilt later.

November 1838 it was reported to be worn out, and Graham recommended that it be rebuilt differently, "According to the specification produced".

The main original features of construction included two horizontal cylinders buried in the boiler at the driving end, in a return to early Trevithick practice, and as Brunton had used in his 1813 'walking' engine. Lagging was not yet very efficient, still consisting of strips of wood laid on the surface of the top half of the boiler, and heat loss was severe, so such a means of keeping the cylinders hot was still thought useful. The main drive was through what in effect were large bell-cranks mounted on a shaft by brackets above the cylinders, and then back down to the front pair of wheels by means of very long connecting rods. These cranks were not very far apart, and the grate was at the driving end, so firing the boiler must have been a very hazardous business; it is not surprising that the engine was fitted with a return flue before long so that it could be fired from the front end, as well as to increase heating surface again (but not until after Rastrick had seen it in 1829). The enginemen called it "Old Elbows", for obvious reasons.

Forward's reconstruction of the valve gear assumes eccentrics on the front axle, the rods coming up horizontally backwards and then through smaller bell-cranks up to gab gear on top of the boiler, which in turn operated the slide valves through a series of rods and cranks to a shaft on each side of the backhead above the cylinders. The lost motion must have been considerable. It is believed that two part cast iron plug wheels were fitted right from the start. The piston rods were guided by parallel motion, the levers of which had to be shorter than mechanically desirable because of the restrictions in geometry due to the limited space available, and it is known that this gave trouble; Hackworth is on record as saying that the packing had to be renewed after every trip, and even assuming that he meant the piston rod packing, as seems probable, this would have been very wasteful in labour and materials, let alone the loss of time involved with the engine out of service. The wheels were coupled together by simple outside rods, in spite of George Stephenson's letter.

One of the most original features of this engine was the attempt to increase the heating surface by means of tubular fire bars connecting the sides of the flue tube and forming the grate, through which bars the water was expected to circulate. Water quality, of course, was very unreliable, and such tubes would very soon choke up with inevitable results; they burned through and burst. There is on record a letter from Harris Dickinson, manager at the Forth Street

Works, claiming that the bars could not burn away since they were filled with water, or if they were choked up there was no water to explode. This letter was addressed to Edward Pease; what he thought is not recorded, what Hackworth thought may well be imagined. Stephensons are believed to have used similar water filled firebars in the two engines supplied to France in March and April 1828, these also proving to be a failure and soon removed.

One minor puzzle is the statement by the Prussians that the chimney was fitted with an eight foot sliding extension when they saw it, the unextended height already amounting to 16-0".

"Experiment" was eventually rebuilt into a Hackworth type engine with outside vertical cylinders, and this may be what John Graham refers to in the 1838 note quoted above, but it may have been earlier. West has illustrated it (Fig.33) with such vertical cylinders, as fitted to "Royal George" and "Victory", although he said clearly in his 1886 paper in Middlesbrough that this was an approximate outline from a published description. Robert Young also said that it had been rebuilt in this form, and there seems to be no reason to disbelieve them. However, a drawing which has gained wide circulation was published by C. E. Stretton, firstly in the old "Railway World" magazine in the 1880's, and then in his book, showing "Experiment" with the Stephenson type of sloping cylinders, but this is almost certainly guesswork and quite erroneous. I have a strong suspicion that Stretton and others at the time had confused "Experiment" with the slightly later "Rocket", an 0-6-0 locomotive which certainly did have such cylinders, derived from "Lancashire Witch". Robert Young says that "Experiment" was sold in 1839, and it is recorded as being sold to T. Hackworth that year for £120. I have also been told that it was taken by Timothy Hackworth in part exchange for No.10 "Auckland" of the 'Tory' class, built at the Soho Works that year, which is probably another way of saying the same thing. In fact, the minutes record that it was sold to Nicholas Downing, Thomas Hackworth's nominal partner in the Soho Works. I imagine that then the engine was rebuilt again and resold elsewhere. Whishaw's list later that year did not include it, and the No.6 was given to "Dispatch" in September 1839.

One other experimental design may be mentioned briefly here. Warren shows a drawing of an engine designed by Stephensons just after "Lancashire Witch" and described as Darlington Engine "Quickspeed" (Fig.34). This was a four-coupled locomotive with diagonal cylinders of the type then in use but with a sun and planet drive instead of the usual connecting rods driving straight on to crankpins. Both Watts and Murray had developed such a sun and planet drive, to overcome the patent granted to Pickard for use of the crank. It could be arranged so as to give a 1:2 rotational drive, and this may account for the name "Quickspeed". Warren says that it appears to be arranged for expansive working, but the valve drive is not shown. There is no record that it was ever built, let alone delivered to Darlington.

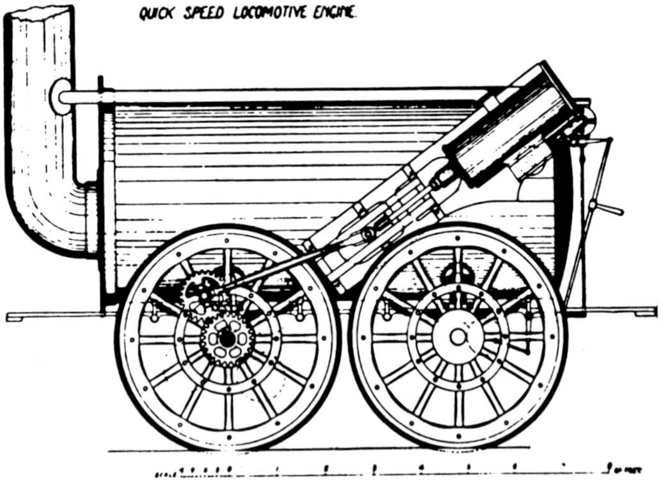

traced from the original (undated) drawing.

Fig.34 R. Stephenson "Quickspeed" locomotive for Darlington.

It seems possible that Stephensons' works No.18, which does not appear in the ledger, might have been reserved for this engine, if it had been built. This comes immediately before the Rainhill Trials "Rocket", works No.19.

Comparisons

The age of the six-coupled mineral engine had come, and Stephensons entered the field as well. No.6 "Experiment" had not been so very successful, and so they went back to more or less established practice. Robert Stephenson was now back in charge, and had been developing the "Lancashire Witch" line, leading to the Rainhill Trials "Rocket" a little later that year, and the "Northumbrian" class. The next locomotive delivered to Darlington was No.7, the first known to be named "Rocket", probably to Hackworth's specifications but of Stephenson design. It was delivered in the autumn of 1829, and according to Robert Young, was waiting for Timothy Hackworth when he returned, disappointed, from the Rainhill Trials.

It was Stephensons' works No.16, cost £550. 0. 0d., and was entered as 1 Loco.Engine on springs with 6 wheels complete, on 25th. September 1829. 1 spare Fire tube etc. and 1 chimney were supplied with it.

The price is the same as that for both "Experiment" and the previous engine.

It was almost certainly developed from the design which Warren called Darlington Engine "A" (Fig.35), which shows clearly the arrangement of the springs. A very good drawing of it, prepared at Shildon in 1831, together with the next two locomotives, is still in existence and fairly well known (Fig.36). This drawing was damaged, and some details are missing; it was also laid out in a rather odd way to fit all three of the engines in, and Warren produced and published a modified version of it with all the engines the same way up and the missing details reinserted, which is very handy, and I have made use of it.

"Rocket" was fitted with six coupled Hackworth type plug wheels of the twelve spoked version, and with the Hackworth developed eccentrics on the rear axle, with long curved rods to clear the dished end of the boiler, going up to the gabs on top between the cylinders. The boiler was fitted with a Hackworth type return flue tube, and was about 10'-0" long by 4'-4" in diameter, very little different to the first engines except for a much increased heating surface area. The driving gear and valve gear was Stephenson type, with Robert Stephenson's diagonal cylinders and much improved valve gear derived from "Lancashire Witch" and the "America", and also as used very shortly thereafter for the second "Rocket", of much greater fame. There was some provision for expansive working by means of a cut-off valve in the main steam pipe, this valve being of mushroom type and operated by a cam on the trailing axle.

There was also a third engine with the name "Rocket", a little later on, an 0-4-0 'Samson' type for another railway.

John Wesley Hackworth has recorded that No.7 was an unreliable engine and tended to start off by itself, Jem Stephenson, its driver, running after it cursing! But I imagine that this sort of occurrence was not so uncommon in any case, valves not being as well designed and steam tight as they had to become later. John Graham says that it was fitted with a wire gauze spark arrestor in August 1832, and others must have been so fitted as well, because in May 1833 he says that these traps have been a failure. The number of fires started in lineside plantations and pasture were excessive, and speed restrictions were introduced in order to try and prevent them. In October 1840 he said that "Rocket" and "Coronation" were not worth repair. No.7 "Rocket" was sold to Binchester & Westerton Collieries near Bishop Auckland, probably at the end of 1840,

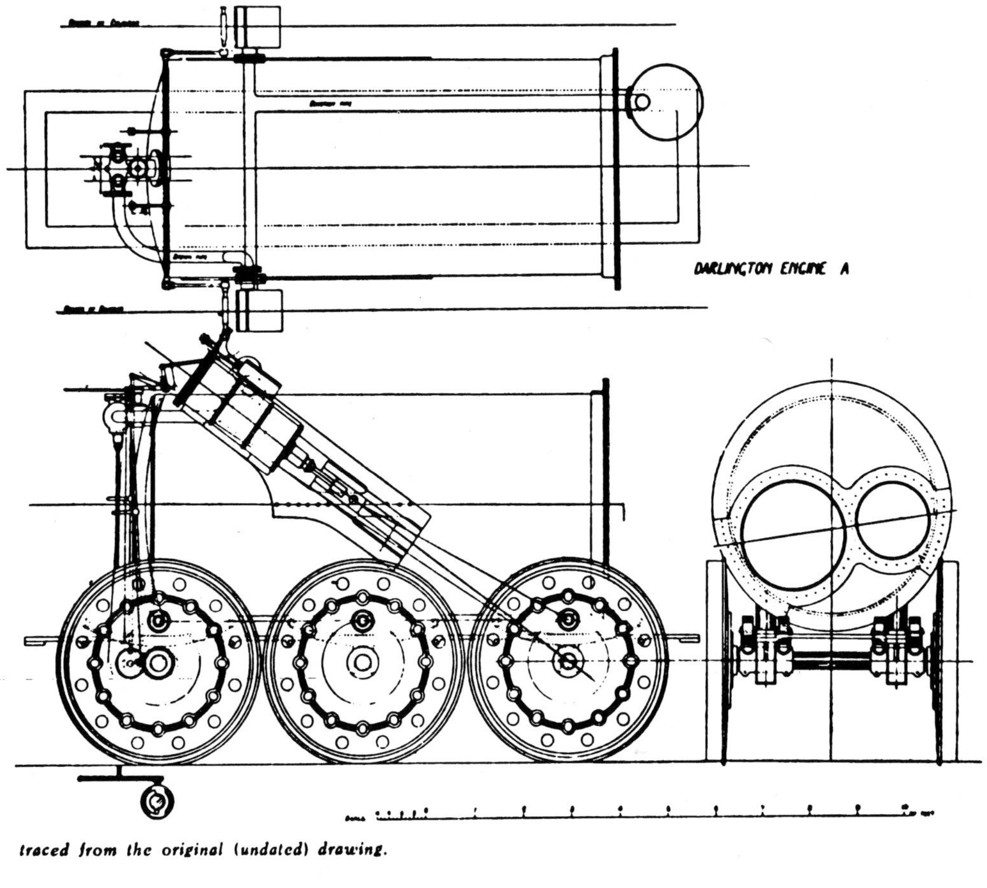

Fig.35 Forth Street Works drawing of Darlington Engine 'A', design basis for No.7 "Rocket".

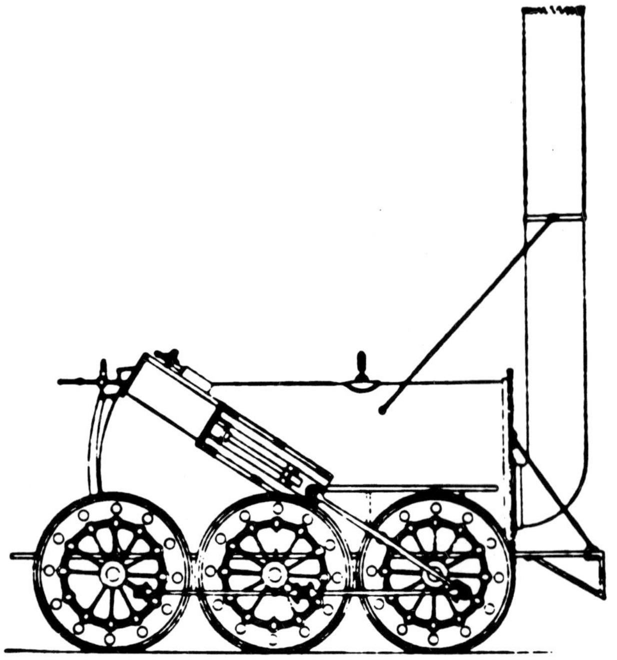

Fig.36 No.7 "Rocket", from J. G. H. Warren's adaptation of the Shildon Works tripartite drawing.

although Craven gives some details of an advertised auction sale, held by Thos. Bowman at New Shildon on March 2nd. 1841, at which the following two second-hand engines were offered:

One six-coupled engine with 14½" x 16" vertical cylinders.

One six-coupled engine with 10½" x 20" inclined cylinders.

The second could have been "Rocket", and the first may have been No.13 "Coronation".

Stephensons are known to have built only one other similar conventional six coupled engine at this time, their Works No.17 for the Baltimore & Ohio Railway, which never reached America. It was almost lost on its way to Liverpool for transhipment, in the 69 ton burden sloop "Fraser" which went down off Peterhead on the 28th. December, 1829, but the engine was salvaged just in time. According to Michael Bailey, the remains were scrapped, and did not include the boiler, which was to have been shipped separately. It was several years before they built another.

The next engine was No.8 "Victory" (Fig.37), whether to celebrate the famous ship or what is not known, as there seem to have been no major battles in 1829. It was a Shildon production and very similar to the modified "Royal George", having a boiler 12'-6" long by 4'-4" in diameter, vertical outside cylinders with slide bar guides, and eccentrics, rods and gabs as fitted to "Rocket". Whishaw said that it had been "remodelled" at Shildon, and as I said in Chapter 3. there is a strong belief that it had been rebuilt from the 'missing' fifth engine, but possibly using a larger boiler ordered from Stephensons in September 1828. It does seem to me, however, that it would have been much easier to have used the new boiler directly for a new engine.

Hackworth was at Forth Street on 19th. September 1828, ordering this boiler, together with a double tube, spare tube, two chimneys and roofs(?), a stay in three pieces and rivets, plus six travelling engine wheel rims hooped with iron and turned. Two such wheels would have been enough had he been going to use the remains of No.5. At the same time he ordered a shaft for the Etherley Incline engine. The locomotive parts were entered in the ledger on 4th. November 1828, at a total of £192. 3.11d. Many other engine parts were booked out to the S. & D.R. that month, including a further four such wheel rims, probably for normal replacements. I can think of no other engine for which this boiler may have been intended, and would have thought that the remains of the boiler from the first No.5 engine were not reusable.

It seems to have been a successful engine, seldom in trouble, was rebuilt in 1833, which is what Whishaw must have been referring to, and went on working until it was replaced by the second No.8 "Leader". No.8 "Victory" was sold to the Wingate Grange Coal Co. in February 1841 for £550 plus a further £50 for the tender, according to John Graham, where it joined "Royal George".

The main disadvantage of these engines with

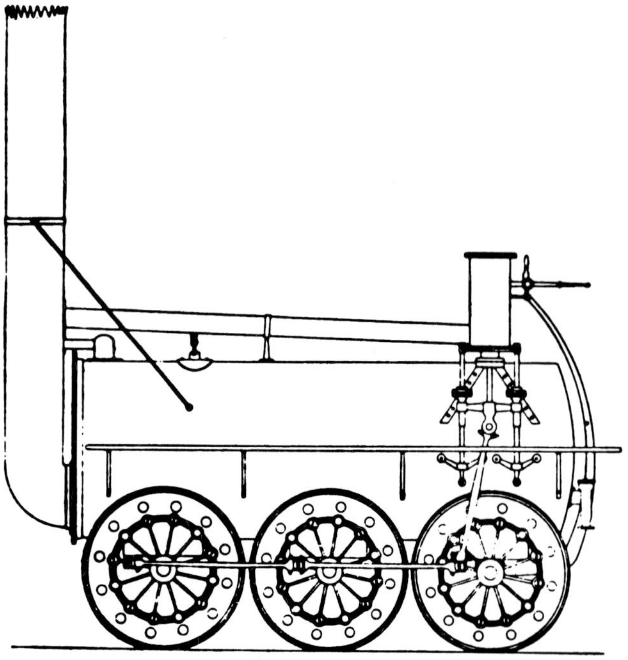

Fig.37 No.8 "Victory" from the same drawing as Fig.36.

vertical cylinders was that the rear (driving) wheels could not be sprung, since the pistons drove straight down on to the cranks, and there was not enough clearance in the cylinders to allow for sprung suspension. As I have said, "Royal George" was fitted with springs arranged on each side across the first two axles and acting as compensating beams. "Victory" may have been also so arranged; there is no way of telling from the drawing. "Rocket", with inclined cylinders, could be and was sprung on all wheels. John Graham mentions later a comparison between "Victory" and the "Director" class, which I shall come to again. This report to the Committee, on December 14th. 1834, refers to "those travelling without a Crankshaft and only 4 of the Wheels on Springs". He then mentions the "Engines on the Victory Plan" in this connection, but, of course, the engine could have been considerably rebuilt by then. Nevertheless, it is an indication that some form of sprung suspension was fitted (see Chap. 5, "Enterprise" etc.).

Then came what is said to have been the first passenger engine on the line, No.9 "Globe" (Figs.38 & 39). The Middlesbrough Branch was about to be opened and the railway was considering changes to the horse drawn coach traffic run mainly by contractors, with the intention of taking over all the passenger traffic themselves and hauling the coaches by steam, more for the sake of orderly operation than for profit, although it was to be a few years more before this came to realization. In February, 1830, the Committee instructed Hackworth to proceed with the construction of this engine, which he had been designing since the year before, and it was ordered from Stephensons in March. Presumably the Shildon Works was fully occupied with other work in hand, and in any case could not have coped with the driving axle, which was cranked to work with inside cylinders; or at least this is what everyone seems to have believed. But take note of the Stephenson ledger extract below. Robert Young maintains that this was the first engine designed with such inside cylinders, and that Stephensons deliberately held back the construction so that they could copy this feature and produce their own 'Planet' class first. This seems unlikely, although there was certainly unexplained delay; in any case Bury was already working on a similar design, which seems to have been the first to run, if not in service, and Gurney had patented the idea and used it in his steam driven coach in 1828. Also, ingenious as the "Globe" was, the cylinders were not at the best end, for various reasons. Young records that one of Stephensons' officials had some very odd remarks to make about the principles involved and the effective length of lever at the cranks, which he seemed to think worked on a different principle to the usual outside cranks. Obviously this did not stop them from taking up the idea themselves.

Fig.38 No.9 "Globe", from the same drawing as Fig.36.

The Stephenson ledger entry, works No.27, on 2nd. November 1830, gives "For 1 - Loco.Engine delivered here. (i.e. ex works, the customer to pay the freight charges) They finding wheels, axles, not fit up and boiler plates for body of boiler. £515. 0. 0d." I assume that this means that the railway supplied the plates for Forth Street to build into a boiler. There is a suggestion that the plates concerned came from another and scrapped engine; this could have been the first No.5. It would have been an unwise thing to do, and if true may explain why the rebuilt boiler failed again a few years later, but little would then have been understood about stresses built in by such occurrences, and the plates would have been reworked to some extent. Where the wheels came from is unclear. Stephensons were supplying wheels all the time to other customers, but there are none shown for the S. & D.R. at the time, so they may have bought them elsewhere. I doubt if they had their own wheelwright for wooden wheels. Even more odd, Stephensons did not at the time supply the axles, one of which was cranked. If they supplied them separately there is no ledger entry to cover it. I would suggest that the railway ordered both wheels and axles from a firm such as Lumley Forge or the Bedlington Ironworks direct, or possibly Kitchings, and assembled the engine at Shildon.

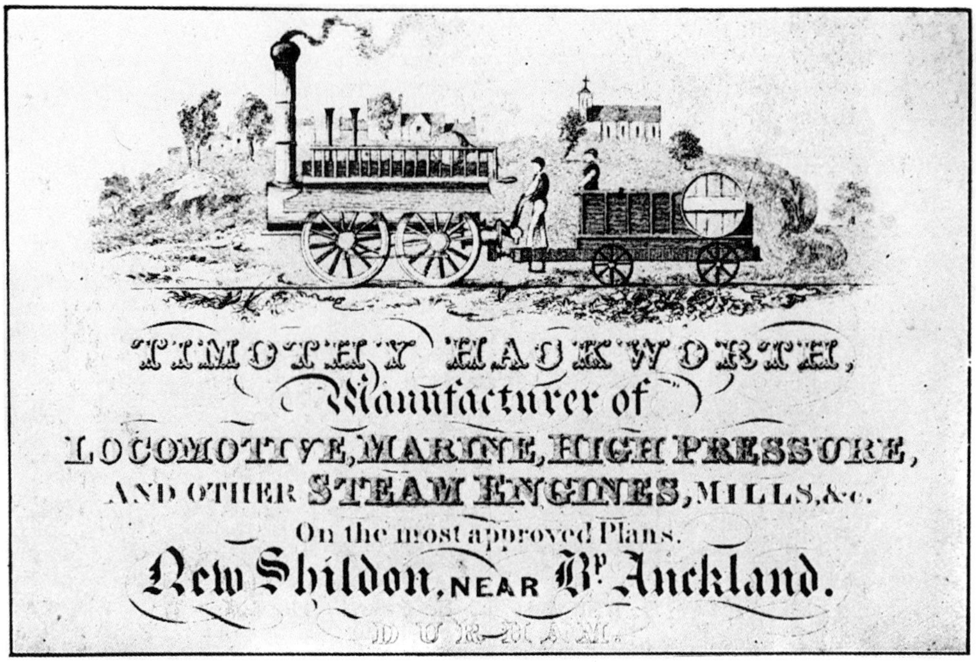

Fig.39 Timothy Hackworth's business card, showing No.9 "Globe".

There are no details of the valve gear, which must have worked from eccentrics on the driving axle, but the drawing of this shaft in Robert Young's book show none; one would have expected them to be between the cranks. However, Young did say that there were no reversing levers in use before this locomotive, an unproveable statement, but it does indicate the type of valve gear that must have been used. Two levers are clearly shown in the locomotive drawing referred to above, and in that by Zerah Colburn, with a regulator lever above. The name of the engine came from its spherical steam dome/dryer, mounted above the boiler at the rear and with the main steam valve behind it. L. T. C. Rolt, however, connects it, in the enginemen's opinion, with the "Globe" tavern at Shildon, their favourite haunt! It had a boiler 9'-3" long by about 3'-6" diameter, fitted with a single flue tube and a series of cross tubes in a spiral formation; Inness says that there were 120 of them to start with. The usual problems beset them and the design was modified considerably, Tomlinson says to a shortened flue tube with 80 x 4'-0" firetubes, and John Graham noted that "Globe" was in for retubing on 11th. July, 1834. Nevertheless, it was a very interesting early attempt at a multitubular boiler; Young says that it was the basis for the 'Galloway' patent boiler, and I remember somewhat similar auxiliary marine boilers not so many years ago.

The four wheels were wooden, presumably for lightness, and maybe also because they were larger than the standard so far used, having a diameter of 5'-0". The whole engine was given a decorative appearance, and Hackworth used it as the heading of his business card (Fig.39). It was said by Young to be much faster than any of the others so far in use on the line, and to have reached 50 m.p.h. on more than one occasion, with great steadiness in running for such a short wheelbase, but John Graham's notes do not confirm this; see the report of the race with No.6 above. Of course, in that incident No.9 was probably hauling chaldron waggons, not coaches, a duty for which it was not designed, and we do not know how many. The engine was delivered in October, with a comment from Harris Dickinson that they considered it to be deficient in steaming power; sour grapes, maybe, but John Graham has recorded one occasion in December 1831, when this engine blocked the Yarm Road at the Urlay Nook crossing and the driver blamed it on shortage of steam. Graham also carried out a survey in September 1833 on the suitability of this engine for 'coaching' duties, together with No.1 and the two 'Planet' type engines, and said that in his opinion the "Globe" was not fast enough; neither, of course, was No.1!

It was ready in time to take the first train into Middlesbrough in late December 1830, the driver being Johnny Morgan, and it carried on with varying success for another seven years. It was in trouble from time to time, with the crank axle broken at Aycliffe Lane in August 1834, and was not included in John Graham's list dated October 7th. 1836, so must have been out of action at the time. In January 1838 it exploded at Middlesbrough through shortage of water, and was never repaired. Strangely enough this explosion is not recorded by Graham. In the 1838 stock list it is shown as broken up.

The explosion took place just outside what is shown on the Middlesbrough 1853 town plan as a goods station, and must have been close to the north end of North Street, as several windows were damaged in the Exchange building, still existing on that corner, and an adjacent ironmonger's shop was also damaged.

There is a report in John Graham's notebooks for March 16th. 1832 that "Carter the Builder had a Roap stritched across the Line at Simpasture Bridge to low for the Engines to pass and the Globe Engine Came up and Broke the Chimney off with it; . . . Damages to be deducted from his acct."

The same thing happened to "William IVth." a little later, with a different builder, one Simpson, at Coddling Bridge, noted April 18th. 1832.

Three other drawings are known, one by Colburn in 1861, one a sketch by R. W. Kidner, published in 1946 and therefore derivative, and one in the Science Museum archives, on the sheet of drawings that I attribute to C. E. Stretton, in App.III. This is quite detailed, but omits the decoration on the splasher valances shown by Colburn and on the Shildon drawing, and the only safety valve shown is a deadweight valve on top of the spherical dome. Colburn's drawing also shows only this valve, whereas Timothy Hackworth's business card and Kidner's sketch both show two tall Stephenson type valves in line on the boiler, close behind the chimney. None of these drawings show springs, but there was no reason for not fitting them, and for the speeds of which this engine was capable it would have been stupid not to have fitted springs.

There appears to be a primitive bar frame, and the Science Museum drawing shows buffers and a coupling chain and hook attached to the front of this frame; none of the others include this, and I suspect Stretton's imagination again.

The tender was wooden and four wheeled, with a large water butt mounted athwartships at the rear, and a coal space in front of it. The driver stood on the tender front platform, there being no working footplate on the engine.

Craven says that the remains of the engine were sold to Wm. Lister with parts of other scrapped engines to be used in building a new four wheeled engine, but the next No.9, supplied by Lister, was the 0-6-0 'Tory' class mineral engine "Middlesbro", in August 1839, and in any case the agreement was modified (see Chapter 5). There is a mention in the minutes of a Lister contract for "Globe" and "Planet" in September 1837, so it is possible that he was operating them at that date. There is also a note of 15th. November 1833, in a Hackworth report, that "he had got the "Globe" engine painted ready for coaching from Darlington to Shildon".

In 1925 the "Locomotive" magazine published a special Centenary Supplement: this did not contain much about the Stockton & Darlington Railway, but it did have a short article on early railway liveries, which mentioned an 'original' drawing of No.9 "Globe" in colour. The chimney and boiler were black, the side valances blue and black with black panels and the decorative features in green and yellow, the steam dome yellow and the wheels blue. E. F. Carter repeated this information in his "Britain's Railway Liveries", in 1952. So far I have been unable to locate this drawing.

George Otley's second Middlesbrough town plan of 1830 has a very nice little vignette in the lower right hand corner, showing a view of the Port Darlington coal drops, Hackworth designed, on the bend of the river which later became the 'Ironmasters' district', with two locomotives apparently waiting to go in, one with eight chaldron waggons and the other with a passenger coach. They are both four wheeled engines, but it is not possible to identify them clearly and no tenders are shown. I imagine that it was prepared long before No.9 was completed, and although Otley was still a servant of the railway at that time, it was in another department altogether, and the vignette was probably done by a junior draughtsman in any case.

In order to assist with the newly developing passenger traffic, the 'coaching' trade, the railway ordered two 'Planet' type engines from Forth Street, no doubt also with the idea of comparing them with "Globe". The first one was entered in the ledger on 23rd. November 1830, and was actually the second of the class built, closely following the one delivered to the Liverpool & Manchester Railway. This was No.10, "Planet", Stephensons works No.28, costing £780. 0. 0d. The second one, No.11 "North Star", works No.36, was booked out on 9th. March 1831, at the same price, so at that point should have been similar to the first. Robert Young very disparagingly says that they were used mainly to drive the Committee members around in a single coach, but it is clear from John Graham's notes that they were in fact used largely for the increasing passenger traffic, and most of the time on the Middlesbrough branch, where they proved to be highly suitable until they finally wore out and were disposed of!

Both seem to have been standard 'Planet' type engines but are listed with differently sized wheels. There is no factual drawing available to show just what they looked like, but Warren is quite definite

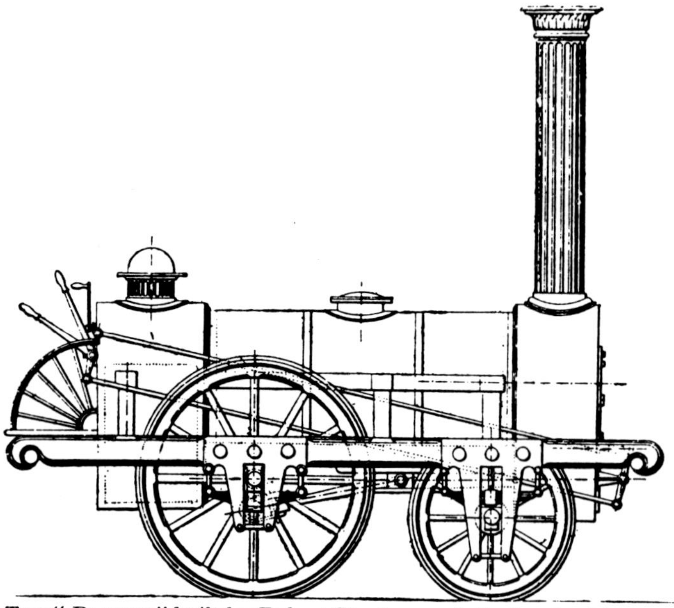

THE "PLANET," built by Robert Stephenson & Co. for the Liverpool & Manchester Railway, 1830.

Fig.40 Stephenson 'Planet' type 2-2-0 with springs below the frames, as illustrated by both Robert Young and R. H. Inness, from Nicholas Wood.

that they were of the type with the frame above the driving axle (Figs.40 & 41). He says also that the second one was of four-coupled type, but both Young and Inness disagree with this, and the 0-4-0 'Samson' type, often spoken of as a 'Planet', did not appear until early 1831, and its price was then £1000. 0. 0d., so there is no doubt about the type concerned here. It is of interest that the replica "Planet" built recently at Manchester by Michael Bailey and John Glithero, which must have been very exhaustively researched, has the springs above the frames. They say, however, that it is not intended to represent the prototype but simply a typical example of the class. The other drawing comes from Nicholas Wood's 1834 edition of his "Practical Treatise on Railroads", so there may well have been two or more different versions built.

The first of the two, No.10 "Planet", not a very original name, since it followed the first of the name, had 5'-0" driving wheels, while No.11 "North Star" had smaller wheels and boiler, according to Inness. He may have used the wheel sizes of those fitted later, and there seems no reason whatsoever for a difference in boiler size when built. Maybe the names were left to the builders, who tended to be rather repetitive; there was more than one "Planet" and at least three different "North Star"s, including the one that got the Great Western Railway on to its feet after the early failures. I have already mentioned the various "Rocket"s.

However, Craven supports Inness about the difference between the two engines, saying that No.10 had 11" x 16" cylinders and 5'-0" driving wheels, while No.11 had 9" x 16" cylinders and was clearly a much smaller engine. All the same, he does say that the driving wheels of No.11 were also of 5'-0" diameter. It is possible, of course, that he took these data from the Inness articles in any case.

De Pambour gives the above numbers in his 1833 list. The number 10 was given to "Auckland" in 1840, by which time "Planet" had disappeared, so this seems to be in order.

The No.11 was given to "Beehive" in 1839, when "North Star" was still on the list; it remained there until 1841, but without a number (see earlier comments). None of the three were numbered in the 1838 list.

No.11 "North Star" suffered bad damage in a collision with waggons in October 1832. No.10 "Planet" was similarly damaged later, and a minute of 19th. February 1836 says that Mr. Nesham was to pay half the cost and the company would pay the rest. Inness says that "Planet" was withdrawn about 1837, and according to Young was then used as a stationary engine pumping water at Shildon, while "North Star" was sold to the contractors for the Great North of England Railway. But Inness reverses these details, and the lists support him, as also does John Graham. The 1838-44 stock valuation lists show "North Star" in use up to 1839, but the next year it was put aside at a value of £50 only. "Planet" had disappeared from the list in 1839, and seems to have been sold to the contractor William Grieves, of Newcastle, together with the later engine No.18 "Shildon", in April that year, following which they passed into the possession of the Great North of England Railway, and they were later sold again in August 1841 to Wm. Lister for £100 each. John Graham's October 1836 list does not include them, following his report a year earlier that they were both in very bad repair, especially the frames. No.11 "North Star" was advertised for sale in July 1837, unsuccessfully, and in December 1837 Graham reports that "North Star" was installed at the Black Boy Branch to pump water, where it stayed until its crankshaft broke in February 1842; temporary repairs were effected, but the original stationary engine was soon to be replaced.

Craven quotes a note dated 8th. November 1833 (apparently in a Hackworth notebook) in which he reports that the "Star" Loco's coke box has failed. This must have been "North Star" at that date. He goes on, "As it is intended to run a Coach or Coaches from Darlington to Shildon . . . , T.H. thinks the Co. will be justified if they put in a copper coak (sic) box & pipes suitable. The "Star" engine would then be fit for the Shildon Coaching Depot. N.B. The engine is

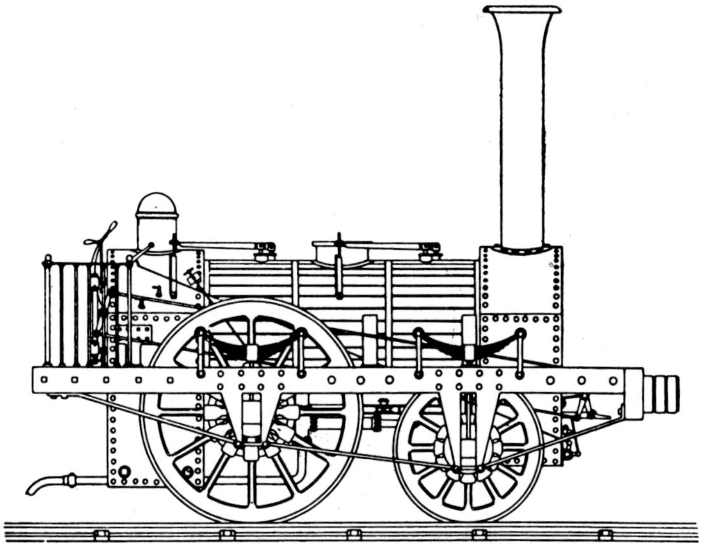

Fig.41 Stephenson 'Planet' type 2-2-0 with springs above the frames, as illustrated by J. G. H. Warren.

at Mr. Kitching's manufactury." Brass tubes for this engine are mentioned in the minutes for that day.

Young also says that the original wooden wheels were replaced in 1835, but does not say with what. George Graham mentions a report by his father in January 1835 that No.10's wooden wheels were in a dangerous state, recommending that they be replaced by cast iron wheels, which apparently was done. He also recommended that No.11 be similarly treated (see remarks about wheel sizes above). From the notebook of an engineer named Adcock at Kitchings' foundry it appears that they worked on both engines in January 1835, repaired the fireboxes, and may then have fitted the iron wheels, of malleable iron with flat spokes.

Data as early condition in general, and approximate.
n.k. = not known. n. = nominal.

No. Name	W.P. psi	Boiler Length Dia.	Surface Area F²	Cylinders Bore Stroke	Wheels Dia.	Wheelbase	Weight Tons Cwt
5. Royal George	52	13'-0" 4'-4"	141	11" 20"	n.4'-0"	2 x 4'-3½"	12 7 + tender not included
6. Experiment (rebuilt 1838?)	45+	13'-0" 4'-0"	n.k.	9" 30"	n.4'-0"	2 x 4'-6" (as 0-6-0)	7.4 light, engine, no tender. 8 8 full
7. Rocket	c.50	10'-0" 4'-4" + 6" (domed end)	n.k.	9" 24" (Inness) 11" 20" (Dendy-Marshall)	4'-0"	2 x 4'-3"	n.k.
8. Victory	c.50	12'-0" 4'-4" + 6" (domed end)	n.k.	12" 22"	4'-0"	2 x 4'-3"	n.k.
9. Globe	n.k.	9'-3" 3'-6"	n.k.	9" 16"	5'-0"	5'-3"	n.k.
10. Planet	n.50	6'-5" 3'-0"	n.k.	11" 16"	L.3'-1" D.5'-0"	5'-0"	c.8 engine only
11. North Star	n.50	6'-0" 2'-9"	n.k.	9" 16"	D.c.4'-8"	n.k.	n.k.

"Royal George" firetube 12'-6" x 2'-1" dia., return tube 12'-6" x 1'-6" dia.

Chapter 5. The Double Tender Classes: 1831-46

The 'Majestic' & 'Director' Classes

Mineral traffic was heavy and lucrative, and there were still not enough locomotives to handle it all. What there were needed constant maintenance, and were frequently out of service. Timothy Hackworth had gained enough experience and confidence to be able to persuade the Management Committee to embark on a programme of expansion. He had two somewhat similar but different in detail designs in view, both developed at Shildon, and got these approved in late 1830. In the case of the first type the parts were ordered both from Stephensons and, for the first time, from R. & W. Hawthorn, whose Forth Bank Works were virtually next door to Stephensons. The parts for No.13 "Coronation" were apparently all produced by Hawthorns, and are said to have been their first locomotive, although the Works No. was 123. Craven believes that Hawthorns actually built this engine entirely themselves, but Inness does not agree.

These parts began to be received during 1831, and construction began at Shildon, where the whole class was assembled, according to Inness. The first of this class was originally called "Bedlington" (see earlier) and given the number 2. This was, of course, the name of the ironworks with which George Stephenson had been closely associated for some time. However, William IV had come to the throne in 1830, and on 2nd. August 1831, the number and name of this engine were changed to No.12 "Majestic", followed by No.13 "Coronation", the only one of the class that has been illustrated, by a sketch of Theodore West's (Fig.43) and by a drawing in the Science Museum (Fig.42), probably based on a Stretton drawing, neither of which can be regarded as very accurate. But they are all we have, and so they will have to do.

Robert Young says that they were called the 'long engines', and they were certainly tending that way. He says also that although they all looked alike externally, the number and arrangement of the boiler tubes varied, but gives no details of these. De Pambour does give some details, but seems to be confused between the two classes (see page 12). Certainly it was not a return tube arrangement, as the English translation states; if West's sketch is to be believed at all they were fired from one end and the chimney was at the other, the driving end, which is believed to have been the normal leading end. The original French edition mentions tubes only, there is no mention of 'returning', and with this word removed his descriptions make more sense. The double tender arrangement was retained, because they were fired at the other end from the driver. Inness, however, gives more details of the boiler arrangements, apparently taken from Whishaw, which seem to have been of a true multitubular straight through type with a long firetube, 9'-0" x 2'-6" dia., with 106 copper tubes of 1 " dia. x 4'-0" long, making a total boiler length of 13'-0" plus a smokebox. However, Craven says that the 'Director' class were similar, repeating de Pambour's error, and this is certainly not true. The heating surface area is variously reported, by Inness as over 300 sq. ft. and by Ken Hoole as 71.1 sq. ft. in the firetube + 194.22 in the tubes, a total of 265.32. The other engines of this class were No.14 "William IVth" (spelling variable), No.15 "Northumbrian", No.17 "Lord Brougham" and No.18 "Shildon".

The cylinders, 14½" bore x 16" stroke, were carried on outriggers on each side of the driving platform, vertical and driving downwards on to a cranked intermediate shaft, from which coupling rods connected with all six wheels. This enabled the engine to be sprung on all three axles, without the problems mentioned in the last chapter arising from direct drive from the piston on to a crank on the wheel, but it introduced other complications, and more bearings to maintain. The piston rods, as shown in the drawings, have crossheads working on slide bars, but Joseph Tomlinson said that they were guided by parallel motion. This is supported by a letter dated July 1839 from Timothy Hackworth to the Committee, saying that the cost of repairs to the "Shildon" had been increased because they (the Committee) had replaced the parallel motion by slide bars.

The wheels were the standard 4'-0" two part cast iron plug type, although Joseph Tomlinson says that some of these engines had wheels with a cast iron boss and wooden spokes (similar to the Bedlington wheels fitted to No.3?). He may not have seen the engines before 1837, so such wheels may have been replacements. All six engines seem to have been in service before the end of the year.

Theodore West's sketch shows a weight fitted at the rear end, "to steady the engine", and there does seem to have been some trouble in this way. Whether

Fig.42 No.13 "Coronation"

this cure worked is another matter. There were many derailments and accidents; John Graham reported a collision between "Coronation" and "Lord Durham" at Urlay Nook in May 1833. "William IVth." was particularly unlucky, it seems. It was derailed at Aycliffe Lane in April 1832 by stones in the points, and hit the depot, causing damage, its safety valve was stolen in June that year at Croft Branch end, it was damaged by a block on the rails in February 1835 and it was derailed again at Aycliffe Lane in July by someone's carelessness in leaving a crowbar standing upright in a block near the line, which had already been passed by at least two other engines, but fell as No.14 passed it, between the first tender and the engine. In September 1838 it needed major repair, and only Lister would quote for the job.

No.13 "Coronation" was badly damaged in October 1833 at Bowesfield Lane by a collision with waggons, and was 'beyond repair' in October 1840. No.18 "Shildon" and No.13. "Coronation" are mentioned together with No.23 "Wilberforce" in November 1833 as 'coaching' engines, not the duty for which they were designed. No.18 "Shildon" was retubed completely in June 1834. No.12 "Majestic" is mentioned in October 1835 as working the Middlesbrough goods service.

Obviously they were not as successful as hoped, and in the 1838-44 stock lists were not valued very highly. No.15 "Northumbrian" and No.17 "Lord Brougham" were broken up in 1838, and the parts used as spares for the others, while No.18 "Shildon" was sold to H. Greaves the same year. The K. Hoole notes call him William Grieves, contractor for the G.N. of E.R., and say that he paid £640 for "Shildon" together with "Planet". Wm. Lister bought them back in August 1841 for £100 each (see earlier). No.13 "Coronation" was the only one still in service, less than nine years since being designed, but

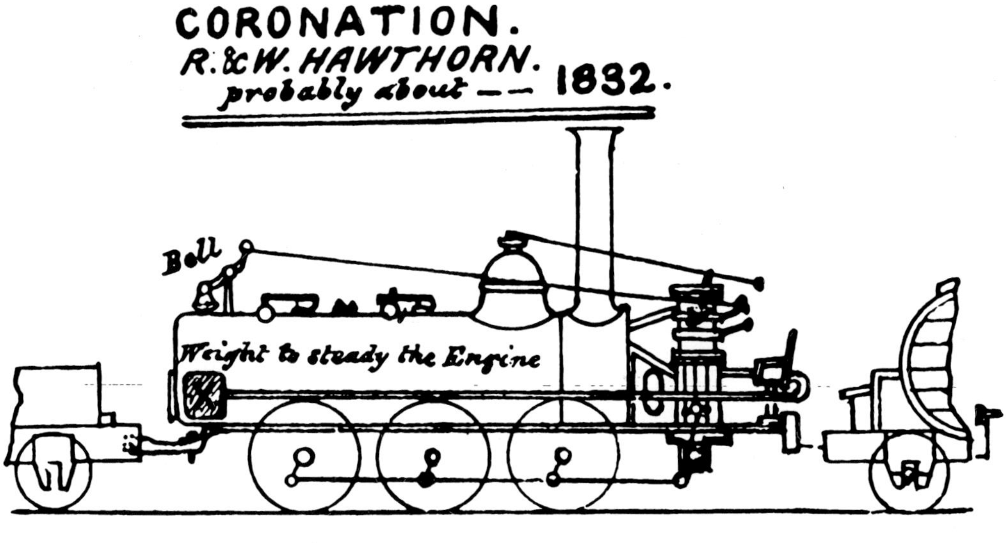

Fig.43 No.13 "Coronation", as sketched by Theodore West.

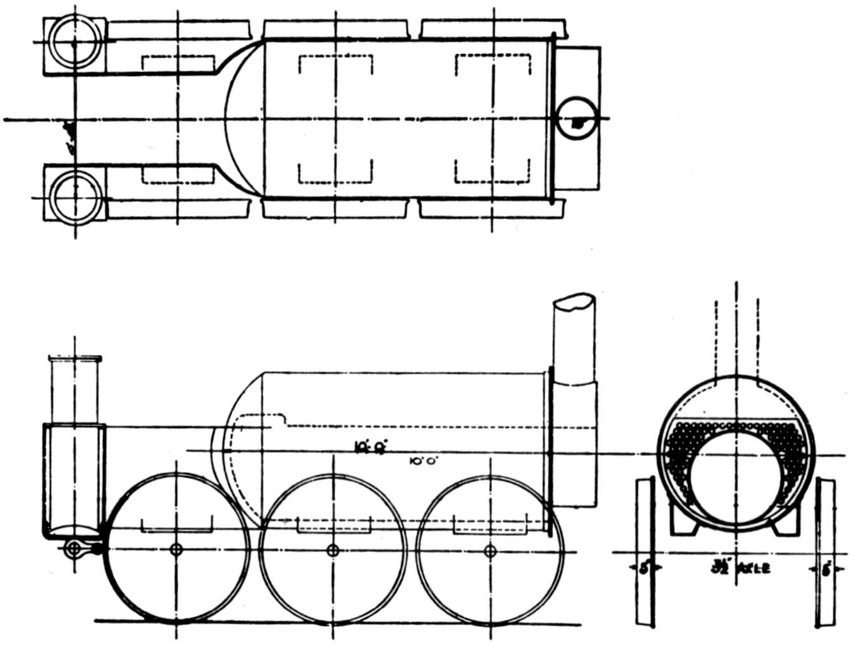

Fig.44 Shildon Works tender guidance drawings for the 'Director' class.

disappeared from the list in 1841, ten years after delivery, sold at an auction at Shildon for £100, apparently to Timothy Hackworth.

It seems from the minutes and Craven's notes that an agreement was entered into in September 1838 with Wm. Lister to take No.12 "Majestic" and No.14 "William the IVth." in part exchange for a new engine, together with the remains of No.9 "Globe", but an offer was received in April 1839 of £560 for the pair as they stood from William Hutchinson, a railway contractor, of Poppleton, and instead Lister was given a contract for two new engines. It is probable that he did take the "Globe" remains.

The other class, which became much better known and lasted a lot longer, was basically a Shildon design (Fig.44), and a drawing dated October 1831 is shown in Warren's book as well as by Robert Young. This drawing was sent to Forth Street for detailing and finalisation, with the result that many people have believed that in fact the class was of

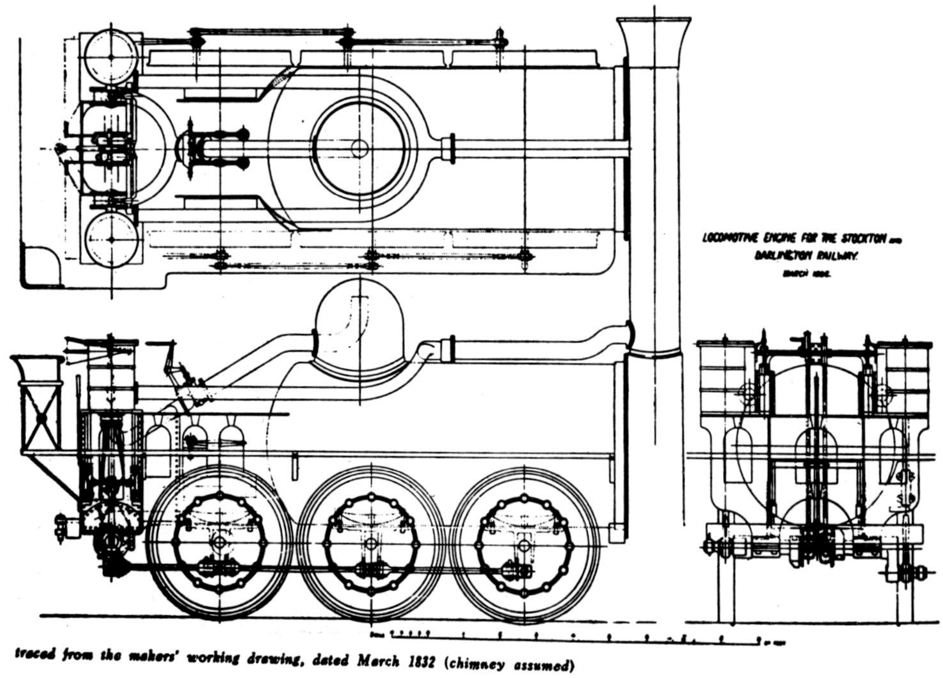

Fig.45 Tracing of R. Stephenson & Co. drawings for the 'Director' class, prepared from the Shildon drawings. Chimney assumed.

Stephenson design. Probably they modified the valve gear, but it is difficult to be sure. Surprisingly, the Forth Street detail drawings (Fig.45) show that there was a return to two eccentric gab gear, two years after the 'Planet' type was introduced, although there was a long lever for some purpose between the eccentric rods.

These engines had multitubular return tube boilers designed by Hackworth, which increased their steaming capacity greatly, and were the first such used on this line. The tubes were arranged round the main flue tube from a chamber at the rear end returning to a horse-shoe shaped smoke box over the firing door; the Shildon drawing showed 91 of them, while de Pambour said 104 return tubes. This enabled the size of the boiler to be kept to a minimum, 10'-0" long by 4'-4" diameter, with a heating surface of 500 sq. ft., making a much more compact engine than the previous class. The chimney could now rise from the centre of the smokebox (Fig.48), instead of being set to one side as with the return flue boilers. The cylinders were again mounted on outriggers, but much closer to the wheels, and there was not a lot of room for the driver, although he had a seat. He would have had to stand in the middle to work the regulator and the valves. The piston rods appear to have been fitted with crossheads working on guide bars. All six wheels were fully sprung, with the bearings working in hornblocks.

Craven's notes of the boiler tubes show great variations between engines, so may refer to the state after later retubing.

Both classes had steam domes, with the main steam pipe taking comparatively dry steam from high inside them, and the regulator valve was fitted in the pipe outside, between the dome and the valve chests.

The first locomotive of this class was delivered in 1832, No.16 "Director", from Stephensons. John Graham says that three new engines were received in September, and the other two would have been No.19 "Darlington" from Hawthorns and No.20 "Adelaide" from Stephensons. No.21 "Earl Grey", from Hawthorns and No.22 "Lord Durham" from Stephensons were delivered later the same year, and No.23 "Wilberforce", the best known of them, from Hawthorns most probably the next year, although John Dixon gives 1832. The three built by Forth Street Works were given the first three works numbers, 1, 2 and 3, in a new series, probably because they had started a new ledger.

The last engine, No.23 "Wilberforce", was well photographed, and two of these (Figs. 46 & 47) have survived, one of them believed to have been taken by John Dixon himself. As a result the class became better known later as the "Wilberforce" class, especially since this engine seems to have outlasted all its classmates. This photograph, showing it without its tenders, was later copied by Bleasdale, who improved it somewhat, the original being rather blotched, and it was used by Joseph Tomlinson as one of the illustrations to his Presidential Address to the Institution of Mechanical Engineers in 1890. It must be one of the very earliest railway photographs known (see also 'Liveries', Chapter 1.). A better known one is that taken outside the first Shildon

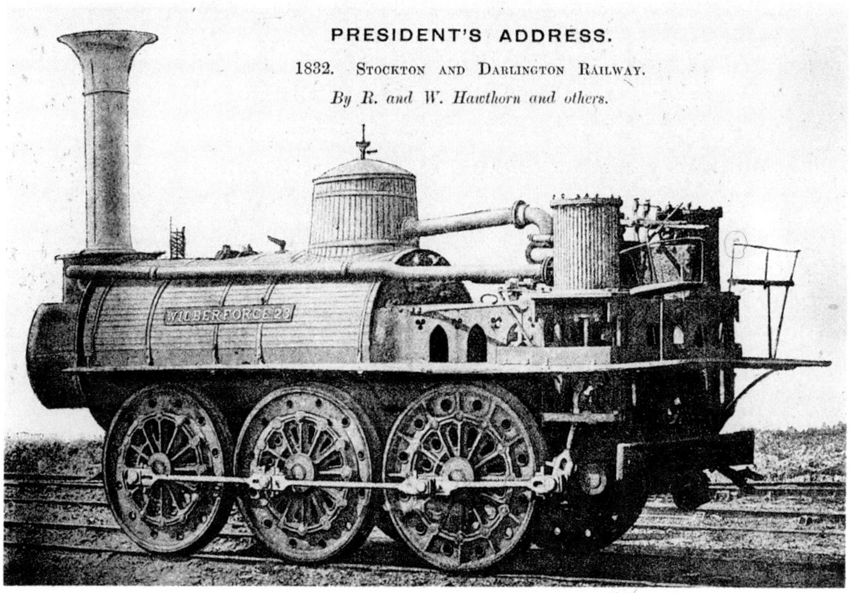

Fig.46 Plate made from early photograph attributed by Zerah Colburn to John Dixon. This was retouched and enhanced, and used by Joseph Tomlinson for his Presidential Address to the I.Mech.E. in 1890.

"Wilberforce," Stockton and Darlington Railway, 1832.

Fig.47 Engraving of No.23 "Wilberforce", by J. Swain of the "Engineer", from an early photograph taken outside the Shildon roundhouse.

roundhouse, built in 1854, and this was also used as a postcard by the Locomotive Publishing Co. Dr. Peter Holmes has published a photograph of Hownes Gill Viaduct just after completion in 1858. showing an engine crossing hauling a train of trucks; it is difficult to be sure, but it could just be one of the "Director" class, which by then can only have been "Wilberforce" or "Earl Grey".

No.20 "Adelaide" survived as a stationary engine for several years, firstly at the Saltburn Brickworks, probably late 1858, and later driving a mortar mill on a site just opposite the new Saltburn Station when the Zetland Hotel was being built, about 1860. There is a well known but not very sharp photograph of this.

The multitubular boiler apparently infringed on a patent of Robert Napier's, and Stephensons pointed this out in late 1831, but after amicable discussion with Napier, they agreed each to continue manufacture without argument; it would appear that in fact Hackworth had probably been first with it, but he seldom bothered to patent anything unless pressed by his son. Marc Seguin, of course, had produced and patented a true multitubular boiler in France, and fitted it to his first locomotive in 1829, probably just beating Robert Stephenson (see Chap.3).

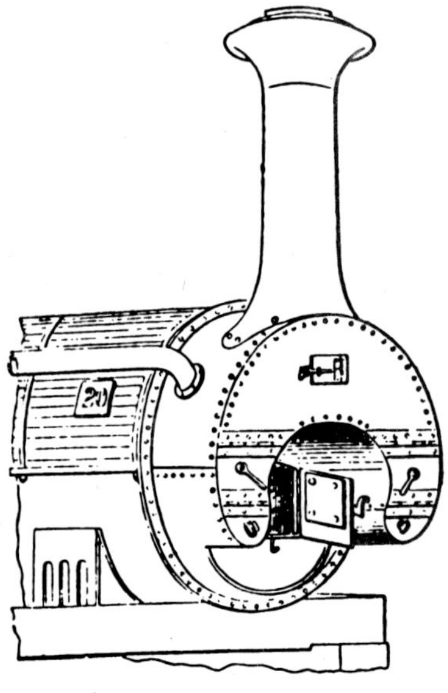

Fig.48 Colburn's sketch of the smokebox of the 'Director/Wilberforce' class.

These engines seem to have had less trouble than the preceding class. No.20 "Adelaide" was derailed at Myers Flatt Battery in October 1833, because of a broken rail, and "tumbled the whole way to the bottom of the battery and lay with her wheels up!". "Adelaide" was derailed again at the foot of the Black Boy Branch inclined plane in July 1835 by wrongly set points. No.22 "Lord Durham" is mentioned above; it is reported in December 1837 with a broken piston and cylinder, and in November 1839 with the flue tube 'down'. No.23 "Wilberforce" had a brass cup (lubricator?) stolen from its cylinder at Stockton Coach station in June 1834; they were having a lot of trouble there with theft and hired a police officer to keep watch. "Wilberforce" was retubed in July 1834, having done 14,000 miles with the last 'one', so it must have been the main flue tube, not the smaller return tubes. "Wilberforce" and "Earl Grey" were in use as 'coaching' engines in 1834/5. The committee complained to Timothy Hackworth in January 1835 about the "disgraceful state when Earl Grey was used Tuesday last for the coaching and merchandize train"! A leak in the boiler of No.23 "Wilberforce" was reported to have delayed a 'coaching' train in mid November 1835.

The "Director/Wilberforce" class was certainly better than the "Majestic" class, and held their value

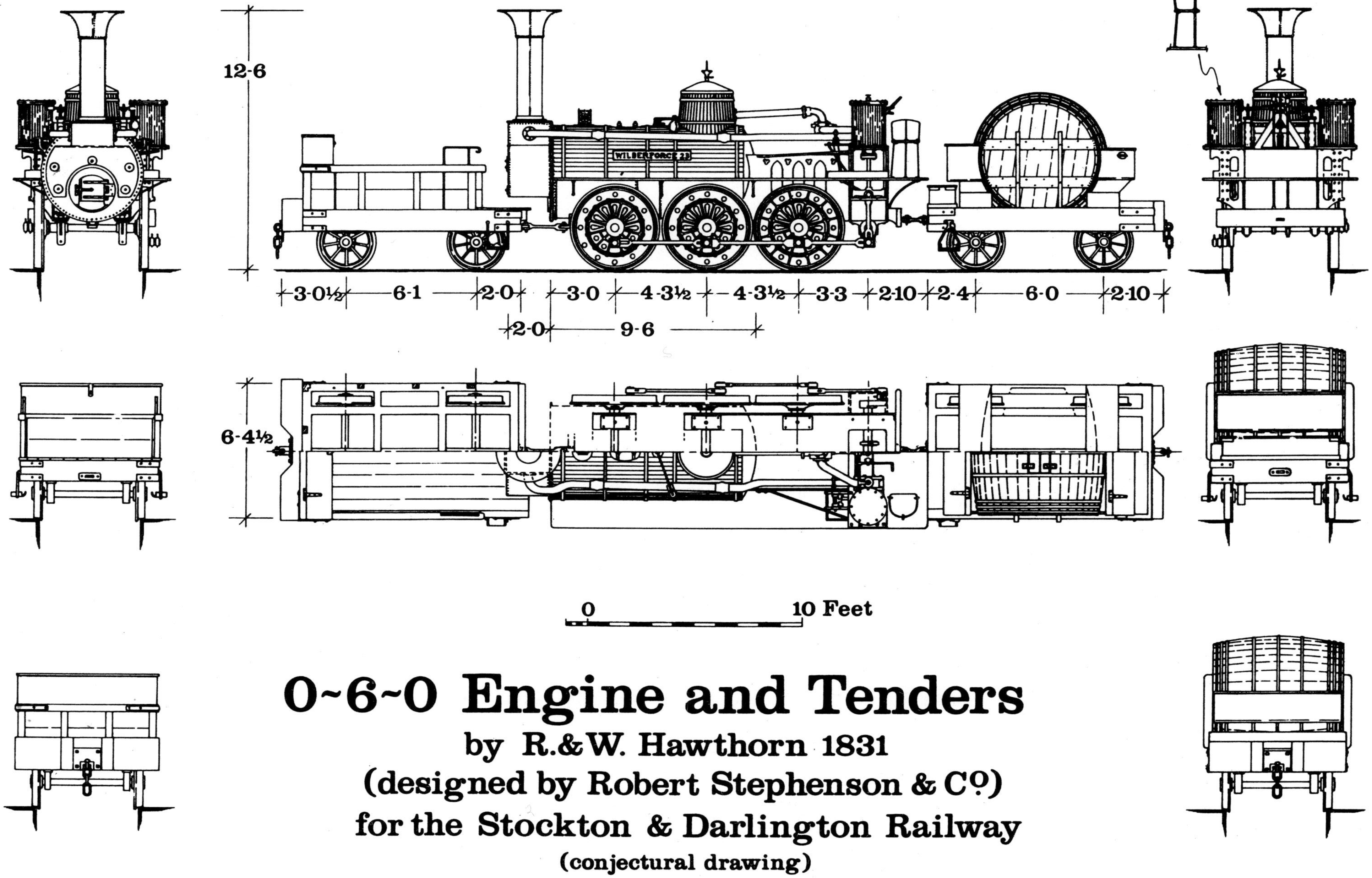

Fig.49 Drawings of No.23 "Wilberforce", by Alan Prior.

quite well in the stock lists. They had been moved to the 'duplicate' list in 1843-4, but some of them were still working in 1848, while "Wilberforce" itself lasted for a long time. There was not another No.23 until 1865. Apparently "Earl Grey" lasted almost as long, since the K. Hoole notes say that they were both scrapped in 1858.

No.16 "Director" was sold to James Bray, contractor, in September 1843 for £400. No.19 "Darlington" was probably sold to John Anderson, another contractor, in March 1856, for £250.

In December 1834 John Graham made a comparison (referred to in the previous chapter) between the "Victory" type of engine, with no springs on the driven axle, and the "Wilberforce" type with all axles sprung, and reported to the Management Committee that in his opinion damage to the track was much less with the latter engines. I quote this report in full, as it seems to me important, and clarifies the question of the suspension of the so-called 'springless' engines to some extent.

"December 14th. 1834. Last week I have paid particular attention to the Engines traveling on Springs throughout and Compaired them with those traveling without a Crank Shaft and only 4 of the Wheels on Springs.

It is my opinion that the Engines on the Victory Plan Traveling at 10 miles per Hour will do as much or more Injury to the Railway as the Wilberforce Kind of Engine will do going at 15 miles per Hour.

J. Graham is further of Opinion that should the Spead not exceed 8 Miles per Hour that the Injury done to the Railway would not be serious."

The only apparent Committee reaction to this was the request to Kitchings early in 1835 to fit springs to the new engine (No.25 "Enterprise"), which was quickly withdrawn when Kitchings advised them of the extra cost, with a rider that they hoped that it would be possible to fit all the engines with springs when it became convenient, or words to that effect.

Of course, it is still not possible to say with any certainty whether "4 of the Wheels on Springs" meant that each of the four was independently sprung, and it is my belief that only two springs were used, arranged as compensating beams over the first two axles.

This report makes the next developments less easy to explain, but de Pambour had made it pretty clear (list, page 6) that cost for cost the earlier engines did as much work as the later and more complicated ones; no doubt this influenced the committee.

"Magnet" & the "Enterprise" Class

An enlarged and improved version of the "Director" class was next on the board, but strangely only one was built, probably because of its cost. No.24 "Magnet", was built at Hackworth's Soho Works, Shildon, in July 1835, according to Inness. Hackworth had opened these works in 1833, as mentioned earlier, and was now not only the senior official locomotive engineer for the line but also a major contractor to it. Since this meant that he was responsible for the maintenance of his own products it was a very good reason to make sure that they were good to begin with.

The drawing of "Magnet" used in the Inness articles in the "Locomotive" magazine may have been taken from an article in the "Engineer" in 1925, but there are other copies, including one in the Science Museum, and I have used one obtained from another source altogether (Fig. 51). This article says that the engine is as rebuilt by Hackworth in 1850, and shows a most peculiar arrangement of boiler and cleading, difficult to explain unless the boiler had been greatly enlarged, or something else hidden under the cleading. The K. Hoole notes suggest that the original boiler had been cut in half and a smaller diameter section inserted, but this does not tally very well with the apparent lengths (the 1840 list is said to state 11'-0" long x 4'-4" diameter, while the drawing shows about 13'-0" long x 4'-0" diameter at the ends, and 3'-0" in between). A better general illustration is the vignette (Fig. 50) used for the heading of the railway's timetables in 1837 and 1838, believed to be of "Magnet", and showing a very similar engine to the previous class, but with a larger boiler mounted further forward. The later drawing shows clearly the built-up plate pedestal for the cylinders, a box like construction very similar to that still in existence on the preserved "Samson" in Canada (see later). Also clearly shown is the parallel motion, of Watt's type, unlike that on "Samson". This drawing is one of the very few showing the warning bell fitted to most of the engines, and what looks like an early stick type pressure gauge at the back of the chimney.

The dimensions of the boiler are not known with certainty, but the 1840 list dimensions should be fairly reliable; Inness and Young both say that it had a 2'-0" diameter flue tube with 110 return boiler tubes. I assume that full springing was fitted; there was no point in using a countershaft drive otherwise. Young gave a very good drawing of the intermediate shaft and eccentric arrangement, with the rods, cylinders and valve chests, showing a four fixed eccentric assembly but little detail of the gabs or control lever (Fig. 8). There is mention of a firebrick wall dividing the fire-tube and furnace; progress towards a separate firebox was being made.

Fig.50 Illustration of No.24 "Magnet" from the heading of the S. & D.R. timetable for passenger traffic, summer 1837-8.

This engine seems to have been employed mostly on passenger duties, which seems strange since it must have been the most powerful locomotive on the line during the first half of its life. Its cost, with tenders, was £1050, and in the 1844 list it was still valued at £1100. In March 1838 it was damaged in collision with waggons near 'Wilson's' branch (colliery). By 1842 it had been transferred to 'merchandize' duties together with No.29 "Queen" (see later), and in those lists (1842/3/4) was not numbered, which may have something to do with the puzzle below. It was one of the engines loaned to the Great North of England Railway for one of the first trains on that line. In 1846 the number 24 was allocated to another engine, and I would have assumed that "Magnet" was then taken out of service. It does not appear in John Dixon's list of 1849, in spite of the rebuilding date of 1850 given in the "Engineer" article. Either it was running as a 'merchandize' engine without a number, or Hackworth bought it as a

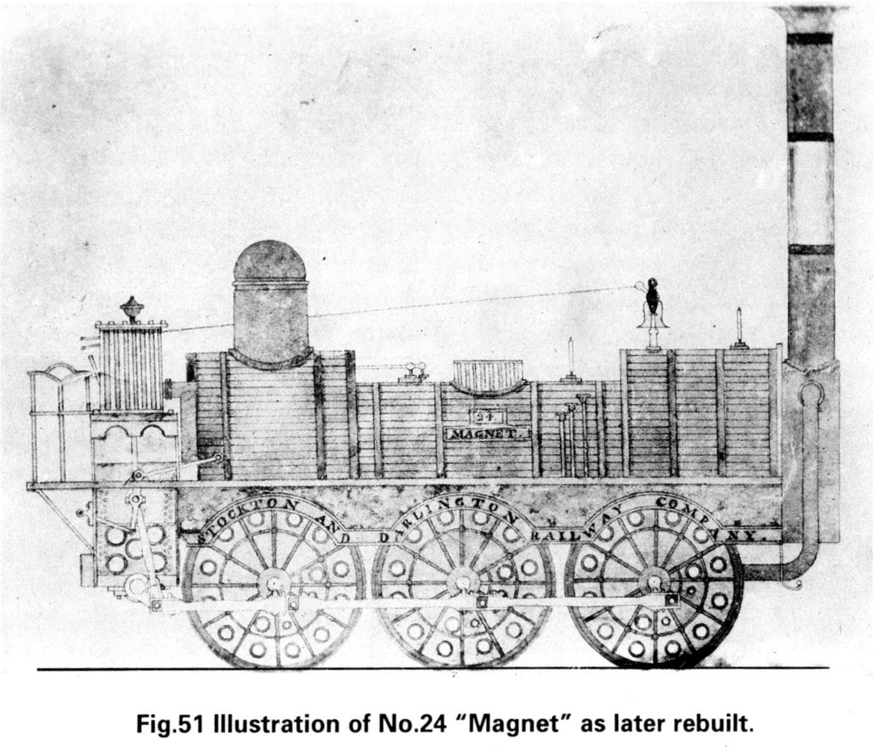

Fig.51 Illustration of No.24 "Magnet" as later rebuilt.

speculation for rebuilding and resale and set it aside until convenient. The K. Hoole notes say that it was sold to N. Downing (nominal partner in the Soho Works management) in February 1846 for £630.

It seems to have had a much shorter life than might have been expected, in view of the average life of the "Director" class, and since there are very few reports of trouble with it. The fact that no more similar were built can only mean that the overall cost, including the cost of maintenance, was higher than could be justified; it is probable that during those years the railway was having a hard time and finding it difficult to make ends meet, with a lot of competition from other railways in the coal trade.

W. & A. Kitching now come into the picture as engine builders; they had been associated with the railway as suppliers since the beginning, and this William (Jnr.) as a shareholder and Management Committee member. William senior had set up as an ironmonger and blacksmith, his shop and forge being situated at the top right hand corner of Tubwell Row in Darlington, opposite the market (interesting to note that Newcomen followed much the same trade; most ironmongers then had to produce their own goods), and they moved to a new foundry at Hopetown in 1831, quite close to where the first North Road Station was built in 1833, and even closer to the later one. It was sited on the land between the S. & D.R. main line and the Darlington Branch to the coal depot. William Lister also had a foundry at Hopetown and bearing its name, and this was taken over by Kitchings a few years later.

There has been a lot of confusion about their first two engines. Robert Young says that they were given the drawings of the "Magnet" to work from; so they may have been, but what they produced was a smaller and simpler 'austerity' version. One of the two Whessoe official histories has another story. It says that Kitchings had a clause in their contract with the railway allowing them to build an engine at contract price should new engines be required. It then says, "This opportunity came in 1834 when one of the engines crashed and in the course of repairs the Kitchings were allowed to strengthen it and make it larger". It was then renamed the "Enterprise". This is not true; that engine was No.2 "Hope", which they rebuilt in 1834-5 as an 0-6-0 very similar to No.8 "Victory", and to which they allotted their works No.1. I have found no record of such a crash at the time, however, but this is not unusual. There was an accident to No.2 "Hope" in November 1835, but this would seem to refer to the rebuilt engine after completion. The price of the rebuilding was £680, and the engine retained its railway number and original name. I have a copy of part of another Whessoe history which states the above correctly, and gives a list of engines built with works numbers and prices, although the first two prices in the list are reversed and contradict the text and the committee minutes.

This rebuilding was drastic - all that was re-used of the main parts of the original engine was four of the wheels with their axles and one tender. It is no wonder that Kitchings considered it to be a new engine.

Kitchings had received an order for a new engine, possibly as early as 1832, it seems, and it is far from clear why this engine took so long to build and deliver, let alone to be paid for. It seems to have been delivered in June 1835, and was named "Enterprise", No.25. There is a puzzling minute of 4th. August 1837 that Kitchings' engine "Enterprise" be purchased at a price of £856 including three tenders, the valuation being attributed to James Kennedy. This probably indicates that up to that date Kitchings, as contractors, owned the engine as well as operated it. The Whessoe history concerned says that the original cost of construction was £900 (see below). As I have said earlier, there was a request from the railway company just before the engine was completed that it be fitted with springs, but the extra cost for this modification was too high, and the request was withdrawn. By 'fitted with springs' they must have meant six single springs, one to each wheel, which would have entailed a countershaft drive. Whether or not they actually had the drawings of "Magnet" for guidance, they certainly followed Hackworth design practice in these two engines.

As a matter of interest, their old ledger indicates that it was originally intended to call this engine "Confidence", and makes it quite clear that in each case work was started in June 1834, giving a completion date for No.2 of March 1835, at a total cost of £745-10-0d., and for No.25 of July 1835, total cost £861-13-3d., including 19/3d. for "1 Large Bell". These figures are for parts and labour only. The name "Enterprise" is first mentioned only on the last pair of pages, the original heading being "Confidence". This cost book contains a detailed and itemised record of all materials and labour for both these engines, and clarifies many points. Having studied it closely, and compared the entries for each engine, it is clear to me that both of them were fitted with two fixed main bearings, and four 'spring' bearings fitted in horns. The springs weighed in total a little over 3 cwt. not including the shackles and bolts, but unfortunately no quantities are given. At that total weight I think that they were probably

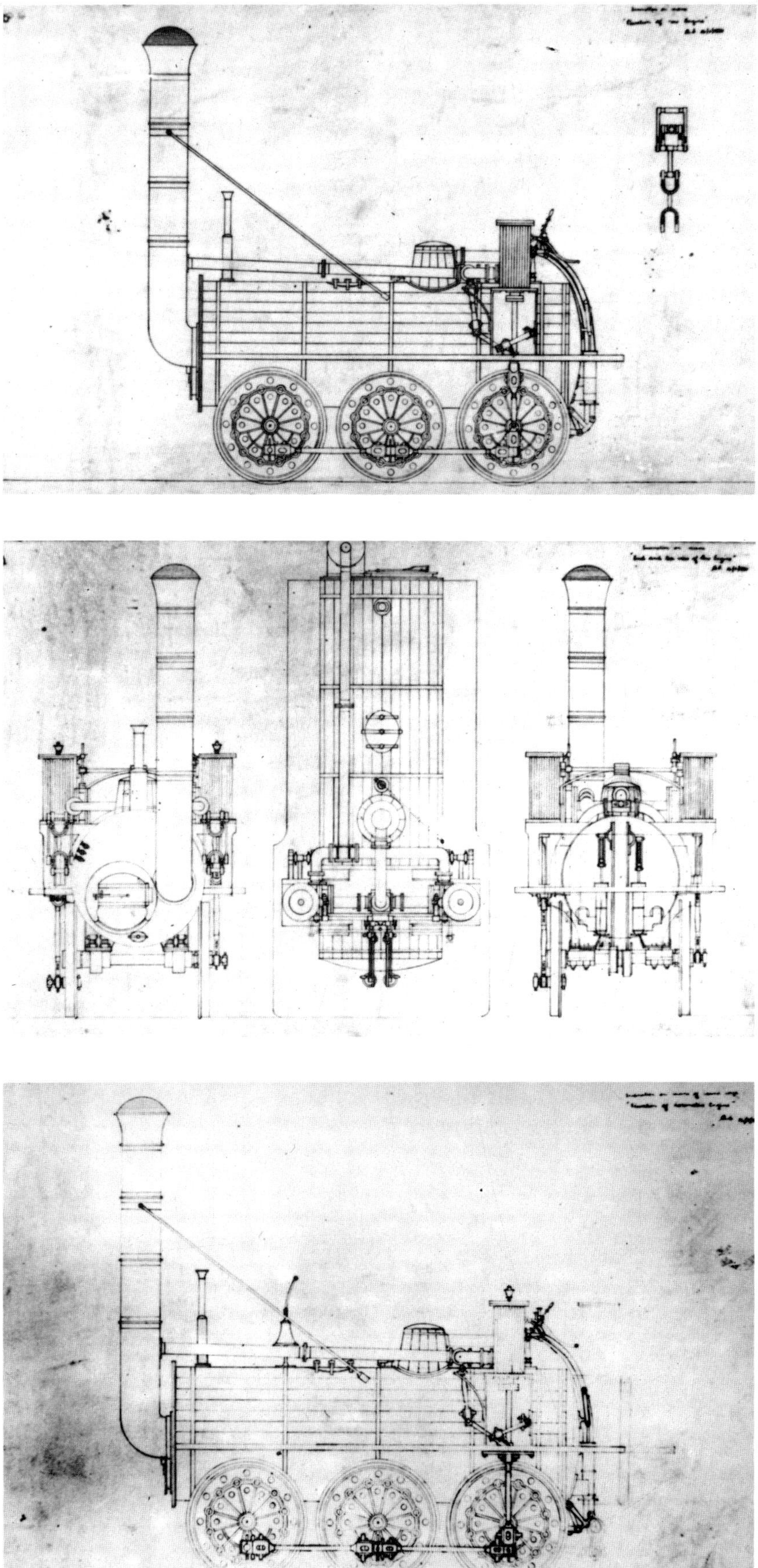

Fig.52 Drawings of a Hackworth type 0-6-0 engine from a Kitching/Whessoe album in Durham County Archives, believed to be either design drawings or conjectural drawings of No.2 "Hope" as rebuilt and No.25 "Enterprise".

arranged in the same way as for "Royal George", as flexible compensating beams (Fig.30), across the first two axles and fixed in the centre.

The main parts for both engines were similar, which meant that only one set of patterns was necessary, saving cost. The boilers came from Newcastle, presumably Forth Street, and each was supplied fitted with flue tube and with one spare flue tube, the weight of one boiler with flue tube being just over four tons. Each engine was supplied with one coal tender and one 'barrel' tender, and for No.2 the original tender was modified to become a 'barrel' tender only. These barrels were quite expensive items, at £7-10-0d. each. The third tender mentioned for No.25 must have been a later addition. It is noteworthy that the grates were fitted with fire-bricks, fire-clay and fire-bars. Spring steel for the 'safety' balances is mentioned, as also is a 'fire-basket' for the rear light. Boiler cleading took 144 feet of 3/4" boards per engine.

Also in the Durham County Record Office is a large album, probably assembled by someone at Whessoe many years ago, of Kitching drawings. Many of these are pages of tinted drawings of waggons, trucks, vans, coaches and mobile cranes, and are probably sales drawings. However, there are three which show an 0-6-0 locomotive of Hackworth type, with vertical cylinders over the rear wheels, and on one sheet pencilled preparations for showing the cylinders mounted further back, over a countershaft. There are two R.H. side elevations, with slight variations in the valve gear, and a sheet of end elevations and cross section. No compensating suspension of any kind is shown in the side views, but what appear to be shackles for springs are to be seen in the end view. These are almost certainly the drawings referred to in one of the Whessoe histories, and must be intended to represent No.2 "Hope" as rebuilt, and No.25 "Enterprise", but the provenance of the drawings is not known, they are not signed, identified or dated, and in my opinion are of comparatively recent origin, possibly this century. I could well be wrong. These drawings (Fig.52) are marked "inscribed on back -" and initialled A.C. 16/3/1950, obviously when they were mounted in the album, each with a further title; "Elevation of new Engine", "End and top view of new Engine", and "Elevation of Locomotive Engine", respectively. As they are pasted firmly to the album pages there is no way of checking this further.

The eccentric rods and other valve gear are similar to that fitted to No.8 "Victory", except that small bell cranks are fitted just where the rods come round the bottom of the boiler. The wheels are c.4'-0" dia. on a 2 x 4'-6" base, the boilers about the right size, steam domes well to the rear and spark arresting cages on the chimneys. The piston rods are guided by parallel motion somewhat similar to that fitted to "Royal George" to start with, but with shorter levers (Fig.30).

A rough list of locomotives built by Kitchings which is held by the Durham County Archives Office begins with "Enterprise", but gives the date as 1833, without an engine number. This list is almost certainly part of the preparations for the history printed later, and may not have been fully checked. Again, a Whessoe letter which I have seen said that the rebuilding was of "an earlier Kitching engine" but there seems to be no record whatsoever of any such earlier engine.

No.25 "Enterprise" was certainly the first locomotive built entirely by Kitchings, but it was far from the last. Their workmanship was acknowledged to be excellent, and they remained associated with the railway for many years. They were, of course, also Quakers. Eventually they sold the Hopetown Foundry to the railway for use as workshops, and moved over the border into Whessoe township, not far away, where they opened another and bigger foundry, somewhere about 1860, and at the same time the surviving Kitching partner, Alfred, transferred the firm to his cousin Charles Ianson. Eventually they became Whessoe Foundry, and the firm still exists as a controls and instrumentation design company.

The minutes record on 2nd. September 1836 that a meeting of the three main 'power' contractors was held, at which they discussed the best type of new locomotive to be built. It was agreed that the specification should comprise a six wheeled engine with 12" dia. cylinders, a simple return iron flue tube and with a weight not more than 11 tons. They agreed to state prices at which they would be willing to supply one or more engines, with time of delivery, and they were to state in what respects these new engines would vary from the engines called "Hope", "Enterprise" and "Victory". It seems obvious that a railway company representative must also have been present. In the event of such an order, the company might take any one of the multitubular engines and dispose of it. It was also stated that no valuation need take place that year.

The next engine built was No.14 "Tees", from Kitchings, in July 1837, at a cost of £900, the same as the stated original cost of "Enterprise". Then No.11 "Beehive" was delivered from the Shildon Works in February 1838 followed by No.12 "Briton", also from Shildon, in June the same year. When delivered they apparently carried Nos.30 & 31 for a while, but received the above numbers in 1839. They are said by Inness to have had no springs, although such a requirement is not mentioned in the specification above. I find it very difficult to believe that this could have been possible, in view of Graham's December 1834 report regarding speeds and track damage. Also by then Hackworth was building engines for other customers which certainly had springs on the first two pairs of wheels. In view of the detailed information in the Kitching cost book mentioned in connection with No.25 "Enterprise" and the rebuilding of No.2 "Hope" I am sure that these three engines were simply a return to the type with vertical cylinders driving straight on to cranks on the driving wheels, making it impossible to fit that axle with springs, and that the other two axles were sprung, so as to do away with the expense and complication of the countershaft drive, as well as a reversion to simpler boilers with return flues only. I may be wrong, as Inness goes on to say that track damage resulted, as one might expect, and time and time again the minutes refer to "the engines without springs", with frequent recommendations that they should be modified. Nevertheless, I think that the track damage may be well enough explained by the hammer-blow of the rear unsprung axle and wheels, and that the phrase was just an inaccurate way of saying "without springs on all wheels".

No.11 "Beehive" was derailed at Darlington coach station within a fortnight of being delivered, by wrongly set points; a wheel was broken and two axles bent.

All four of these engines, together with No.2, were declared to be no longer economic in September 1841; a very short life. Tenders were then asked for from the contractors for rebuilding them. "Enterprise" and "Briton" were rebuilt by Timothy Hackworth in 1842 at agreed costs of £524.10.0d. and £543.10.0d. respectively, but when it came to payment this was changed to £500 each, the remainder being set against the £100 bid by Hackworth at the Shildon auction mentioned earlier, at which he bought No.13. "Coronation", so he seems to have gained a little overall. The two rebuilt engines were delivered in September that year, with "Enterprise" being renamed No.8 "Leader" and "Briton" becoming No. 12 "Trader", which explains why these two engines vanished from the annual valuations; there was not another No.25 until late 1845.

The other two were also rebuilt, at the Shildon Works, No.11 "Beehive" costing £550 and completed at about the same time as the two above. It was renamed "Bee". No.14 "Tees" is a little more difficult. There is a Shildon Works Committee minute of September 1844 saying that No.14 and No.6 ("Dispatch") were to be completed by the beginning of December, which may indicate a rebuilding, or simply repairs. Certainly "Tees" was rebuilt at the Shildon Works later, in 1848 (Fig.67). John Dixon said that it was rebuilt into a 'firebox' engine, which is explained below.

No more such were built for the railway. One of the engines built by Hackworth for other owners, was the locomotive "Samson", which with two others was exported to Nova Scotia in 1838 after the "Enterprise" class was built, and which is now preserved in Canada. This engine is both well documented and illustrated (Figs.53 & 54); several photographs are available and it has also been the subject of an engraving in the "Engineer". It is my belief that these three engines were similar in most details to the "Enterprise" class, if one can call it a class; no intermediate shaft, no springs on the driven axle, possibly leaf springs on the other two axles and with the same box mountings as "Magnet" for the cylinders, although with different parallel motion, of Scott Russell type (Fig. 55). As preserved, "Samson" has a casing fitted over the firebox end, apparently forming a sort of smokebox with a door over the grate, a development of the earlier 'horseshoe' shape, but this may be a later addition. Quite recently I received two drawings, thanks to two interested friends, found in the Ken Hoole collection at Darlington Railway Museum, and, I find now, repeated in the Laurie Ward collection, one of which shows "Tees" as rebuilt, dated 18/5/48, the other drawing being of No.26 "Pilot", dated 10/5/52, both marked S.W.Co. (Figs.66 & 67). This drawing of "Tees" explains what Dixon meant by a 'firebox' engine, and shows many other interesting details. The boiler is clearly a straight through true multitubular type with a proper firebox and a smokebox with cleaning door, 12'-6" long over tube-plates by 4'-6" diameter over the cleading. Buffers are fitted at the front end, on a beam secured to horns from the front wheel mountings. There is no main frame; all the wheel mountings are still riveted to the boiler, the second and third mountings being hornblock type, with a bar between the two, but there are no details of the front bearings to go by. Since the cylinders are at the front end I would expect that that axle was also sprung. Between the first two pairs of wheels there is a steam brake working upwards, a large wedge shaped wooden block forming the brake itself, with a small steam cylinder above on the boiler side, similar to the brakes fitted to the "Commerce" class (see later). The cylinders are at a

Fig.53 Locomotive "Samson" as preserved in Canada.

shallow angle as in the later 'Tory' class, and driving on to the third set of wheels. The valves are straight forward slide valves above cylinders, operated by the Shildon link gear (Fig.9) with a lever at the side of the driving platform, counterweights and a weigh-shaft, and a single crosshead on the piston rod working on a slide bar, this arrangement and the steam pipe layout all being similar to the later engines of the 'Tory' class, but at the front end. Another smaller lever operates the cylinder drain valves from the driving platform. The regulator is not shown.

Here we have a very much improved engine, explaining the high value of £1950 in John Dixon's 1849 list. It was still on the list in 1869, in N.E.R. days, and the next engine bearing this number was built in 1871. These two drawings, however, appear to be copies of older drawings, and their provenance is not

Fig.54 Engraving of locomotive "Samson", from the "Engineer".

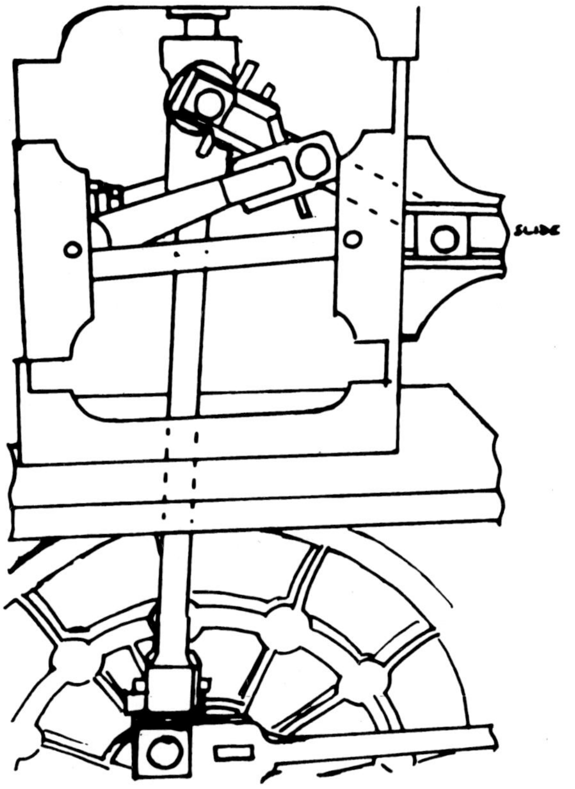

Fig.55 Parallel motion valve gear fitted to locomotive "Samson"

definitely known, so although they seem authentic, some doubt must remain. It is almost certain that they are by R. H. Inness

Apart from "Samson", the only other survivors of those days are No. 25 "Derwent", of course, a rather later engine, and the remains of a locomotive known as "Braddyll", at the Hackworth Museum at Shildon, built c.1835/7 by Hackworth for the South Hetton Coal Co. and in my opinion rebuilt later, and perhaps the "Albion", also preserved in Canada, one of the second batch built for the Albion Colliery in Nova Scotia in 1839, attributed to Hackworth but possibly by Fossick & Hackworth and of a very different design.

The "Tory" & "Miner" Classes

A committee minute dated 19th. January 1838 says "More coal engines to be got"!

The 'Majestic' class had not proved very satisfactory, and although the 'Director' class gave good service, and did not damage the track so much, the reciprocating effect of the vertical cylinders must have caused some problems of stability at the higher speeds now being attained. There had been a speed limit of 6 m.p.h. for mineral trains in the early days, since the cast iron waggon wheels did not last very long at higher speeds, but this was frequently exceeded. Indeed, John Graham pointed out in 1835 that this limit would delay the passenger and parcels trains, which were on a much faster schedule, and it cannot have continued for so many years.

Higher speeds, then, were becoming necessary, and heavier mineral trains to cut costs and fight competition needed more powerful engines, with fully sprung suspension to protect the track. Chaldron waggons were being replaced by bigger trucks. It was necessary to change the basic design somewhat, and Hackworth decided, probably with some regret, to use the Stephenson arrangement of sloping cylinders with slide valves on top, which still gave easy access for maintenance. Designs and specifications were prepared (by the Shildon Works Co., according to John Dixon) and tenders called for from a number of local builders.

The first orders were placed in February 1838 and in November the first of the new class, No.15 "Tory", was delivered to service from Hackworth's own Soho Works, soon followed by No.17 "Whig" from Kitchings. As a class they varied considerably; the first few had plain return flue boilers while the later ones had the Hackworth return tube boiler developed from the 'Director' class; a drawing exists of this last development as applied to No.8 "Leader" (Fig.58).

In addition the actual size of boiler seems to have varied from one to another, and Inness says that the wheel diameters varied between 3'-9" and 4'-4", most, of course, having the standard 4'-0" wheels. So there can have been very little in the way of standard spares, and very limited interchangeability. Even the cylinder sizes varied, from about 12" bore by 18" or 22" stroke up to 16" by 24", and the weights varied between 15 and 20 tons total for engine and tenders in working order. One imagines that the load to be hauled would have had to be chosen or adjusted to suit whichever engine happened to be available for service! Most of the earlier engines were rebuilt later, some more than once.

However, in spite of their differences, they had a general overall similarity. One, No.4 "Stockton", has been thought to have been different, mainly because of the sketch (Fig.56) by Theodore West which shows the cylinders at the front end, as became the practice in the later 'Miner' class. This engine was the only one built for the S. & D.R. by Timothy's brother Thomas after he had joined the firm of Fossick & Hackworth at Stockton, a firm which also built marine engines and was much later to become Blairs, a famous name in that field. Thomas had entered into partnership with Fossick in 1839, after his few years working for Timothy as manager (and partner) at the Soho works, and was still there while "Tory"

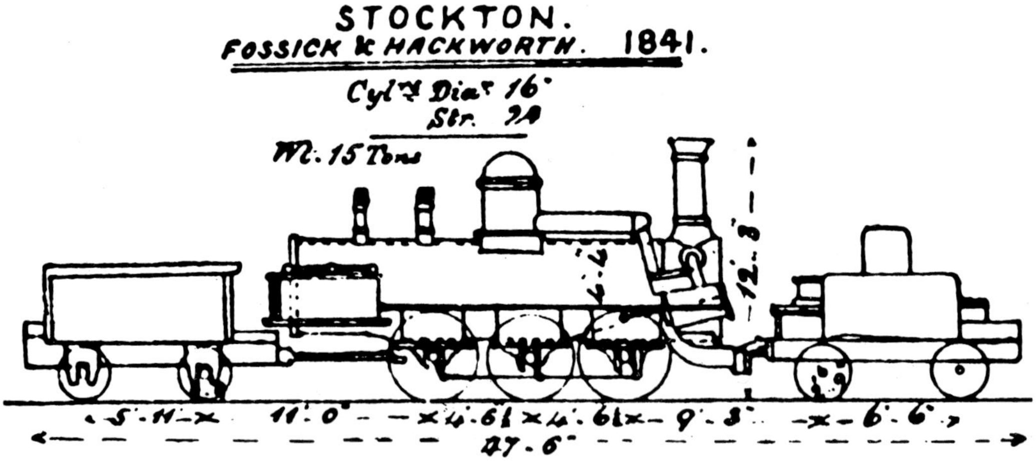

Fig.56 Theodore West's sketch of No.4 "Stockton"

was being built. Fossick & Hackworth continued to make locomotives of similar type for many years. Another difference about this sketch of "Stockton" was the boiler. West depicts it with the water tender at the chimney end and the coal tender at the other end, so it looks as if the engine was fired and driven from the same (rear) end, and it seems obvious that it was a straight through boiler of multitubular type. Because of this, some authors have listed "Stockton" separately, but in fact the sketch almost certainly shows it as later rebuilt, and I believe that when built it was very like the others. No doubt since they had the tenders they just used them. It was certainly altered in June 1851, when it was fitted with "Fossick's Patent" (coke burning tube) at a cost of £355. Fig.57 shows it as rebuilt, and follows West's sketch closely. I am following Inness in including it in the 'Tory' class, more than one of which were rebuilt in a similar way.

I have said in the last section that all four of of the 'Enterprise' class were rebuilt into the 'Tory' class, and shall treat them accordingly, but there is not enough information regarding No.14 "Tees" at this time to give many dimensional details, except where they can be lifted from the drawing (Fig.67).

Most of this class had the later form of four fixed eccentric gab gear, probably all of them up to No.22 "Alert", built in time to be fitted with the 1847 'hook' type of link motion (Fig.9), which was certainly fitted to "Tees" when rebuilt in 1848. When they were reboilered later the valve gear was often changed at the same time, but not always, as is to be seen on the "Derwent" (see below). They were built at at least six different works between 1838 and 1848, including

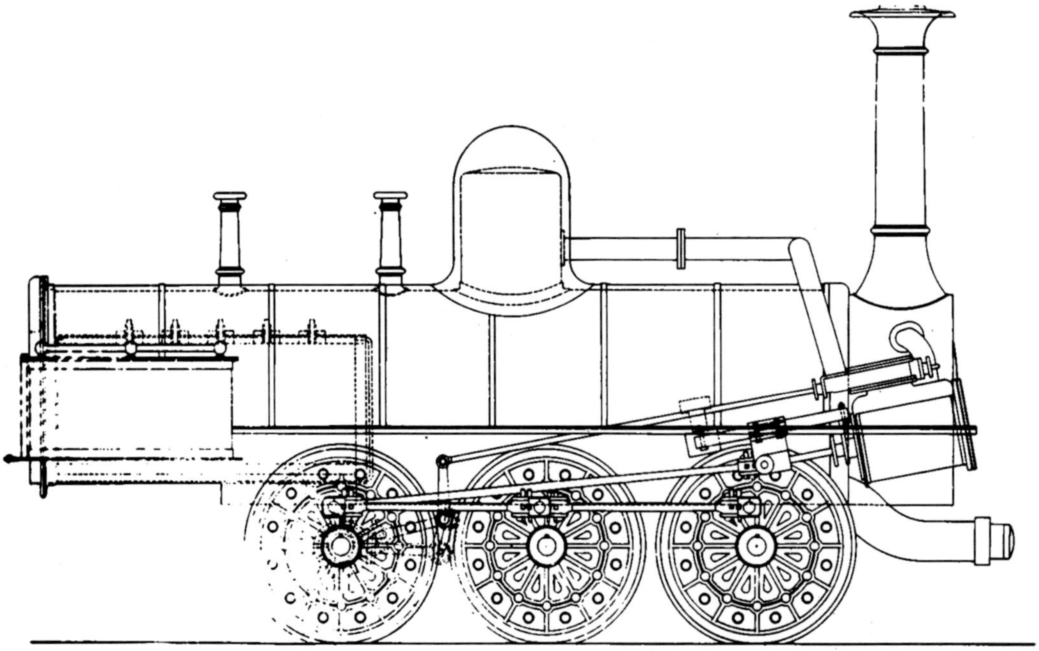

Fig.57 Drawing of "Stockton" as rebuilt.

Shildon Works, the Soho Works, Kitchings, Wm. Lister, Neasham & Welsh, and Fossick & Hackworth. Inness gave few dimensions, but Craven offers more and I shall list these all at the end of the chapter. In general the original boilers seem to have been about 14′-0″ long x 4′-4″ in diameter, with minor differences, and the flue tubes 13′-0″ long x 2′-0″+ dia. with return tube 13′-0″ x 1′-8″ dia.

No.15 "Tory", the first in service, was delivered by Hackworth in November 1838. Six months later it suffered a boiler explosion at Preston (Stockton, Yarm Lane) and John Graham says that the flue tube and boiler end blew out and went 50 yards! This engine was followed by No.17 "Whig" from Kitchings in March 1839. In August 1839 Hackworth delivered No.10 "Auckland" and Lister delivered No.9 "Middlesbro′" (Fig.59), followed by No.6

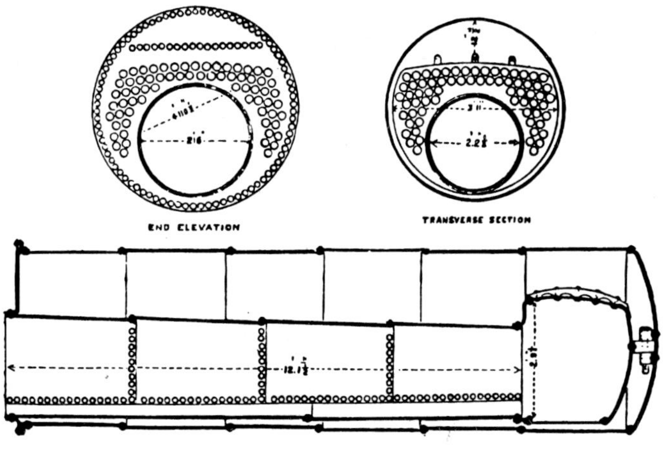

Fig.58 Drawing of the boiler for No.8 "Leader".

"Dispatch" from Hackworth. "Dispatch", according to Craven, was fitted with a multi-tubular firetube boiler, presumably of return tube type, and so was Kitchings' "Pilot"; these were the first of the class so fitted. Inness says that No.10 "Auckland" had a multitubular boiler, and gives details, but I believe that this would have been after rebuilding later. Craven, no doubt, relied on the 1840 list.

No.18 "Etherley" came from Lister in February 1840, No.26 "Pilot" from Kitching in June and No.28 "Ocean" from Lister in August. The three Lister engines reverted to return flue boilers. No.4 "Stockton" had originally been ordered from Timothy Hackworth, but was subcontracted to his brother's firm of Fossick & Hackworth and delivered by them in July 1841, followed by No.5 "Hopetown" from Kitchings in August. Unfortunately John Graham did not record any further deliveries of this class, but Craven has given some more, and for all of them the year of entering service is shown in Appendix IV, following the text.

No.27 "Witton Castle" came from Neasham & Welsh in May 1840, No.7 "Prince" from the Shildon Works Co. in March 1842, No.8 "Leader" from the Soho Works in April 1842, No.12 "Trader" from the same works in September 1842, No.16 "Stanhope" from Kitchings in April 1845, No.25 "Derwent" in November and No.24 "Skelton Castle" from them in May 1846. No.22 "Alert" did not appear until January 1848. Of the other two rebuilt engines, No.11 "Bee" was delivered from the Shildon Works Co. about September 1841, and No.14 "Tees" presumably in 1844. No details of which kind of boiler was then used are available.

No.7 "Prince" blew up at Crook in 1859, killing one man, and was in collision with No.123 "Belle Vue" in April 1862 at Sunniside Bank Foot. The damage is not recorded. "Leader"'s wheels were examined in July 1840, for an unstated reason. John Graham did say in June 1841 that they were good engines, and perhaps discipline was better.

No.12 "Trader" was sold to the Peases' West Collieries, but blew up soon after, in 1868, and was then replaced by No.25 "Derwent", (Fig.65) which had been working on the railway since 1845. "Derwent", the sole survivor now, was similar to "Leader" (Fig.60) but built by Kitchings, one of a pair with No. 16 "Stanhope" (Fig.62). The "Derwent" is now, of course, to be seen at the North Road Station Railway Museum at Darlington, in near original condition, still fitted with four fixed eccentric gab gear (Fig.13), and with the two feed pumps below the rear end, driven from eccentrics on the centre axle, which were a feature of the class. One point to be noticed is that the cylinder on the right hand side has been repaired at some time in the past by fitting a flange which is bolted to the cylinder itself, replacing the original flange cast with the cylinder, which must had suffered serious damage, and a plain plate cover is fitted instead of the original cast cover, still to be seen on the other side. To judge by what can be seen in various photographs, this must have happened a long time ago.

According to the "Railway Engineer" magazine, "Derwent", while working at Crook for the colliery, was still capable of hauling 60 empty chaldron waggons up the Sunniside incline, or 80 loaded waggons on the level. I have been told that this is unlikely, but I am not arguing. There was also a spell at the construction of the Waskerley reservoir. Pease and Partners presented the engine to the N.E.R. in 1898, when she joined No.1 "Locomotion" at Bank Top Station. She was still able to be steamed in 1925,

Fig.59 Old photograph of No.9 "Middlesbro'", attributed to John Dixon by Zerah Colburn. The print here is by Bleasdale, but may be a copy.

Fig.60 Photograph of No.8 "Leader"

Fig.61 1875 photograph of No.26 "Pilot", disguised as No.10 "Auckland"

Fig.62 Old photograph of No.16 "Stanhope".

Fig.63 Press photograph of No.25 "Derwent" in steam at the 1925 Centenary.

Fig.64 1875 photograph of N. E. R. No.1033, ex No.33 "Shildon", as rebuilt with later standard boiler, and with tenders as they would have been originally.

when she formed part of the Centenary procession (Fig.63), in second place after the Hetton Colliery engine, and is said to have reached 12 m.p.h. on trials the day before

This engine was taken to Newcastle in 1881 for the Stephenson Centenary celebrations, and again in 1887 for the Queen's Diamond Jubilee. It is difficult to see why, as it had nothing at all to do with either George Stephenson or Newcastle-upon-Tyne!

I wanted to include some details of No.25 "Derwent's" boiler tubes, but no record of them was to be found. Kitchings' cost book for this engine, covering a total building period of from May 1842 to December 1845, has no details of tubes. However, Michael Bailey and John Glithero have had a look inside, and given me all the details, which I have included in the list at the end of the chapter. Mr. Bailey has also told me that this engine has been fitted with a grate end of Adamson patent type, of 1850's or later date.

A well known photograph of No.9 "Middlesbro'" (Fig.59), which Colburn attributed to John Dixon, shows clearly that she was fitted with a return flue boiler. Bleasdale copied this photograph, as he did that of "Wilberforce". Ken Hoole told me that he considered it to have been a fake, taken as a posed shot after the engine was out of service, but Colburn, in his 1864 edition, said that at the time "Middlesbro'" was in use on the South Durham & Lancashire Union Railway. However, she was supposed to be scrapped in 1865, according to Craven.

No.26 "Pilot" was borrowed by the Great North of England Railway to run as the pilot engine to the opening train on the Darlington to York line on the 29th. May 1843. It was rebuilt by the Shildon Works in 1852, with a peculiar high extension to the steam dome (Fig.66), and the Shildon sliding link valve gear mentioned earlier, with the cylinders still at the rear end and buffers shown at that end as well, but although the reversing lever is clearly shown at that end, there is no platform for the driver, who must have had to stand on a tender. This engine is said by Ahrons to have been disguised as No.10 "Auckland" for the 1875 Jubilee celebrations (Fig.61) in a letter to the "Locomotive" magazine in August 1925. He said that this was because the organizers had wanted to include an S. & D.R. engine of Hackworth build in the exhibition, and there wasn't one left. "Pilot" had been bought by the Consett Water Co. (one of Bouch's positions was as an engineer to this firm) in 1868/9, and it was borrowed from them, cleaned up and renumbered as the Hackworth built No.10 for the occasion. Robert Young used the same photograph in his book, and mistakenly captioned it No.10 "Despatch", but he made several other errors in his description of renumbering at the time. "Pilot" was lucky to have survived so long, as a committee minute of 30th. October 1867 authorized that it be broken up at the Shildon Works.

The photograph (Fig.61) confirms most of the details in the drawing referred to, but shows a decrepit tender with a platform which goes under the buffers on the rear of the engine, and unless there was a plate between the buffer horns, there is nowhere for the driver to stand on the engine itself, so why the buffers? The same caveat must apply about this drawing as about that of No.14 "Tees" as rebuilt (Fig.67).

Several other of the older engines on display in 1875 had the numbers heavily overpainted, and doubts exist about the authenticity of some of them.

No.28 "Ocean" became No.13 later (see earlier note in Chapter 1) and then No.12 on the 1868 list; No.12 "Trader" has vanished from this list, as also have No.9 "Middlesbro", No.26 "Pilot" and No.27 "Witton Castle". This latter engine, the only one of the class built by Neasham & Welsh, of East Portrack Lane, Stockton, is said to have survived until 1875, but there was certainly another No.27 built in 1867, so it may have been sold and survived as a colliery or industrial engine somewhere. It was supposed to be scrapped in 1866 according to Craven. Most of the class lasted in service until the late 1860's. There are a couple of other puzzling points about Neasham & Welsh; James Lowe says that "Witton Castle" had a countershaft drive, which seems unlikely. The John Graham notebooks contain an entry for October 12th., 1838, saying, "Inspected Neasham & Welshes Locomotive Engine, and finds the Dementions Sent to be Correct; but she is not an Engine he could recomend the Company to buy on account of the Weight upon 4 Wheels and the Height she is mounted would be very injurious to the Railway." Lowe does not mention this one at all. It is possible that it was this earlier engine which had a countershaft drive, but not "Witton Castle".

Most of the class were rebuilt, some more than once. I have mentioned that No.26 "Pilot" was 'remodelled' as John Dixon calls it, at Shildon Works in 1852. Nos.13 "Ocean" and No.18 "Etherley" were also 'remodelled' there in 1842 and 1843, No.4 "Stockton" 'altered' in 1851, No.17 "Whig" and No.24 "Skelton Castle" 'remodelled' in May and June 1853, "Tory" being altered at about the same time to a return multi-tubular at a cost of £647-19-0d. (Craven says £450!). No.10 "Auckland" was rebuilt at about the same time. In most cases the boiler was changed to the multitubular return tube type. All of

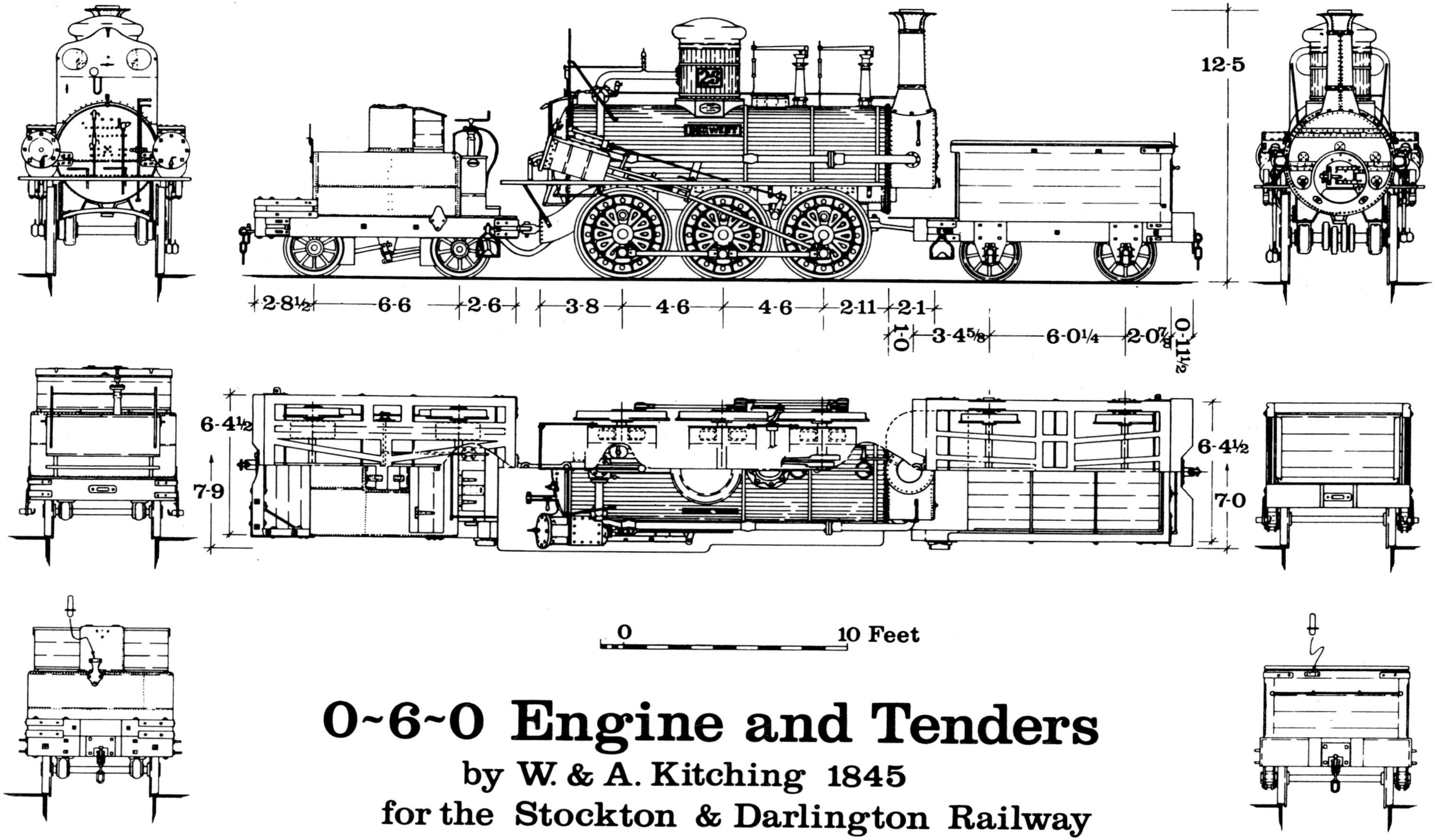

0~6~0 Engine and Tenders

by W. & A. Kitching 1845
for the Stockton & Darlington Railway

As preserved at North Road Station, Darlington.

Fig.65 Drawings of No.25 "Derwent" as preserved, by Alan Prior.

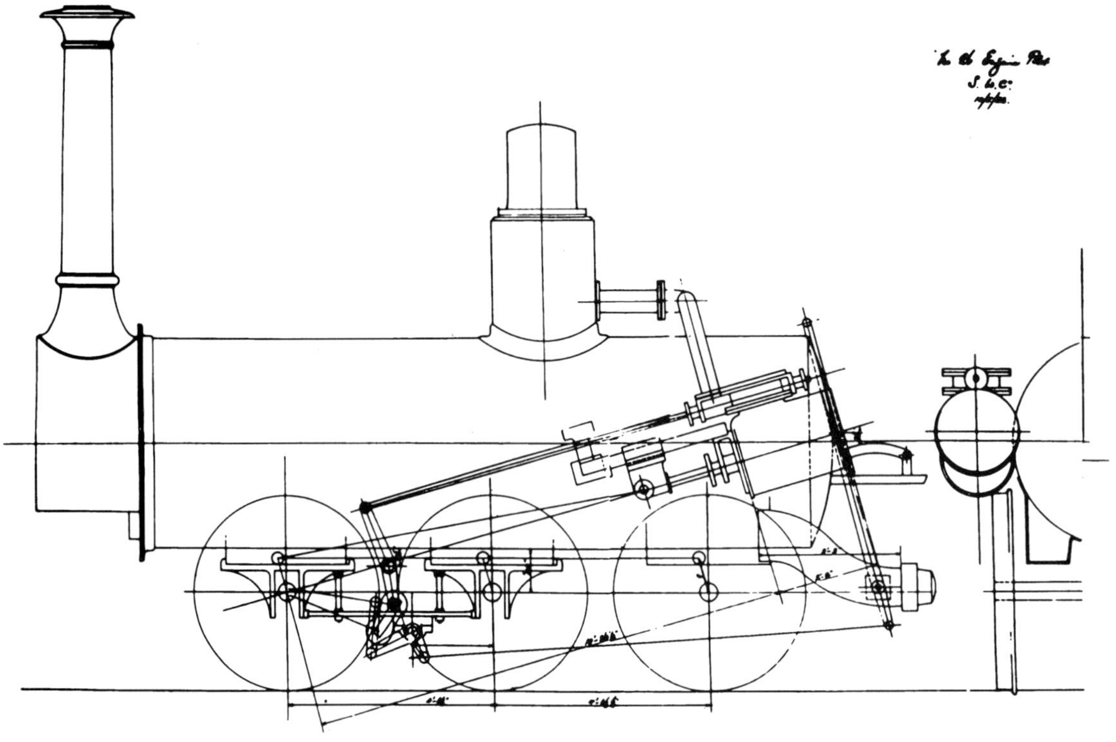

Fig.66 Drawing of No.26 "Pilot" as rebuilt in 1852.

them had their cylinder dimensions changed later. The only ones of which there seems to be no record of such rebuilding are No.27 "Witton Castle", No.5 "Hopetown", No.7 "Prince", No.16 "Stanhope", No.25 "Derwent" (of course!) and No.22 "Alert". I am not including the already rebuilt engines in this.

The trouble is that there is too much conflicting information on rebuilding. I have a list of valuations and repairs from the P.R.O. covering the years June 1850 to June 1857, and many of the dates disagree with those from other sources as given above. It is obvious that many of these 'remodellings' were very

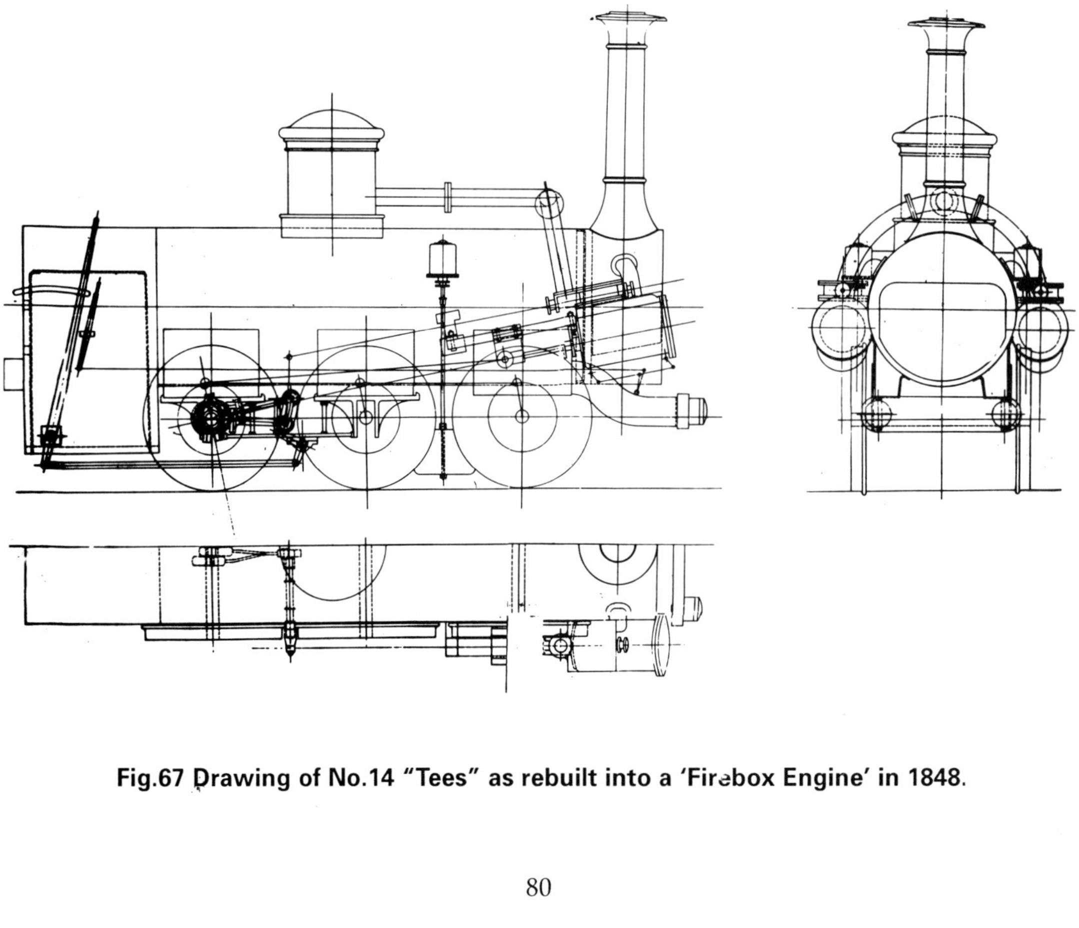

Fig.67 Drawing of No.14 "Tees" as rebuilt into a 'Firebox Engine' in 1848.

extensive; in the case of No.17 "Whig" it says, "In fact, very little of the old engine left". But as I have said before, there would be no change in the No. in the stock book to complicate things. I have listed these data where possible.

As I said above, Kitchings' cost book for works No.12, which was the "Derwent", has no details of tubes. It does have a page devoted to boiler materials and labour, which covers nearly 4 tons of plate for the boiler shell and a bit over 13 cwt. for the tube, singular, obviously the main firetube. This book is shared by the costs for works No.7, "Raby Castle", and there 100 brass tubes are clearly listed. This proves nothing in itself: the wheels and axles for No.12 are not mentioned either, and I do not believe that this book was used as the main ledger for costs, probably just a daybook, or a pocket book. However, heresy as it may be even to suggest such a thing, it seems to me that a strong possibility exists that the engine was built with a return tube boiler and reboilered later with the multitubular return type.

They were followed very soon by an improved version called for by William Bouch; one cannot call it a new design. This was the 'Miner' class, larger and a little more up-to-date, but very similar in most ways. They had the cylinders at the front end, but still in the sloping position, the boilers were of greater capacity; Inness says that they were fitted with copper fireboxes and that they had 'Howe' (Stephenson) link motion with four eccentrics. Craven says that the boilers were much larger, but their external dimensions were very similar to those of the 'Tory' class. The first orders were placed with the Shildon Works Co. in September and November 1844. Six were built in two groups of three, and they did seem to be all much the same in their major particulars. The cost of the second three is given as £1800 each, and that for the first three cannot have been much different. The boiler was 13'-0" long by 4'-4" diameter, fitted with 213 iron tubes, a grate area of 10 sq. ft., and had a total heating surface of 1363 sq. ft.; the cylinders were 15" bore by 24" stroke, and the steam pressure was 75 p.s.i. The boilers were still of the return tube type, and two tenders were needed, having a water capacity of 2 tons, and the total weight loaded was 22 tons 7 cwt. Craven says that the intended weight was about 18 tons, but does not say whether this included the tenders.

Neither of these two classes had main frames; the wheel brackets were riveted to the boiler barrel, as was the drawbar at each end. The wheels were of the standard two part cast iron plug type. They were delivered; No.29 "Miner", June 1845; No.30 "Wear" the same month; No.31 "Redcar", September 1845; No.32 "Eldon", February 1846; No.33 "Shildon" the same month, and No.34 "Driver" in March 1846. All of them except possibly No.31 "Redcar" were rebuilt into 'firebox' engines later. John Dixon omits "Redcar", but Craven includes it. This must mean straight through multitubular boilers as fitted to No.14 "Tees" when rebuilt, but according to the "Railway Engineer" magazine No.33 "Shildon" (Fig.64) was later fitted with a Smith's tubular boiler, possibly the same thing. The rebuilding dates were 1848 (Nos.30, 31 & 33) and 1849 (Nos.29, 32 & 34), while No.30 "Wear" was again rebuilt in 1858 with a new firebox and tubes at a cost of £400. They worked well, and nearly all lasted into the 1870's, "Shildon" reaching 1877.

It is believed that after rebuilding two tenders continued to be used, although it is obvious that since the driver and fireman were now at the same end, a single tender only would have been necessary. This may have been so as to use existing tenders until more were built, or a misinterpretation of the well known photograph of this engine from the 1875 album (Fig.64) which shows the tenders in their original position, an obvious piece of deception. West's sketch shows one tender only, in the conventional position.

A further three such were ordered in March 1846; these were to bear the names "Times", "Gypsy" and "Graham". In August the order was put into abeyance, and when reactivated only two engines were built, of quite another type (see Chapter 7.) The intended name "Gypsy" was never used.

The 'Tory' and 'Miner' classes seem to have been fitted with steam whistles, which were then coming into general use, but there is no evidence that earlier engines had them, in spite of George Graham's claim that John Dixon invented the steam whistle. They had been in use in ships for some time before this. There is a report in John Graham's notebooks dated November 20, 1835, that he had been to inspect steam whistles fitted to the engines of the Newcastle & Carlisle Railway, and that it "is not so good a thing as was Represented to be he therefore cant Recomend it". Colburn attributes the invention of the cup whistle to a workman named William Stephens at the Dowlais Iron Works in 1833, and says that Sharp, Roberts then took it up in 1835. I believe that the organ pipe whistle was earlier.

If one examines the photographs available, it is apparent that the whistle was fitted in a different position on almost every engine, which may indicate that they were not fitted originally.

There is an interesting report from John Graham minuted on 18th. June 1841 that "since the engines have been supplied with water at the Yarm Branch, and the extra tank carried by them removed, the loads have all been increased, to 32 wagons for the 5 engines with vertical cylinders, to 36 wagons for the 12 with oblique cylinders and to 40 for the 5 engines with multitubular boilers.

Data as early condition in general, and approximate.
n.k. = not known. n. = nominal. a. = actual.

No.	Name	W.P. lb/in² (p.s.i.)	Boiler length	Dia.	Surface Area F²	Cylinders Bore	Stroke.	Wheels Dia.	Wheelbase	Weight Tons Cwt.
2/ 12.	Bedlington/ Majestic	prob. c.50	13'-0"	3'-10"	300+ (Inness) 265.32 (Hoole)	14½"	16"	4'-0"	2 x 4'-6"	c.20 total working.
13.	Coronation	"	"	"	"	"	"	"	"	"
14.	William IVth.	"	"	"	"	"	"	"	"	"
15.	Northumbrian	"	"	"	"	"	"	"	"	"
17.	Lord Brougham	"	"	"	"	"	"	"	"	"
18.	Shildon	"	"	"	"	"	"	"	"	"
16.	Director	60	10'-0"	4'-4"	500	14½"	16"	4'-0"	2 x 4'-6"	c.20 total working.
19.	Darlington	"	"	"	"	"	"	"	"	"
20.	Adelaide	"	"	"	"	"	"	"	"	"
21.	Earl Grey	"	"	"	"	"	"	"	"	"
22.	Lord Durham	"	"	"	"	"	"	"	"	"
23.	Wilberforce	"	"	"	"	"	"	"	"	"

Majestic class: Straight through multitubular;
Fire tube 9'-0" long x 2'-6" dia. 104 or 106 x 4'-0" long x 1¾" copper smoke tubes.
Tenders; four wheeled, 500 gallons water in leading tender, 30 Cwt. coal in trailing tender.

Director class; Multitubular return tube, Flue tube 8'-0" long x 2'-6" or 2'-4" dia. at firegrate end, 2'-0" dia. at combustion chamber, 91-104 x 1¾" copper tubes (Ahrons states 52 - 70 tubes).
Tenders; four wheeled, 600 gallons water in trailing tender, 2 tons coal in leading tender.

Craven notes - Whishaw:	Boiler	Cylinders	Firetube	Ret. tubes	H.S. f²	Weight
No.19 Darlington	9'-9" x 4'-3"	14½" x 16"	9'-6" x 2'-5"	89 x 1⅜"	R.T. 326.14 F.T. 86.5 Total 412.64	11¾ tons gross 10¼ " net

Nov. 1840 list:

Hawthorn engines:

No.19 Darlington	9'-6" x 4'-4"	14¾"	9'-0" x 2'-3"	88 x 1¾" x 6'-6"
No.21 Earl Grey	" "	"	" "	" " x 7'-0"
No.23 Wilberforce	" "	"	" "	100 " x 6'-6"

Stephenson engines:

No.16 Director	Boilers, Cylinders & firetubes as Hawthorn	80 x 1¾" x 6'-5"
No.20 Adelaide	" " " " " "	88 x " x 6'-11"
No.22 Lord Durham	" " " " " "	80 x " x 6'-6"

No.	Name	W.P. lb/in² (p.s.i.)	Boiler length Dia.	Surface Area F²	Cylinders Bore Stroke.	Wheels Dia.	Wheelbase	Weight Tons Cwt.
24.	Magnet	prob. 60	Nov.1840 list 11'-0" 4'-4"	n.k.	15" 16"	4'-0"	probably 2 x 4'-6"	n.k.

As Director class but fitted with 110 tubes 7'-6" long x 1½" i.d. (Inness)
" " " " x 1 7/16" (Whishaw)

Nov. 1840 list gives 100 tubes 7'-6" long x 1¾".

Tenders probably as Director class.

Rebuilt 1850(?) at Soho Works.

No change in wheels or wheelbase, boiler c.13'-0" long x 4'-0" dia. at ends, 3'-0" in middle.

25.	Enterprise	c.50 Craven:	13'-0" 4'-3" 13'-6" 4'-4"	n.k.	11⅛" n.k. 11" 20"	4'-0"	probably 2 x 4'-6"	n.k.
			Firetube 13'-0" x 2'-1" dia., return tube 13'-0" x 1'-3" dia.					
		Rebuilt 1842 as No.8 "Leader", Tory class, Soho Works.						
30./ 11.	Beehive	" Craven:	" " 13'-6" 4'-4"	n.k.	" n.k. 11⅛" 20"	"	n.k.	n.k.
			Firetube 12'-6" x 2'-1" dia,, return tube 12'-6" x 1'-6" dia.					
		Rebuilt 1842 as No.11 "Bee", 'Tory' class, Shildon Works.						
31./ 12.	Briton	c.50	13'-0" 4'-3"	n.k.	11⅛" n.k.	"	n.k.	n.k.
		Craven details. as No.11 "Beehive"						
		Rebuilt 1842 as No.12 "Trader", Tory class, Soho Works.						
14.	Tees	"	" "	n.k.	" n.k.	"	n.k.	n.k.
		Craven details as No.25 "Enterprise", except cyls. 11⅛" bore.						
		Rebuilt 1844 into Tory class, Shildon Works.						
		Rebuilt 1848 as "firebox" engine, Shildon Works.						

Original dimensions probably similar to rebuilt No.2 "Hope" as 0-6-0.

No details of tenders. Return flue boilers as built.

"Tees" fitted with straight through multitubular boiler when rebuilt 1848.

Scaled from drawing; boiler 12'-4" long x 4'-4" dia., wheelbase 2 x 4'-6",
crank throw = 12" = 24" stroke, bore c.15", w.p. probably as Tory class.

No.	Name	Ordered	Builder /works	Delivered	Price £	Rebuilt	Builder /works	Cost £	Disposal
15.	Tory	16-2-38	T.H. /Soho	23-11-38	1020	early 1853 & 1857	S.W.Co.	647-19-0 (J.D.) 450 (Craven)	Rep.1869
17.	Whig	16-2-38	Kitching	15-3-39	1020	April 1842 June 1853	 S.W.Co.	? 650	 Rep.1870
10.	Auckland	11-1-39	T.H. /Soho	23-8-39	900 + Experiment	early 1853		500	Rep.1870
6.	Dispatch	12-4-39	T.H. /Soho	Aug. 1839	1040	1844 1851		? 200	 Rep.1869
9.	Middlesbro'	7-9-38	Lister	23-8-39	955				Cut up 1858? Rep.1865
18.	Etherley	7-9-39	Lister	7-2-40	955	Jan.1843 1850	S.W.Co. S.W.Co.	? ?	 Rep.1869
26.	Pilot	22-2-39	Kitching	5-6-40	1150	Sept.1852		300	Sold 1868/9 Rep.1868
28/13/12	Ocean		Lister	Aug.1840	955	June/July 1843 1853	S.W.Co. S.W.Co.	? 300	 Rep.1870
27.	Witton Castle	5-7-30	N. & W.	15-5-40	975				Cut up 1866? Rep.1867
4.	Stockton	28-8-40	F. & H.	9-6-41	980	early 1851 (Fossick's Patent) & 1857	?	355	Rep.1870
5.	Hopetown	28-8-40	Kitching	16-7-41	980				Rep.1869
7.	Prince	26-12-40	S.W.C.	March 1842	1330				Blew up 1859 Rep.1869
8.	Leader (ex No.25 Enterprise)	rebuild	T.H. /Soho	April 1842	900 + Coronation				Scr.1868 Rep.1869
12.	Trader (ex No.12 Briton)	rebuild	T.H. /Soho	30-9-42	900 + Coronation				Sold 1867/8 Blew up 1868
11.	Bee (ex No.11 Beehive)	rebuild	S.W.C.	Sept.1842 May (J.D.)	550		S.W.Co.	?	Rep.1867
14.	Tees	rebuild	S.W.C.	1844	550	May 1848	S.W.Co.	?	Rep.1871
16.	Stanhope	20-12-44	Kitching	1-4-45	1160				Rep.1870
25.	Derwent Now preserved	27-2-45	Kitching	Nov. 1845	1140				Sold 1868 Rep.1869
24.	Skelton Castle	27-2-45	Kitching	May 1846	1140	early 1852	S.W.Co.	230	Rep.1870
22.	Alert	12-11-47	Kitching	Jan. 1848	1400				Rep.1869

Data approximate. Listed in order of building.
n.k. = not known. n. = nominal. Rep. = Replaced. Scr. = Scrapped. J.D. = John Dixon.
Boiler type; R.F. = Return Flue; M.T. = Multitubular Return Tube; M = Single Multitubular.

In each case line 1. gives data from R. H. Inness, line 2. from the Hoole/Craven notes and the 1840 list of dimensions as far as that applies; i.e. up to November 1840. Details of tubes are given with length and dia. of fire tube first and that of return tube or tubes second.

No.	Name	W.P. lb/in² (p.s.i.)	Boiler length	Boiler Dia.	Type	Tube details	Cylinders as built Bore Str.	Cylinders later reb. Bore Str.	Wheels n.dia.	Wheelbase
15.	Tory	(1) 60/70	n.k.	n.k.	R.F.		12" 22"	16" 18"	4'-0"	2 x 4'-4½"
		(2)	13'-6"	4'-4"	"	13'-0" x 2'-1"	12⅝" 18"			
						13'-0" x 1'-8"				
17.	Whig	(1) "	n.k.	n.k.	R.F.		12" 22"	16" 18"	"	" " "
		(2)	14'-0"	4'-4"	"	13'-9" x 2'-2"	12½" 18"			
						13'-9" x 1'-8"				
					M.T.	14'-4" x 2'-6"				
						103 x 1 13/16" x 12'-0½"				
10.	Auckland	(1) 70	14'-6"	4'-4"	M.T.		12" 22"	16" 18"	"	" " "
		(2) all as No.15 "Tory"			R.F.					
6.	Dispatch	(1) 70	prob. as No.10		M.T.		14" 22"	16" 24"	"	" " "
		(2)	13'-4"	4'-4"	"	12'-6" x 2'-3"				
						70 x 1⅜ x 10'-0"				
9.	Middlesbro'	(1) 60/70	n.k.	n.k.	R.F.		12" 22"	16" 18"	"	" " "
		(2)	14'-0"	4'-4"	"	13'-9" x 2'-2"				
						13'-9" x 1'-8"				
28/13/12	Ocean	60/70	n.k.	n.k.	n.k.		14" 20"	16" 18"	"	" " "
18.	Etherley	60/70	n.k.	n.k.	n.k.		14" 22"	15" 24"	"	" " "
26.	Pilot	(1) 70	prob. as No.10		M.T.		14" 20"	16" 18"	"	2 x 4'-6½"
		(2)				12'-0" x 2'-4"				
						72 x 2" x 10'-1"				
27.	Witton Castle	(1) 70	13'-6"	4'-4"	M.T.		14" 20"	16" 18"	"	2 x 4'-6"
		(2)				12'-7" x 2'-2"				
						80 x 2" x 10'-1"				
4.	Stockton	70	17'-3"	4'-4½"	n.k.		15" 24"	16" 24"	"	2 x 4'-6½"
5.	Hopetown	70	n.k.	n.k.	M.T.	similar "Pilot" ?	15" 20"	15" 24"	"	2 x 4'-6"
8.	Leader	70	n.k.	n.k.	M.T.		14" 22"	16" 24"	"	2 x 4'-6"
12.	Trader	70	prob. as No.8		M.T.		14" 22"	16" 24"	"	2 x 4'-4"
7.	Prince	70	n.k.	n.k.	M.T.		15" 18"	16" 24"	"	2 x 4'-6"
16.	Stanhope	70	c.14'-3"	4'-6"	M.T.	As No.25.	15" 24"	15½" 24"	"	2 x 4'-6"
						11'-8¾" x 2'-5½" o.d.				
25.	Derwent	70	as No.16		M.T.	99x1⅞"o.d.x11'-9½"	15" 24"	15½" 24"	"	2 x 4'-6" *
						691·5 sq. ft. total H.S.				

No.	Name	W.P. lb/in² (p.s.i.)	Boiler length	Boiler Dia.	Type	Tube details	Cylinders as built Bore Str.	Cylinders later reb. Bore Str.	Wheels n.dia.	Wheelbase
24.	Skelton Castle	70	as No.16		M.T.		15" 24"	16" 18"	"	2 x 4'-6"
22.	Alert	70	as No.16		M.T.		15" 24"	15" 24"	"	2 x 4'-6"

No.	Name	W.P. lb/in² (p.s.i.)	Boiler length	Boiler Dia.	Type	Tube details	Cylinders Bore Str.	Wheels n.dia.	Wheel-base
11.	Bee	70	n.k.	n.k.	?R.F	14" 16" 16" 18"	14" 22"	4'-0"	2 x 4'-6"
14.	Tees as rebuilt	"	n.k. 12'-6"	n.k. 4'-6"o.d.	?R.F. M	15" 22" n.k. n.k.	16" 24"	"	" " "

* as measured by Alan Prior from No.25 "Derwent"; Inness gives 2 x 4'-4½".

Tubes etc.; No.27; Flue tube 2'-5" dia. at firegrate - 1'-10½" at Comb. Chamb., + 92 x "2 dia. tubes.
No.10; 103 x 12'-0" long x 1¾" tubes.
No.8 ; 69 x 2" tubes.
No.4 ; Firebox cylindrical copper, 6'-9" long x 3'-8½" dia. + 150 x 10'-11" long x 1⅞" tubes.

Data as early condition in general, and approximate.
n.k. = not known. n. = nominal. a. = actual.

No.	Name	W.P. lb/in² (p.s.i.)	Boiler length	Boiler Dia.	Surface Area F²	Cylinders Bore	Cylinders Stroke.	Wheels Dia.	Wheelbase	Weight Tons Cwt. in W.O.
29.	Miner	75	13'-0"	4'-4"	1363	15"	24"	4'-0"	2 x 4'-5½"	22 7 engine.
30.	Wear	"	"	"	"	"	"	"	"	"
31.	Redcar	"	"	"	"	"	"	"	"	"
32.	Eldon	"	"	"	"	"	"	"	"	"
33.	Shildon	"	"	"	"	"	"	"	"	"
34.	Driver	"	"	"	"	"	"	"	"	"

Copper firebox 3'-11" high, 3'-5" long, 3'-1" wide, 213 x 1⅞" x 14 W.G. Low Moor iron tubes.
Front tender 700 gallons water, rear temder 2 tons coal.
Rebuilt into 'firebox' engines later, 1848/9, no details.

Chapter 6. Passenger Traffic: 1837-60

No.9 "Globe" and the two 'Planet' type engines had been the S. & D.R's first passenger locomotives, joined later by "Magnet" and others from time to time. Now it was time for more.

Stephenson had brought out the 2-2-2 'Patentee' type, an enlargement of the 'Planet' class, in 1833, and soon everyone was building similar engines. Hackworth's version was a very up-to-date looking locomotive, based on one built the year before for the first Russian railway, with firebox, smokebox and a multitubular boiler, ash pan below the firebox, and a tender that at last began to resemble the later form, with fully sprung wheels. The engine, of course, also had fully sprung suspension with hornblocks, and probably buffers on the front beam, as Graham shows it. The accepted price was £1400, for delivery on 1st. March, 1837, but there were delays.

No.26 "Arrow" was delivered eventually in May 1837. It had double frames with 5'-0" diameter driving wheels and 3'-6" diameter leading and trailing wheels of spoked type, materials not known but probably cast iron, although by now malleable iron wheels were coming into use.

There is a sketch of it by Theodore West, and a better version based on this and published in the "Engineer" of January 1880 (Fig.68), there attributed to a Mr. Graham (probably George). Its appearance is more likely to be evident from a drawing of the Russian locomotive, a little larger but very similar, built in 1836 (Fig.69). "Arrow"'s boiler had a diameter of 3'-4" and the length seems to have been 7'-9" over the tubeplates, with 135 x $1^1/_2$" tubes according to Inness, these being 8'-0" long. This was probably the i.d., as other sources give $1^3/_4$" tubes. The cylinders were between the frames at the front, and apart from the fact that gab gear was still being used, it all looked almost modern. Similar looking engines were still in use at the turn of the century. The bogey of boiler explosion was well guarded against; the illustration (Fig.69) shows two Stephenson type column valves and one deadweight lever valve. The regulator appears to have worked in a horizontal plane just above the firebox, and what looks like a water gauge glass is visible on the backhead.

The cylinders had unusual dimensions, 17" bore by 9" stroke at first, although it seems that Hackworth's original specification stated 14" x 18" cylinders. Maybe there was some intention to reduce piston speed and bore wear; apparently the Liverpool & Manchester Railway had been carrying out some such experiments, and Hackworth would have heard of it. It would seem that this arrangement was modified some years later by means of a lever linkage which doubled the crank throw without changing the piston stroke - a very strange device. Tomlinson said that this could not have been Hackworth's idea, and attributed it to a Darlington grocer! At the same time the cylinder bore seems to have been increased to 21" or 22" diameter (21" in the 1840 list), which is understandable in view of the reduced mechanical advantage of this stroke increasing lever, although the crank throw must have been increased accordingly. Why the connecting rods then fitted are shown as bent, goodness only knows. The engine was also fitted, at least at first, with a 'traction increaser', a set of small 'jockey' wheels which could

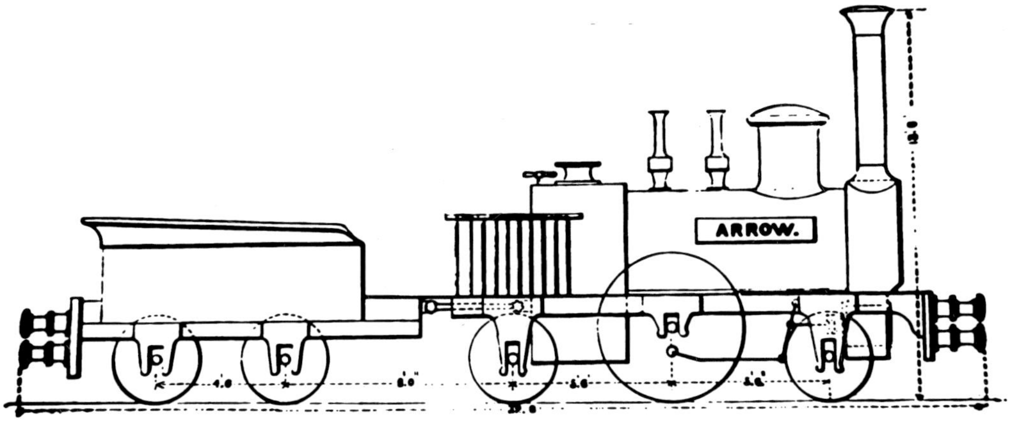

Fig.68 Sketch of No.26 "Arrow", by Mr. (George?) Graham, published in the "Engineer".

be brought into contact with the tyres of both driving and trailing wheels so as to connect them temporarily if slipping occurred.

The engine gave trouble from the outset, and failed time and time again, with demands from the committee for explanations and reassurance. A new cokebox was fitted and changes made to the cylinders and crankshaft. It was not until August 1838 that the engine was behaving sufficiently well for Hackworth to be paid the balance of the price, which he received in September. In December that year Hackworth again had to work on the engine, and this seems to have been when the larger cylinders were fitted. No further major trouble seems to have been experienced, but it must have cost far more than the contract price one way and another. The Shildon Works had a go at it as well, costing a total of £44-11-8d. there alone.

Joseph Tomlinson, who had served his apprenticeship at Shildon, was very critical of the engine in his Presidential Address to the Institute of Mechanical Engineers in 1890, but nevertheless, it seems to have given reasonably good service in the end, and was capable of high speeds for the time. A Mr. Birkbeck, who had also worked at Shildon, when contributing to the discussion after Theodore West's paper to the Cleveland Institute of Engineers in 1886, mentioned in Chapter 1, said that it frequently attained 60 m.p.h. on express passenger service!

The engine certainly had more than its share of troubles. It collided with a train of waggons standing near Whessoe Lane in July 1837, when hauling a 'coaching' train, luckily without any injury to passengers. The coke box was reported as "done" in February 1838 (see above). The driving wheel tyres were reported as worn out in October the same year. The crankshaft broke in September 1841. In August 1838 "Arrow" and "Sunbeam" were being used on alternate trips to compare the fuel consumption, but the results were not noted. It was one of the engines to have axle weights checked in December 1838. "Arrow" and others were giving trouble with wheels and tyres in January 1839, and John Graham noted with some exasperation that he considered "Arrow" and "Queen" to be a nuisance and not worthwhile. It was later numbered 51, when the 'coaching' engines received numbers again in 1846/7, but according to Craven did little work after that, and was finally scrapped in February 1864, fetching a total of £188-13-5d. for the metal and the value of the wheels, etc. It had been written off and taken in for scrap by the Shildon Works in December 1858.

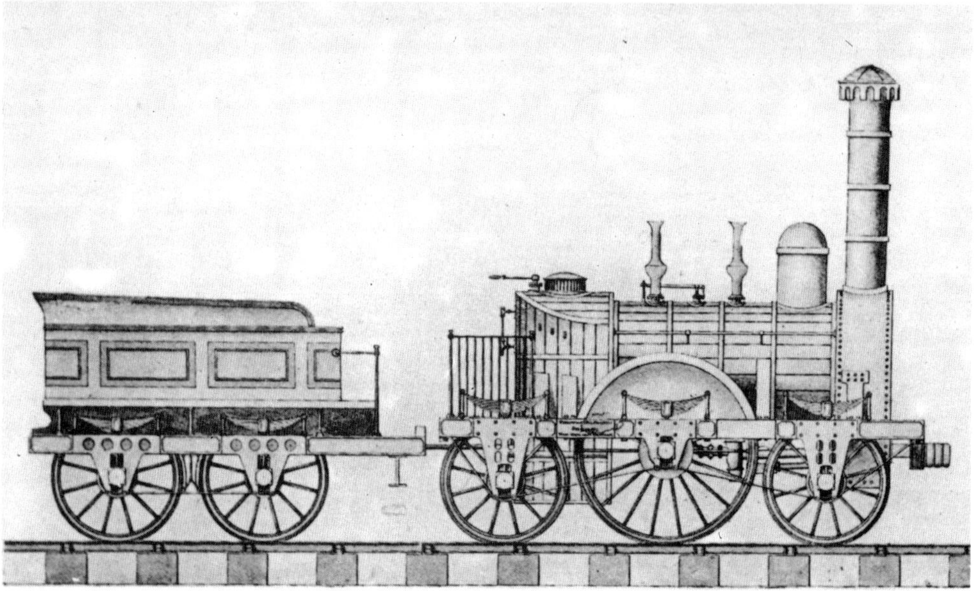

Fig.69 Drawing of Hackworth's engine for Russia, 1836.

The next engine, in numerical order but built in 1836, was "Swift", a peculiar machine from R. & W. Hawthorn. Inness says that it was No.27, but it may have been No.26 for a while before "Arrow" came into service. The works No. is 199, and can be read on the plate shown on the G.A. It was four-coupled but the cylinders, 10" bore x 16" stroke, were mounted vertically between the wheels at boiler top height each side, and it drove through an intermediate shaft which carried the eccentrics and was connected on each side to the centre of the coupling rod.

This rather odd arrangement was more common in early days than I can understand; Crampton used it for tank locomotives, and others used it as well. I find it difficult to see just what advantage there was, while the disadvantages of additional bearings are

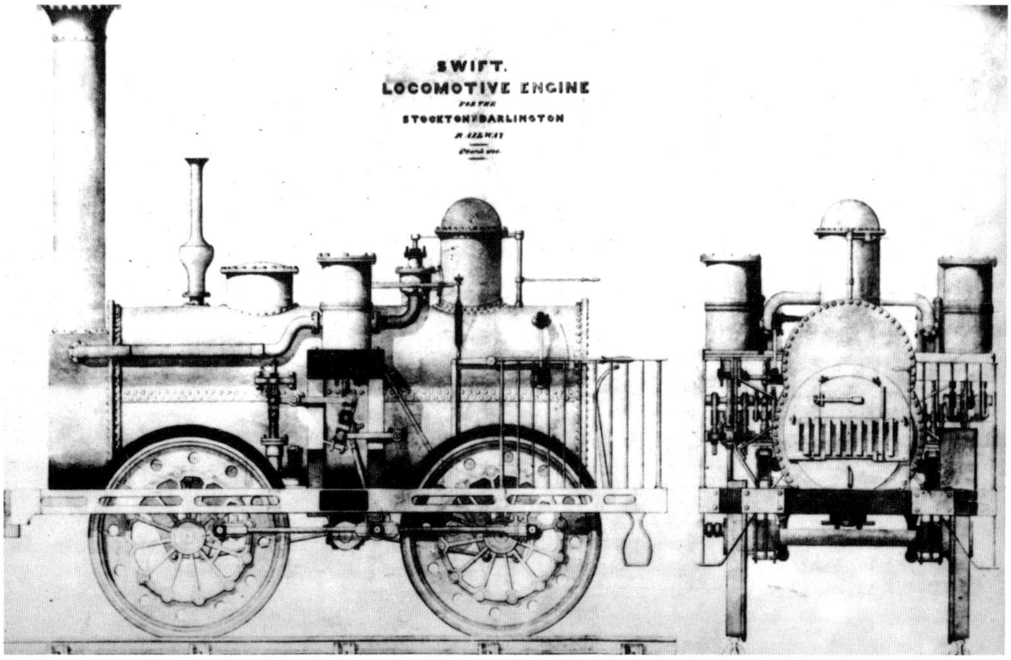

Fig.70 Works drawing of No.27 "Swift", once on display in the foyer at R. & W. Hawthorns' Forth Bank Works.

easy to see. Suspension, of course, as in Hackworth's 'Majestic' and 'Director' classes, but what else? The piston rods were guided by parallel motion of Watt's type, and the feed pump was driven from an extension of one of the parallel motion arms. Several drawings exist, and the best (Fig.70), a works G.A. used also by Inness, gives a cross section which shows that the original boiler section was formed of two intersecting circles forming a figure eight. It was a fairly light engine, of under eight tons (plus water, I assume), and out of date when received, so may have been a bargain offer. Ahrons says that it had a single flue, gab type valve gear with two loose eccentrics and that the weigh-bar shaft passed right through the boiler inside a transverse tube. Woods, according to Craven, gives it a firetube 4′-8″ long x 2′-5″ dia. with 102 x 1⁵/₈″ x 5′-0″ long tubes, which does agree with the boiler length of 10′-0″. The better drawings show a light perforated plate frame. and the fittings include a water gauge glass, the regulator valve being mounted just in front of the dome, with linkage to a horizontal handle behind the dome.

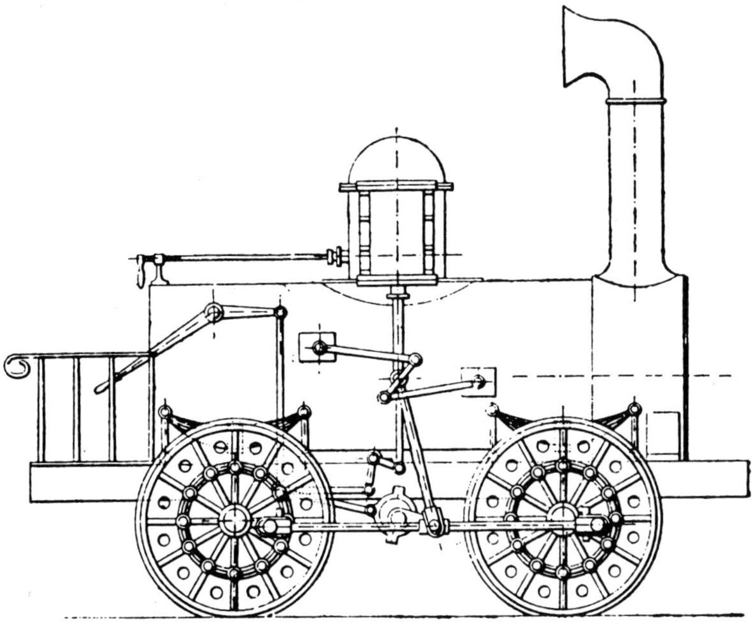

Fig.71 Line drawing of No.27 "Swift", published in Deghilage's "Origine de la Locomotive", 1886.

Theodore West did an earlier and more detailed drawing of it in 1880, before his large charts, but on them he shows "Swift" as having a four wheeled tender with a water barrel mounted athwartships, as on the earlier engines. This engine does not seem to have given much satisfaction.

The wheels seem to have been of an unusual design variation on the Hackworth two part cast iron plug type, with twelve spokes open in the wheel centres. The G.A. shows them of 4'-0" dia., and Inness uses this, but Wood said that they were 4'-7" and Theodore West that they were 4'-6". They could easily have been increased in size later, there is plenty of clearance.

Fig.72 J. S. MacLean's decorative drop capital, showing No.27 "Swift" as rebuilt.

It was mentioned by John Graham in May 1837 as having a flat on one wheel, and the wheels and tyres were giving trouble in January 1839. The minutes record that in November 1836 "Swift" was "throwing fire into the Tourist Coach" (later No.7 and certainly not 1st. class!). Lister was instructed to remedy this, and an article in the "Engineer" in June 1881 states that a revolving cowl was fitted at the chimney top, as shown in Deghilage's drawing (fig.71). I have been told that"Swift" was shown as No.26 in an 1839 list (see above), but it does not appear at all on the 1839 stock valuation list, nor the 1838 or 1849 lists, come to that. I have also been told that it was actually owned by Mr. Wm. Lister, one of the main subcontractors, and was never in fact the property of the S. & D.R., which may explain its omission from these lists, and even the fact that it was there at all; Lister may have got it very cheaply. "Arrow" was said to be No.27 in the same list; another small puzzle. As a 'coaching' engine it was probably not numbered at the time, in any case.

According to J. S. MacLean it was sold in 1839 to the contractors building the Great North of England Railway, together with two others (No.18 "Shildon" and No.10 "Planet"), and was much modified; he has illustrated it (Fig.72) as then rebuilt, with the parallel motion replaced by crosshead and slide bars, and with a different boiler with a conventional round section. In fact the G.N. of E.R. bought "Shildon" themselves, possibly not until construction was complete and the contractor had finished with it, but "Swift" then disappeared. The Deghilage drawing shows an apparently round boiler but with the parallel motion still fitted, so rebuilding may have been in two stages.

Hawthorns also built No.28 "Sunbeam", in 1837. This was a straight forward copy of a Stephenson 2-2-0 'Planet' type with 12" x 18" cylinders, double frames, 5'-0" driving wheels, 4'-0" leading wheels, 8'-0" long boiler of 3'-2" diameter with 104 x $1^{3}/_{4}$" x 8'-4" tubes, with firebox and smokebox. John Graham describes how he and W. Lister went to Hawthorns to inspect it, which suggests that it was another bargain offer, not built to S. & D.R. order, and in fact Ahrons says that it was one of four engines built for the New York, Boston and Providence Railway, of which only one was delivered to America. Craven contradicts this, saying that only three were built, none went to America, and the other two went to France for the Paris - Versailles Railway. The price was £1320, a little less than "Arrow", but obviously better value. It was delivered on the 18th. August, a week after they had seen it.

In December 1837 a tender wheel broke, and the next day the engine was derailed at Darlington. Its driving wheel tyres were reported as being worn out in October 1838, and it was one of the engines giving wheel trouble in January 1839. A new copper firebox was ordered from Hawthorns and fitted in March 1838. In the first half of 1853 new cylinders were fitted, new firebox, smokebox and 104 secondhand tubes, at a cost of £150. I mentioned above the fuel consumption tests in August 1838. The axle weights were checked together with those of "Arrow" and "Queen" in December 1838. In spite of all this, it was said to have given very good service, remained in use until 1863, and was sold in 1864. The number may have been 29 in May 1838, from a Hackworth note. It became No.43 in 1846/7 (fig.73). It does not appear in the stock valuation lists until 1840, when it was shown with a 'coaching' tender. Although bought by the Railway, it seems to have been operated by Wm. Lister most of the time until the Railway took back all the engines. Whishaw gives a few other details.

Kitchings built two quite different passenger engines at about this time. No.29 "Queen" (Fig.74), an 0-4-0, was built in late 1836 and delivered in February 1837, according to Inness. However, John Graham reported in November 1838 that "on Monday Last the New Coach Engine called The Queen was tried with the Coaches, the Engine works well She is Strong and well made and finished; but She is too heavy for the light Rails especialy on the Middlesbro Branch J G would therefore recomend a third Pair of Wheels to be placed under the Foot Board". One of the Whessoe official histories gives the completion date as 1838, and Ahrons agrees with this.

The cylinders, inside the frames and inclined, were 12" x 18" and the wheels at first 4'-6" dia., the boiler 8'-6" x 3'-8" dia. with 86 x 2" x 8'-10" tubes; the price paid was £1470 including the four wheeled tender.

"Queen" had paddlebox like splashers, and in my opinion this was the cause of one of Theodore West's errors, which I shall come to shortly. As mentioned above, the axle loading proved too heavy, and in December 1838 the frames were lengthened and a pair of 3'-0" trailing wheels were added, the cylinder dimensions being increased to 14" x 18" at the same time. The axle weights were then checked again. The wheel tyres gave trouble, as mentioned above, and again in February 1839, this being blamed on the switches being too short, but apart from this it seems to have been quite successful. The engine was derailed at the south end of the Tees suspension bridge in June 1839, caused by the inside rail on the curve being too high. It was later renumbered 40, presumably when "Miner" took the number 29, and by 1842 it had been transferred to 'merchandize' duties. John Dixon's 1849 list gives it a value of £1200, and dates it at 1838. Further rebuilding was carried out in April 1847, and it ceased work in 1850, when it was set aside at Stockton with others. It failed to find a bidder at auction in January 1856, and was sent to Shildon for scrap in December 1858. There was not another No.40 until 1866.

One of the engines sketched by Theodore West in Sheet 1 (Plate III) of his Stockton & Darlington set, was an 0-4-0 called "Comet" and attributed to Hawthorns in 1835. It has a very similar look to "Queen" above, as built. However, the only "Comet" owned by the S. & D.R. was from Kirtley, in early 1841, and it was a 2-2-2. Hawthorns did build an 0-4-0 called "Comet", which had a very similar appearance to the West sketch, but this was for the Newcastle & Carlisle Railway.

Fig.73 Swain's engraving of No.43 (ex No.28) "Sunbeam", from an early photograph.

"Raby Castle", a 2-2-2, was ordered from Kitchings in February 1838, and delivered in October 1839, at a cost of £1480. Inness says that its number at first was 30, and I have used this, but since when it was received all the 'coaching' engines lost their numbers it is far from certain that in fact it ever ran with that number. It was later renumbered 49. This was an up-to-date looking engine, of conventional double framed type with inside cylinders, and Tomlinson has given a good sketch of it (Fig.75). The driving wheels were 5'-0" diameter, leading and trailing wheels 3'-6", and a boiler 8'-0" long x 3'-4½" with a heating surface of about 700 sq.ft., according to Inness, but Craven differs a little. The tubes were apparently 102 x 1¾" x 8'-3" long. The valve gear was four eccentric gab type. It gave good service, but this was marred by a fatal accident at the Mandale Road crossing, when it ran down a waggoner. A charge of manslaughter was brought against the driver, Matthew Appleton, which seems to have been proved, and a deodand of £1400 was raised against

Fig.74 Drawing of No.29 "Queen", originally published in the "Engineer".

the engine, a legal penalty that was rare even then. John Dixon valued it at £1300 in his 1849 list, and it was withdrawn in June 1858. Inness says that the cylinders were 12" x 18", later increased to 13" bore, but an early drawing, used also in the Whessoe official history (Fig.76) shows them as 13" x 20". John Graham's only note on this engine was that the firebox failed in January 1840. There is an entry in the minutes for the 16th. October 1840 that an account was to be prepared from the original report of an accident involving the "Arrow" and the "Raby Castle", but no details were given.

The Whessoe history quotes from Herapath's Railway Journal in 1841; "The Raby Castle is a beautiful engine for her size and weight and second to none that I have yet seen. She is a steady, active, neat, tight little engine capable of generating more steam than she has the weight to apply it to".

Craven says that she was withdrawn in June 1858 and sold to the Shildon Works Co., from where she was sent to the Soho Works to drive workshop machinery until 1877.

An engine was bought from John Hague of London, in July 1839, following an advertisement in the "Railway Times" of April 1839, an 0-4-0 similar to the four-coupled 'Planet' type. The price paid was £1000 without tender. A tender was bought separately at the same time for £40. John Graham says that it was inspected before purchase by G. Scott, probably the foreman mentioned later, in June, and that it appeared to be suitable. There is a note in Timothy Hackworth's small stores book that in July 1839 it was fitted with a safety valve, the wheels retyred and four new stays fitted between boiler and frame. Its

Fig.75 Sketch of No.30 "Raby Castle, by W. W. Tomlinson.

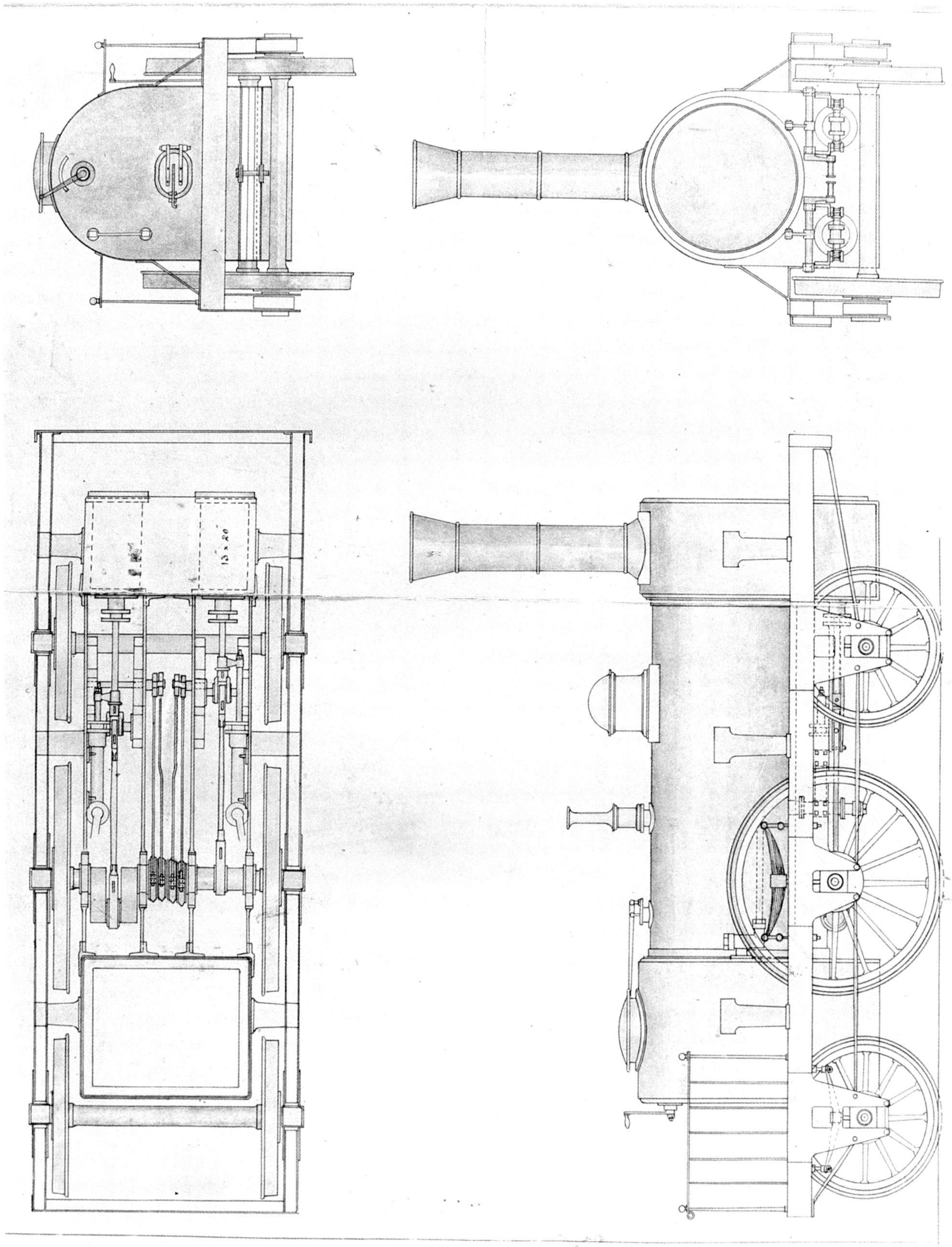

Fig.76 Kitchings' arrangement drawing of No.30 "Raby Castle",

wheels were reported as failing in December 1841, and new ones were ordered from Stephensons, but it was still there in 1849, when it was valued at £800. There is little other mention of it.

No.42 "London" (Fig.77) seems to have been conventional enough, a tender engine with 10" bore inside cylinders, 4'-6" wheels and a boiler 7'-0" long x 2'-4" diameter. Probably it carried no number until 1846/7. Other dimensions from the 1840 list are 11" x 16" cylinders, tubes 71 x 1 3/4" x 7'-4". It lasted until

1853, when it was sold to the Cumnock Iron Co. for £390. Little else seems to be known about this Mr. Hague, or if he built any other locomotives. David Baxter suggests that he was only an agent, in which case who built the engine? Possibly the Rennies, who were active in London at the time? Whishaw gives Hague's address as Cable Street, and describes the wheels as of Losh's design, the inside cylinders as 10" bore x 18" stroke, with two tubes of 3" diameter and 72 (return?) tubes of 1¾" diameter.

It was used mostly on 'coaching' duties. James Lowe, in his "British Steam Locomotive Builders", says that Hague was an early refrigeration pioneer, and gives quite different dimensions: cylinders 11" x 16", wheels as above, firebox 'D' shaped and 2'-7½" long x 3'-3" wide x 3'-7" high, boiler 7'-0" x 2'-7", 71 copper and brass tubes 1¾" x 7'-4"; tender wheels cast iron 2'-10½" diameter with two wooden brake blocks, 491 gallon water tank. It had gab type valve gear.

Fig.77 Theodore West's sketch of No.42 "London"

Craven has a story that it was possibly intended for the London & Greenwich Railway, ordered by their old board of directors, but that this board was ejected and the order repudiated by the incoming engineer, one Miller.

Hackworth designed and built a small passenger engine at the Soho Works, delivered April 1840, at a cost of £1320 plus £200 for the tender, which seems high. No.41 "Dart" (Figs.78-80), an 0-4-0 with inside cylinders but of long boiler type, and still using gab gear. The number would have been allocated later, as for "London". There is a puzzling letter from Hackworth to the Management Committee dated 19th. February 1839, in which he offers them an engine similar to No.28 "Sunbeam", but obviously this is not what happened. Strangely, Craven's extracts from the minutes describe it when delivered as a 'coal' engine. The wheels were of 4'-5" diameter. It was a very useful engine and lasted into N.E.R. days, being rebuilt in the late 1860's and renumbered 1041, in which guise I believe it took part in the 1875 celebrations.

Ahrons says that the cylinders were 13" x 16" and that the wheels were 4'-6" diameter. This is supported by the 1840 list, which gives the boiler as 8'-2" x 3'-3", with 122 x 1¾" x 8'-7" tubes, with a conventional firebox. A larger boiler seems to have been fitted in the late 1860's when it was rebuilt (Fig.80). Its final fate is not known. This must have been the last engine that Hackworth built while still officially the line's chief locomotive engineer, as he retired in May 1840, to devote his energies to his own business at the Soho Works. So far as I have been able to see, he built only two more engines for the S. & D.R., "Leader" and "Trader" (q.v.), in 1842.

Graham visited Leeds in August 1840 to inspect the Leeds & Selby locomotive "Rodney", and recommended it, but the S. & D.R. would not pay the price required and the deal fell through.

The name of William Bouch comes up in September 1840. Craven says that he was sent to the Haigh Foundry, Wigan, to inspect an engine offered for sale at a low price. He reported that "The work was unexceptionable", but the Power Committee decided that the weight was too high, and refused it. Craven thinks that it would probably have been an 0-4-2 goods engine, finally purchased by the Manchester & Leeds Railway, but Lowe does not list it.

Then we come to No.52 "Comet", mentioned above. On October 9th. 1840 John Graham reported that as instructed he had visited Leeds and Warrington to inspect two engines for sale. The engine at Leeds, another Leeds & Selby loco. named "Hood", was old and worn out, and rejected, but the one at Warrington was new and "J.G. considers the Cheapest he has Seen offered for Sale". The builders were Kirtley & Co., of Dallam Foundry, Warrington, a famous name in railway history. This engine was tested at their works satisfactorily early in November, and was to be forwarded shortly. Inness says that it was received in January 1841, but Craven makes it later, saying that it was put on the line on 19th. February and satisfactory. The name "Comet" was allocated in December 1840, and the engine became No.52 in due course. The price asked was £1200, and the engine must have been almost completed when Graham saw it in October.

It was a passenger engine of 2-2-2 type, outside framed, 13" x 18" cylinders, 5'-0" driving wheels (Inness and Lowe both say 5'-6"), 8'-6" x 3'-6" boiler with 118 x 1¾" x 8'-9" tubes. An illustration (Fig.81)

Fig.78 Early photograph of No.41 "Dart"

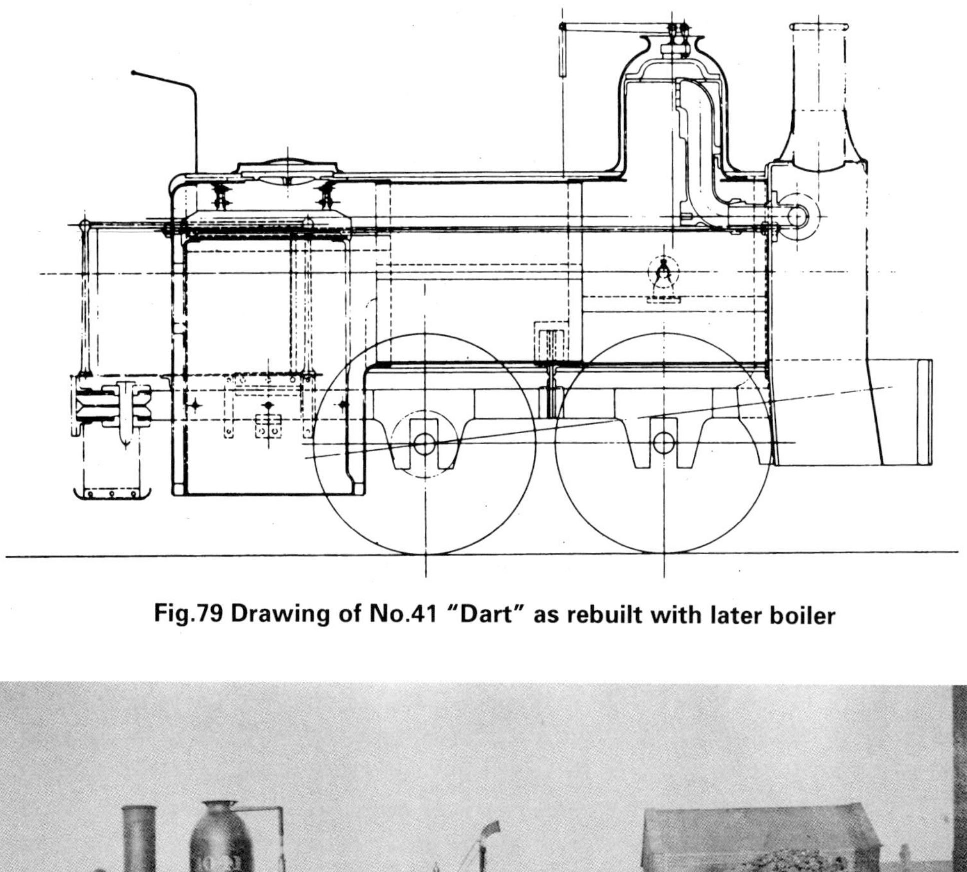

Fig.79 Drawing of No.41 "Dart" as rebuilt with later boiler

Fig.80 1875 photograph of No.1041 "Dart"

has recently come to light which shows clearly that it was a 'Patentee' derivative, probably built under licence. New 14" bore cylinders were fitted in the first half of 1856, 118 tubes put in and the firebox patched at a cost of £200, and in 1858 a new firebox and 120 tubes were fitted. Craven says that the value in 1865 was still £603, but the engine was still on the 1868 N.E.R. list at a value of £1450. The next No.52 was a long wheelbase 0-6-0 in 1875, and Lowe says that "Comet" lasted until then.

On the 23rd. October 1840, Graham reported "That the new Engine from Sharp, Roberts & Co. was tried yesterday and works Satisfactorily". This engine was named "Manchester" (Fig.82), and in due course became No.53. This firm became Sharp Brothers a few years later. The price was £1390; obviously again an engine made to another order and not delivered, so at reduced price, and ready for use, their works No.94. It was a 2-2-2 of their standard design, sandwich framed, with 12" x 18" cylinders, 5'-0" driving wheels and an 8"-0" long x 3'-3" diameter boiler,with 89 x 2" x 8'-4" tubes.

These engines, widely known as "Sharpies", were derived from the Stephenson 'Patentee' type, and Craven says "perhaps the most consistently successful of its period on the main line railways". John Graham reported that the crankshaft broke in May 1841, and new cylinders were made for it at the Shildon Works in October 1845. Craven says that it remained on the active list until about 1860, and that it was ordered to be broken up at the Shildon Works in August 1866.

Laird Kitson & Co. of Leeds offered an engine to the Railway in June 1841, 12" x 18" cylinders, about 13 1/2 tons, for £1365 plus £215 for the tender, but the offer was turned down. Obviously tenders are getting dearer, probably because they now have to carry both coal and water, and because the construction is better.

Sharp Bros. and Stephensons were both asked to tender for a new 'coaching' engine in August 1841, this engine to be like the "Manchester". Sharps' price was £1340 and Stephensons' £1400, but the latter was accepted, possibly influenced by an offer to let the Railway try out a similar engine first, and also remembering Pease involvement with this firm. In 1842 (Inness says October, John Dixon and John Graham both say April, while Craven gives the actual date from the minutes as 15th. April), Stephensons delivered this new 2-2-2 engine, outside framed, with a 'Gothic' domed firebox and Salter type safety valve mounted on top of it (Fig.83).

This drawing, by Inness, shows the engine fitted with the later standard four buffer arrangement, and Theodore West's 1880 drawing of "Swift" also assumes that these were fitted: the photographs of "Dart" show these but as fitted at rebuilding. The subject merits a small digression; most of the earlier engines are shown with dumb buffers or beams, the 'coaching' engines having single buffers, but all these drawings were made later and it is very difficult to be sure when the double buffer practice began. Ken Hoole has some notes on the matter, but does not give a date for the start, saying only that

Fig.81 Old painting of No.52 "Comet"

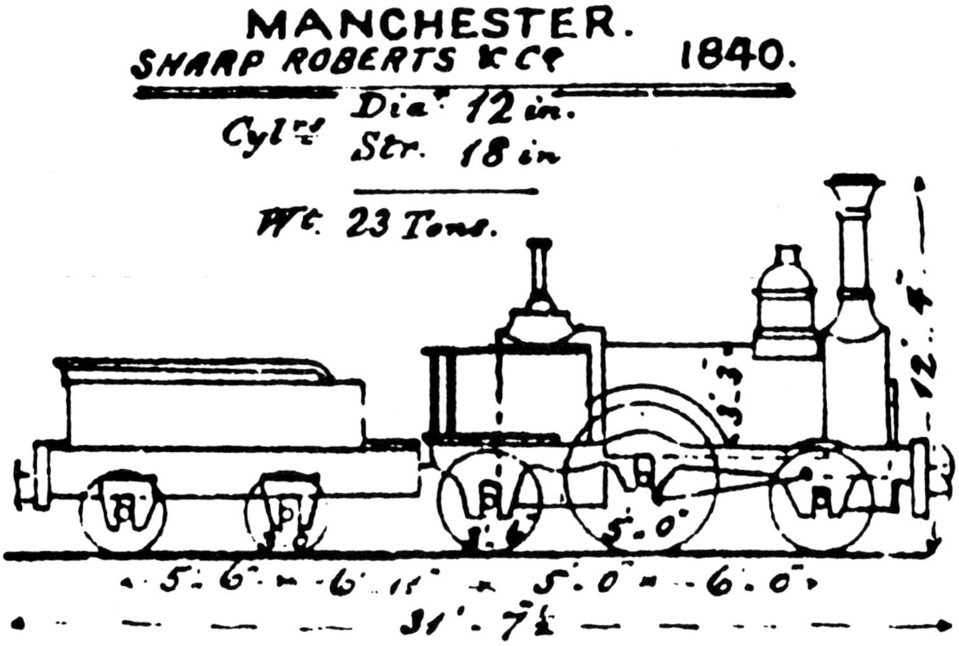

Fig.82 Theodore West's sketch of No.53 "Manchester"

they were not fitted to passenger engines built in the mid 1860's, and were abandoned generally in the 1870's. However, the final classes of long boiler 0-6-0 mineral engines built in 1875 were fitted with buffer beams designed to carry double buffers, whether they were ever fitted or not, and No.1275 is still so provided. It may be noted that the Kitching general arrangement drawings for the "Pierremont" class of 1855 show the upper pair of buffers but only a dumb buffer beam arrangement for the lower ones, while the earlier drawing for "Raby Castle" of 1839 shows a dumb upper beam only. The two sets of buffers were fitted as follows; those for chaldron wagons at 2′-1″ above rail level and 3′-7″ apart, and those for standard rolling stock at 3′-3″ above rail level and 5′-8″ apart.

Returning to the engine in question, it was named "Tyneside" and in due course numbered 54, had 6′-0″ diameter driving wheels, the biggest so far on this railway and for some further time to come, leading and trailing wheels of 4′-6″ dia., inside cylinders of 13″ bore x 18″ stroke, a boiler 8′-11″ long x 3′-5″ diameter , with Stephensons' link motion with box type links according to Inness. If so, it must have been one of the first locomotives so fitted. The boiler pressure was now up to 90 p.s.i. The works No. was 354, and the actual price paid was £1475, maybe including some extras.

Fig.83 Drawing of No.54 "Tyneside"

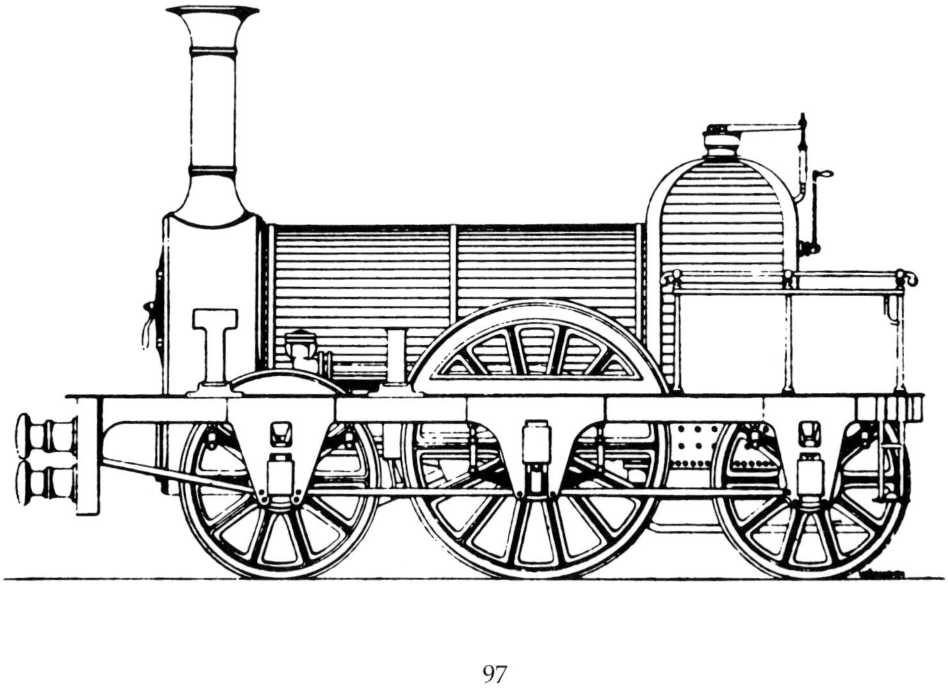

In the first half of 1853 new 14" x 20" cylinders were fitted, new firebox and wheels, and 135 tubes, at a cost of £840. This engine was still on the list in 1868 at a value of £1450, and Inness says that it lasted until 1870. Neither Warren nor Lowes mentions it.

John Graham had inspected another "Tyne Side" in May 1839, a second-hand engine from the Clarence Railway, but said that its steaming was very poor and rejected it.

Another passenger engine was designed by William Bouch soon after he joined the Shildon Works, at an agreed cost of £1380. This was No.50 "Meteor", of which several illustrations exist, the best known showing a hotel in the background which may well have been the Temperance Hotel at Saltburn (Figs.84 & 85). This engine, another 2-2-2 with outside frames, and inside cylinders, had 5'-0" driving wheels and Stephenson link motion. Ahrons says that it was designed "after the pattern of a 'Sharp'", and that the cylinders were 12" x 18" when new. It was delivered in June or July 1843, and lasted until the 1875 Jubilee celebrations, after which it was scrapped. It is mentioned at the foot of the July 1843 stock valuation list, but had not then actually been taken into stock, although valued at £1580, was not valued in John Dixon's 1849 list, and in the 1868 list had a value of £1450. It was 'remodelled and enlarged' in the first half of 1856 at a cost of £883, and again rebuilt in 1857 with 14" x 18" cylinders at a cost of £500.

The first of these photographs was probably taken before its final rebuilding, but possibly will not have shown the engine in its original form. A later photograph, used in souvenir albums at the 1875 Jubilee, shows the engine bearing the original number 50, whereas almost every other S. & D. engine present seems to have had the N.E.R. 1xxx added, whether in working order or not.

Fig.84 Early photograph of No.50 "Meteor", probably at Saltburn

An article in the "Engineer" in 1879 says that the "Meteor" had triple frames, two inside and one outside on each side.

One of John Graham's last reports, in December 1840, advises that the 'coach' engines be fitted with brass feed pipes instead of leather, so that the feed water could be heated with waste steam. This was followed in January 1841 by a Shildon Works Committee minute saying that all the coach and merchandise engines were to be fitted with new telescopic brass feed pipes similar to those now attached to the "Manchester"

The Bury Engines

The Stockton & Darlington railway bought several Bury engines, in two batches, all of them second-hand.

Edward Bury, of Liverpool, was the senior partner in a firm which became eventually Bury, Curtis & Kennedy. The works were known as the Clarence Foundry.

He was also the Superintendent Engineer of the London & Birmingham Railway, which was under construction in the late 1830's, opening in 1838. Robert Stephenson was the civil engineer in charge of the building of this line, one of the greatest pro-

Fig.85 Engraving of No.50 "Meteor", by J. Swain of the "Engineer" magazine, from a photograph published in connection with the 1875 celebrations

jects of the time. The James Kennedy concerned, originally Bury's foreman, had been the Stephensons' foreman at the Forth Street Works at the time when the first locomotive for the Stockton & Darlington Railway was being built, in association, it is believed, with Timothy Hackworth. Ahrons attributes most of the early design work on the Bury engines to Kennedy. He also carried out the S. & D.R. 1837 stock valuation.

His basic design covered a very light locomotive with bar frames instead of the more usual plate type; these bars being of rectangular section, welded and riveted together, and attached to the boiler by brackets at smokebox and firebox. The firebox was of so-called 'Haystack' design, with a high rounded dome which usually carried the safety valves and whistle. The firebox itself was vertical and cylindrical under this dome, and the grate was 'D' shaped. Originally the firebox was of iron, but later became made of copper, dome and all. The cylinders were inside the frames at the front, and Bury was probably the first engineer to use such a design for a railway locomotive, adapted from Gurney's steam road coach patent of 1827, in spite of Stephenson and Hackworth claims, but the design was not ready in time to be the first to run in service. He used two types of wheel arrangement, 2-2-0 for general and passenger use (Fig.86), and 0-4-0 for goods use (Fig.87); the wheelbase was short and his engines could cope easily with short radius track curves. Later he added a pair of trailing wheels. He tended to use larger driving wheels than were then usual, up to 6'-0" diameter.

The wheels were spoked, the spokes consisting of round bars welded to brackets which were riveted to the rims, and secured to the hubs by riveting inside the hollow hub, I believe. I have no knowledge of the original type of valve operating gear, but it must have been of gab type at first; a drawing exists of the valve gear of a London & Birmingham engine before 1842 which shows four eccentric gab gear.

His first engine was under construction in 1829, but the crank axle broke, and the engine had to be rebuilt. Because of this he was unable to submit an entry for the Rainhill Trials in October 1829. It is not known what tube or flue arrangement was used for the first engine, but the second engine, "Liverpool", is said to have had convoluted flues with a bellows mounted under the tender. Ahrons says that the rebuilt engine had a multitubular boiler. Bury supplied an early engine to the Liverpool & Manchester Railway, and most of the early engines on the London & Birmingham Railway later.

Many other engine works built Bury engines under licence, and at least two examples are still preserved. His design was taken up by Norris Brothers of Philadelphia, and the bar frame principle was used for the standard American type for many years, with the leading pair of wheels changed to a four wheeled bogie.

I am grouping all the seven S. & D.R. Bury engines together for convenience, although there were in fact ten years between the two batches.

The first batch was purchased at the end of 1844. They were second-hand, and from the Midland Counties line (by then part of the Midland Railway) according to Inness. They comprised M.R. Nos.74, 75, 90, 94 & 95, and George Graham describes vividly how he and a foreman named Scott (see earlier) went to Derby and brought four of them back in October 1844, mostly in the dark and over unfamiliar metals, with one engine in steam but leaking badly and hauling three others. I assume that the fifth engine was used for spares.

The first two had been built by Fairbairns of Manchester. M.R. No.74 became No.44 "Sun", and was a 2-2-0 of the early type, with 12" x 18" cylinders, 5'-6" driving wheels and gab gear. G. H. Daventry, however, said that this engine was built by Jones, Turner & Evans of Newton, Lancashire, for the Midland Counties Railway, who had named it "Sunbeam". Craven supports this, saying that it was M.C.R. No.4, built by this firm in April 1839. David

Fig.86 Photograph of an early Bury 2-2-0 built for the Liverpool & Manchester Railway

Baxter has suggested that it may well have been overhauled by Fairbairns before delivery to the S. & D.R., but George Graham's account seems to rule this out, and John Dixon's 1849 list says clearly "Fairbairns of Manchester". Craven considers this an error. In any case, there was already a "Sunbeam" (No.28/43) in the list, and so the name had to be changed. Craven says that this engine first appears in the Shildon small-stores book on 8th. November 1845, and may have been refitted in the interval.

M.R. No.75 "Gannymede", built by Fairbairns in August 1840 as M.C.R. No.41, became No.45, and was a four-coupled engine with 12" x 18" cylinders, 5'-0" wheels and gab gear. Both had four wheeled tenders, and were used at first mainly on passenger work. They seem to have given good service; eventually they were used for shunting duties around Shildon, and "Sun" was withdrawn at the end of the 1860's, while "Gannymede" was listed in the N.E.R. 1868 stock list (£1180) and lasted several more years, being scrapped in 1875. Craven says that No.44 "Sun" was fitted with new 13" cylinders in the first half of 1855, ran regularly until mid 1859, and then now and again until 1865, was then written down to £350, and was replaced in April 1868, when a new 0-6-0 mineral engine bearing this number was built. No.45 "Gannymede" received new cylinders of 13" bore, frame and boiler plates, to a total of £110, in the year ending June 1854, and continued working until November 1873.

The next two, becoming No.46 "Antelope" and No.47 "Unicorn", had been built in August 1840 by Edward Bury's Clarence Foundry at Liverpool, although the name of Rothwell has also been mentioned. They were both of 2-2-0 type with 12" x 18" cylinders, 6'-0" driving wheels, the usual Bury domed fireboxes and bar frames. Craven says that when delivered they had similar 5'-6" wheels to the

Fig.87 Engraving of No.1089, ex 89 "Huddersfield", by J. Swain of the "Engineer" magazine, from a photograph taken in connection with the 1875 celebrations

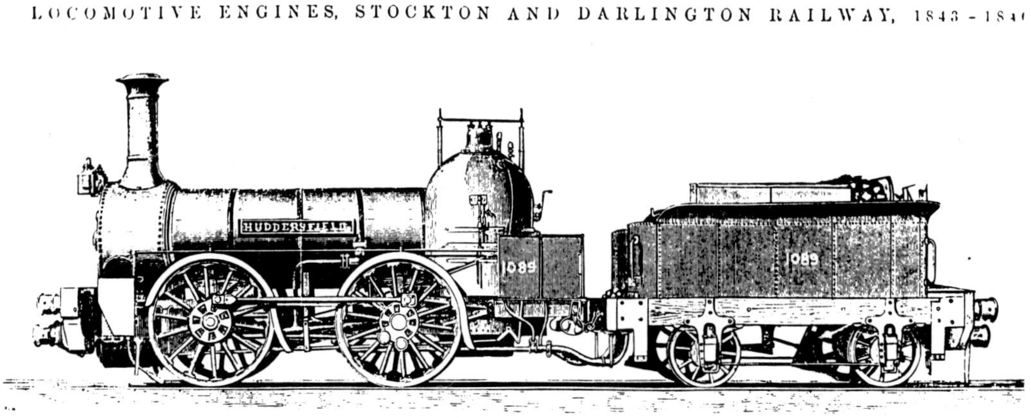

others, and that the larger wheels were probably fitted at Shildon. Their M.C.R. Nos. were 14 & 15. They did not last as long as "Gannymede" although used on similar duties to the previous pair; by 1869 "Antelope" had been scrapped, and "Unicorn" was still on the list at £1130. Craven says that in 1849 No.45 was rated as a 'second-class Mineral engine', while the other three were still 'coaching' engines. Nos.44 & 47 were in the spare stock list in July 1850.

"Sunbeam" was the only one to have its name changed; the other three kept their original names, and as passenger engines received their S. & D.R. numbers in 1847. Some at least of these four engines were used on passenger service between Middlesbrough and Redcar, and George Graham says that salt water was used in their boilers there. It has to be said that there are so many different accounts of the original source, the builder, that is, of these four or five engines that one can believe pretty well what one likes, although I believe that John Dixon's 1849 list and the N.E.R. 1869 list may be relied on, and these are what I have used for my own list. Craven may be right about "Sun/Sunbeam", but I have no way of checking this.

Three more second-hand engines were bought in November 1854, all apparently of the same type, four-coupled with 15" x 24" inside cylinders, 5'-0" wheels and four wheeled 900 gal. tenders. They were of Bury make, built at the Clarence Foundry in 1845/6, and had come from the Lancashire & Yorkshire Railway, previously the Manchester & Leeds, carrying Nos.154, 155 & 156. According to Craven, the first two had been named "West Riding Union" and "Cleckheaton", and the third was not then named. Two of them became No.87 "Fryerage" and No.88 "Deanery", while "Huddersfield", according to Inness and others, retained its name, which must have been given later, and took No.89. There is a well known photograph of this last engine carrying the N.E.R. No.1089 at the 1875 Jubilee (Fig.87). They were used mostly in the Shildon area on short mineral runs and shunting, and lasted until 1882 for "Fryerage", 1871 for "Deanery" and 1877 for "Huddersfield". The "Railway Engineer" gave their wheel size as 4'-9" diameter, but it may have been changed.

It has been said that these Bury engines were popular, easy to handle, and so light that, when derailed, as happened not infrequently, they could almost be lifted back on to the lines!

Mainly 0-4-2's & 2-4-0's

In Chapter 1 I referred to the locomotive "Active" which appeared later. This was built by Kitchings in 1844, their works No.11, an 0-4-2 type, and became No.48. It is likely that this wheel arrangement followed the necessity to give the "Queen" trailing wheels because of original high axle loading. Inness does not mention this engine at all, rather strangely, since he does include No.55 "Wolsingham", built a little later and very similar. West has given a sketch of "Active" (Fig.88) in which he shows it to have 5'-6" driving wheels and a 3'-6" diameter boiler, it looks otherwise like a shorter version of "Meteor" (with a different wheel arrangement, of course). He gives the weight as 20 tons 17 cwt. but does not say whether this refers to the engine alone or the complete engine and tender. Craven says that the coupled wheels were of 5'-0" dia. and the trailing wheels 3'-6". He gives the price as £1200, and thinks that it was ready made, not ordered. The boiler was 8'-6" x 3'-4" and contained 108 copper tubes 2" dia. x 9'-0" long. The valve gear was of four eccentric gab type. It had a four wheeled 773 gal. tender.

It was rebuilt in June 1853, at a cost of £600, the coupled wheel diameter being increased to 5'-7½" and the trailing to 4'-0", with cylinders of 14" x 17", and the valve gear changed to link motion. It was broken up in 1871.

"Wolsingham", built by Kitchings in May 1847, is listed by Inness as having 4'-6" driving wheels and 3'-0" trailing wheels, with a boiler about 9'-0" long. It was rebuilt later with a higher pressure boiler and bigger cylinders, and ran until 1879. I know of no illustration of it. John Dixon's 1849 list gives a cryptic note alongside the entry for this engine, ('for Cockermouth'), without any explanation, but it was possibly an engine built for the Cockermouth & Workington Railway, offered to the S. & D.R. and bought by them. Kitchings built two similar engines for the Whitehaven & Furness Junction Railway in 1847, so it was probably their own design. Craven gives more details: he refers to a letter from John Dixon dated 27th. February 1847, offering this engine to the S. & D.R., as he understood that one would be required for the Wear Valley line, and he had had an application for it from the Maryport & Carlisle Railway. Presumably he was acting as an intermediary of some kind. This letter also states 4'-8" coupled wheels, 13" cylinders. Craven also gives the price as £1350 plus £250 for the tender. The inside cylinders were 13" x 18", 4'-6" coupled wheels (in spite of the above) with 3'-6" trailing wheels, boiler 9'-0" x 3'-4¼" with 99 x 2" tubes 9'-5" long. This engine was also 'remodelled' at Shildon Works in 1852, with 5'-4½" coupled wheels and 14" x 18" cylinders, and link motion valve gear as fitted to "Active". There was not another No.55.

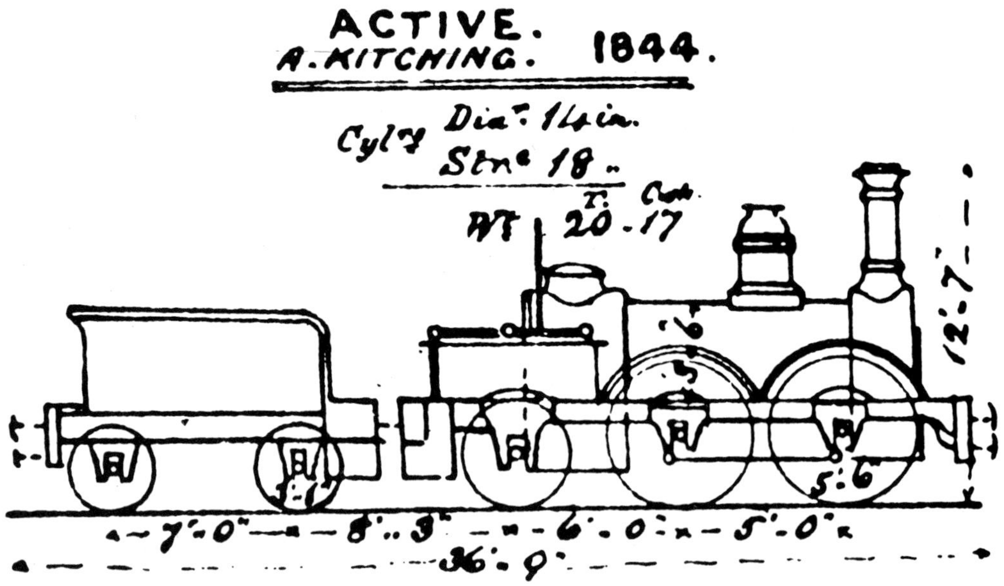

Fig.88 Theodore West's sketch of No.48 "Active"

In the 1868 stock list "Active" was valued at £1400 and "Wolsingham" at £1350, which may confirm the difference in size. As far as I am aware no further engines of this wheel arrangement ran on the Stockton & Darlington Railway. No.55 "Wolsingham" was probably rebuilt later at Gateshead as a 2-4-0 with 5'-6" driving wheels, 16" x 20" cylinders and 100 psi working pressure.

In 1847 two more passenger engines were put in hand by the Shildon Works Co. which call for some comment. They were 2-4-0 types, not remarkable in itself; it became a standard for passenger locomotives for many years on this and many other railways. No.38 "Rokeby" (Fig.90) and No.39 "Ruby" were both delivered in June 1847, of long boiler type with a very short wheelbase, boiler 13'-9" long x 4'-2" diameter, 158 x 2" tubes, and working at 100 p.s.i. The steamdome and the small dome over the firebox are both shown in the first drawing as having hexagonal covers, and are reminiscent of the later dome fitted to the 'Commerce' class, as also is the driving gear (see Chapter 7). The cylinders, $14\frac{1}{2}$" x 24", were outside, and it was thought necessary to keep them in as close as possible. However, the connecting rods are not looped as on "Commerce", but the crosshead is slotted in a built-up assembly which serves the same purpose; the drawing makes this clear. As the wheels revolve, the front wheel crankpin moves up and down in this slot, and the coupling rod works outside of it. The guide bar and crosshead assembly is enormous for the size of engine, and covers most of the front driving wheel. The actual connecting rods are short, although they drive on to the rear axle. One can only guess at the reasons for this very odd arrangement; my guess is the very restricted clearance between tracks in the Shildon Tunnel, which later had to be made single track. But why make these engines outside cylindered at all? Craven has very little to say about them.

Another innovation is that the frames were of bar type, possibly copied from the Bury engines, and apparently took far more man-hours in preparation than they should have done, all having to be draw filed. The valve gear was of Stephenson link type but operated through a rocking lever to each valve. The rocking shaft also drove the feed pumps. Inness has shown a steam brake on each side, working downwards between each pair of driving wheels. The weight, with a six wheeled tender, was about $24\frac{1}{2}$ tons in working condition, the tender carrying 1500 gallons of water and 3 tons of coal. As passenger engines they seem not to have been very successful, and were soon put on to short haul mineral and shunting work. When rebuilt in 1860 (Figs.91 & 92) they were given larger cylinders, conventional crossheads and guide bars and a rudimentary cab, and at the same time they were used to try out an experimental form of William Bouch's chimney feed water heater (Fig.89) which became known as "Bouch's Coffee Can". This had provision for either direct cold feed or for heating it through the casing. Later a well was added, sunk into the smokebox, and a gauge-glass visible from the cab.

An article in the "Engineer" for February 1885 gives a rather inaccurate engraving of "Rokeby" carrying the number 194, which is quite wrong. Inness's drawing is much better. Another puzzle is that after "Ruby" was scrapped in 1871 the number 39 was transferred to "Rokeby", according to Inness, and it then ran until 1875. However, the photograph (Fig.91) of it as rebuilt shows it still carrying No.38, so I assume that it was taken before 1871.

In 1847 another engine building works began to supply locomotives to the railway. This was the firm of Gilkes, Wilson, of the Tees Engine Works, Middlesbrough. Both these gentlemen had a close association with the railway, both became mayors of Middlesbrough, and there are still both a Gilkes Street and a Wilson Street in the town. They were both Quakers. They and their successors Hopkins, Gilkes supplied 97 locomotives to the Stockton & Darlington Railway, more than any other one contractor over the years covered by this book.

Edgar Gilkes had been an engineer at Shildon Works in 1839, and was sent to Middlesbrough to take charge of the repair works on the corner of Grey Street and Lower Commercial Street when these works came into Pease ownership. Isaac Wilson was a member of the Management Committee and became the father-in-law of Joseph Beaumont Pease. He also ran the Middlesbrough Pottery next door to the engine works and really part of the same group of buildings. Both the pottery and the engine works had been the property of Richard Otley, himself an early employee of the railway and the planner of the new town of Middlebrough, and his partners. When they fell into bankruptcy in January 1843 they were bought up by Joseph Pease (son of Edward). Gilkes and Wilson leased this repair works from him in

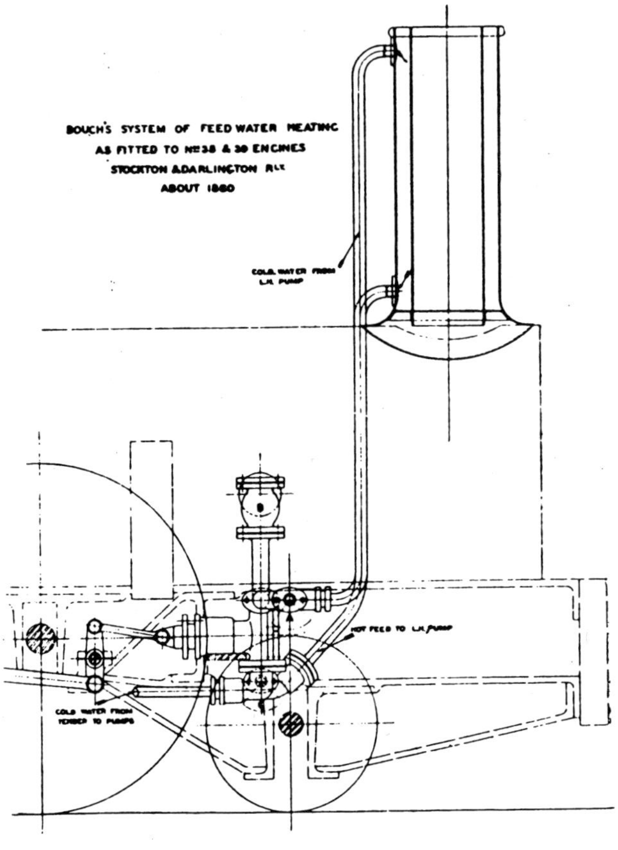

Fig.89 Drawing of Bouch's chimney feed water heater arrangement

1844 and called it the Tees Engine Works. I have no doubt that the Pease family retained a financial interest in the business, and it was obviously a part of the railway empire, although nominally independent. In 1865 they amalgamated with the Teesside Iron Works, not far away, and became Hopkins, Gilkes, while Wilson remained a director. I have a handwritten note on a list of locomotives which refers to it as Gilkes, Pease & Co., but have never seen any other confirmation of such a title, although the iron works became Wilson, Pease later. They fell into severe trouble in 1879, just before the Tay Bridge disaster, with which they were involved, and passed through several other ownerships including that of the Pease family again, and the works were used as the Westgarth marine engine works, then became the property of the Head Wrightson Machine Co., while the iron works and most of the land was bought by the original Teesside Bridge Co. before they moved to North Ormesby. William Bouch was a partner, and when he died left his holding to brother Thomas, not an entirely beneficial inheritance at the time.

Gilkes, Wilson took an order in 1847 for twelve 2-4-0 long boiler type locomotives for the Newmarket & Great Chesterford Railway, which could not afford to pay for more than six of them. The Stockton & Darlington Railway took the rest. They were delivered as follows; No.65 "Stephenson" in April 1850; No.66 "Priam", November 1847; No.67 "Orion", April 1848; No.68 "Brunswick" in July 1850, No.69 "Clarendon", November 1850 and No.70 "Alarm" in February 1851. There is another puzzle here. John Dixon's 1849 (and up to 1854) list clearly says that No.65 was named "Newmarket", and this is the only engine for which he does not give a precise date of delivery. Is it possible that it was in fact delivered first, in early 1847, and the date of April 1850 was when it was renamed "Stephenson"? Certainly the N.E.R. 1868 list appears to support Inness, giving the later name and date. Dixon may have meant only that Newmarket (the railway) was where the engine came from, but in all other cases he gives the known name at the time. My list is a précis only, not the full document, so there may be an error.

Again Craven has another story. He says that the first two were on loan from the order for the N. & G.C.R. mentioned above until May 1848, when it was agreed to purchase them at a price each of £2275. These two engines were at first numbered 56 & 57, and a minute of 17th. May 1848 complains that "the gauge of the two engines last placed on the line" was not right, naming them as "Priam" and "Alarm". What then happened was that when the next two, No.65 "Stephenson" and No.68 "Brunswick", were

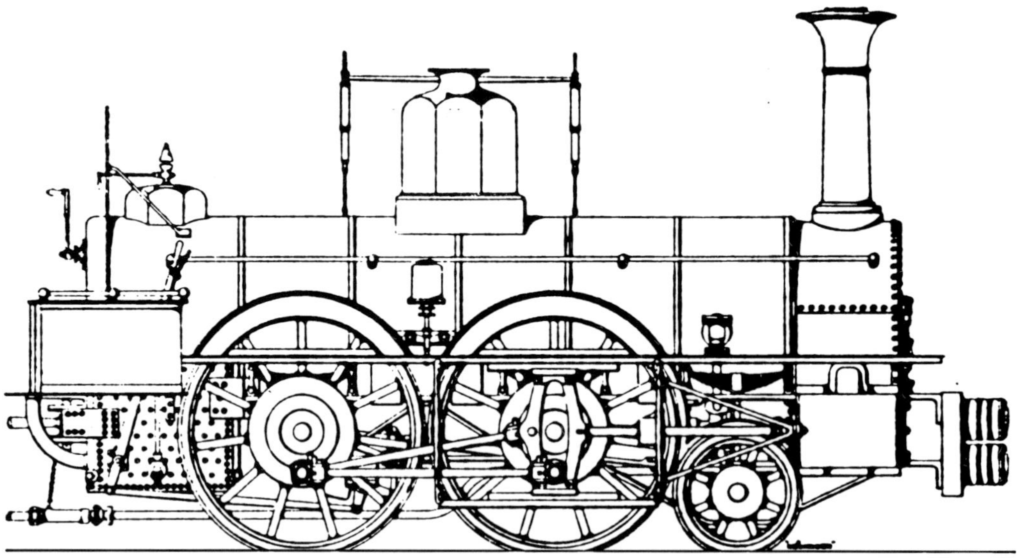

Fig.90 Drawing of No.38 "Rokeby" as built

Fig.91 Photograph of No.38 "Rokeby" as rebuilt, with slide bars

Fig.92 Drawing of "Rokeby" as rebuilt and renumbered 39

S & D R. No 67 'Orion'
Built by Gilkes, Wilson & Co
1847

Fig.93 Drawing of "Priam" class locomotive as built

Fig.94 Photograph of a locomotive of the "Priam" class, numbered 1066, in 1875

Fig.95 Photograph of No.1058, ex 58 "Woodlands", at North Road, Darlington

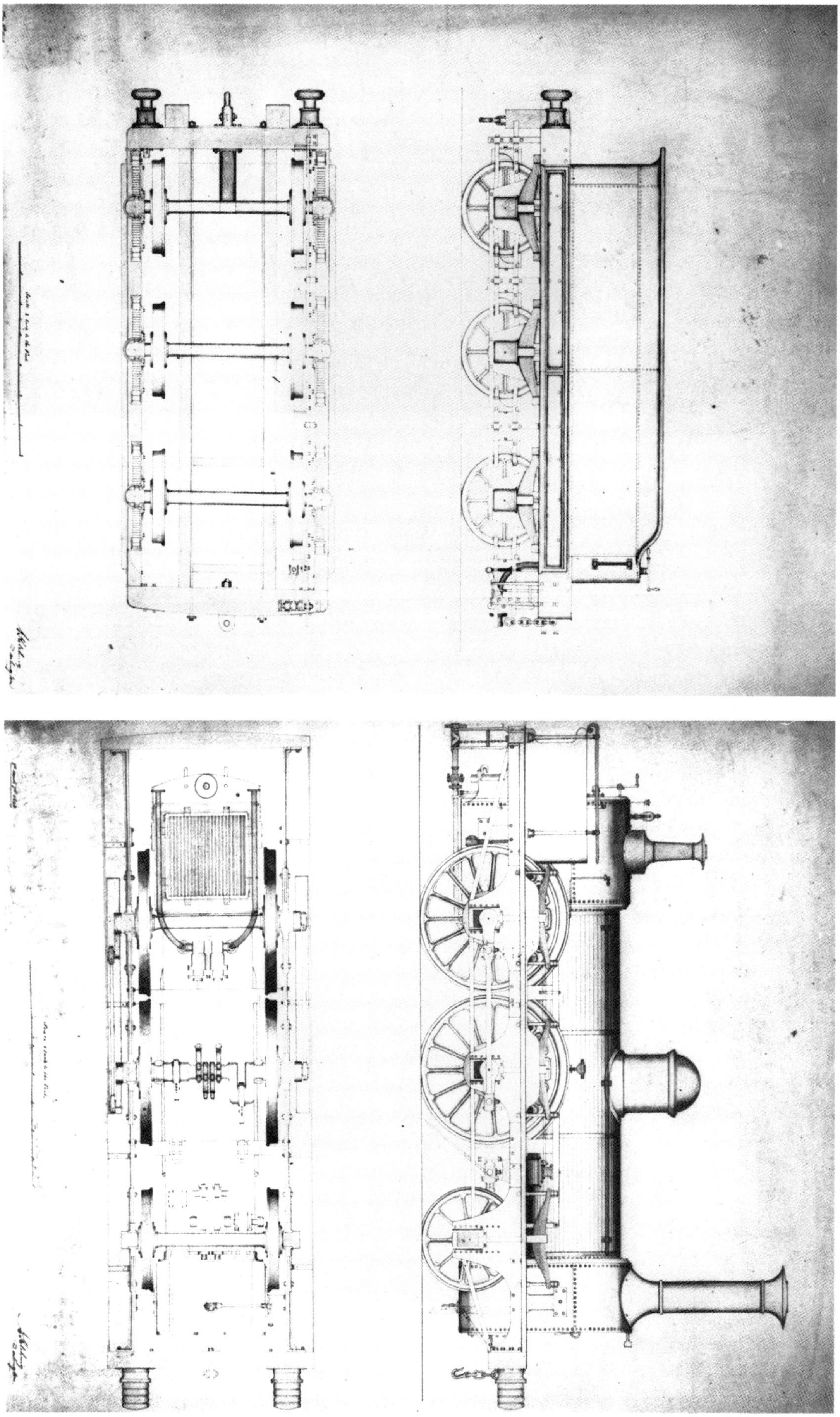

Fig.96 Kitching Works general arrangement drawing of No.58 "Woodlands" & No.59 "Hallgarth", with tender

purchased in 1850 the first two were renumbered 66 & 67, "Priam" keeping that name and "Alarm" becoming "Orion". Then Dixon was recommended to buy two more at a cheaper price, "against depreciation of stock", at £1900, which was apparently also the price of the second pair, these last three being No.69 "Clarendon" and No.70 "Alarm". I would like to see further proof of all this.

They had no cab to start with (Fig.93). They were outside cylindered (15" x 22") engines of conventional type, said by John Dixon to be to Stephensons' patent, with built up spoked coupled wheels of 5'-6" diameter, leading wheels 3'-8", boilers 14'-0" long x 3'-6" diameter, fitted with 127 x 2" tubes giving a heating surface of 941.6 sq.ft. plus firebox 74.4 sq.ft., at 95 p.s.i. The total weight of engine and tender was 44 tons 7 cwt. in operational order. The valve gear is not specified but would certainly have been of Stephenson link type. Dixon says that No.67 "Orion" had a different tyre thickness.

They gave good service, mainly on mineral traffic although designed for passenger work, and all lasted into the 1870's. No.66 "Priam" was on show at the 1875 Jubilee celebrations as N.E.R. No.1066, in what is said to be almost original condition except for the added cab and the S. & D.R. buffers (Fig.94). Or was it, as has been hinted, another of the class made to look like "Priam", and if so, why?

An article in the "Engineer" for February 1880 includes an engraving of No.1066, based on this photograph, but also gives a set of dimensions almost all differing from those given by Inness and repeated above or used in my lists. The driving wheel diameter is given as 5'-0" only and that of the leading wheels as 3½ ft., the boiler 13'-10" long x 3'-8" diameter, of ⅜" thick Low Moor plate, copper firebox 3'-7" high, 3'-6" long, 3'-5" wide, 127 iron tubes, 14'-3" long, 2" dia. x 14 W.G. Total heating surface including firebox 1061 sq.ft., pressure 95 p.s.i. Two inside frames were fitted, and the cylinders are given as 15" bore x 22" stroke, with Stephenson link motion and two feed pumps. The total wheel base is given as 11'-0". The tender was wooden framed, but otherwise much as described by Inness, and the weights also conform closely to those I have listed. The speed is given as 30 m.p.h. This drawing shows sand pipes, but they may well have been fitted later.

The built-up wheels referred to above were constructed, fabricated as we would call it now, with cast iron bosses and X-section spokes made up of 'T'-bars placed back to back, probably cast into the hubs or bosses, and also into the rims, if these were of cast iron. This has to be done one stage at a time, probably into the rims first, since these would shrink more during cooling, and coating the ends of the spokes with a borax solution as flux before pouring the iron. Inness refers to "welded wrought iron type", so the attachment of the spokes to the rims may have been by hammer welding, the spokes slotted in first. Hammer welding was possible on a fairly large scale, but needed very large furnaces and steam presses. I am aware that flame welding of a sort was in use by the middle of the century, but doubt if the Tees Engine Works used it.

A new design of more powerful inside cylindered 2-4-0 passenger engine was brought out, and two of these were built by A. Kitching in 1848, No.58 "Woodlands" (Figs.95 & 96) in August, and No.59 "Hallgarth" in October, at a price of £2350 each. The design was probably Bouch's. They had 14" x 20" cylinders, 5'-2" driving wheels of built-up spoke type as above, and double frames with four bearings for the crank axle. The boilers were 12'-0" long (12'-3" according to Craven) by 3'-8" diameter, with 129 x 1⅞" brass tubes, working at a pressure of 120 p.s.i. Two short stroke feed pumps were fitted between the frames in front of the leading coupled axle, driven by eccentrics on the driving axle. Sanding gear was fitted. The firebox was of copper. The overall weight of engine and six wheeled 1054 gal. tender in working order was 40 tons 13 cwt. Originally no weatherboard or cab was fitted, according to Ken Hoole and confirmed by this drawing. Later a standard S. & D.R. cab was fitted with a much taller dome and a stovepipe chimney. These two engines established a standard design which remained virtually unchanged except for increase in size until the 1860 4-4-0's came into service, and they gave good service themselves. "Woodlands" appeared in the 1875 celebrations carrying the N.E.R. No.1058, and after working for some time on the Barnard Castle and Middleton-in-Teesdale branch was scrapped in 1877, while "Hallgarth" lasted until 1881. Craven says that the price paid was £2350 each including tender.

They were both fitted with new cylinders and valve gear in 1853.

Another similar but larger engine was built by Kitchings, their works No.23, and offered to the Railway for £2100 in March 1851. It was accepted, and delivered at the end of April. I assume that the price included the six wheeled 1104 gal. tender. This was No.71 "Hackworth", after which there was a gap in the building of this type for a few years. This engine had driving wheels of 5'-9" diameter, and cylinders of 15" x 20" or 21". The boiler had a length over the barrel of 9'-0" according to Inness, which should mean that the 12'-0" given for the earlier pair

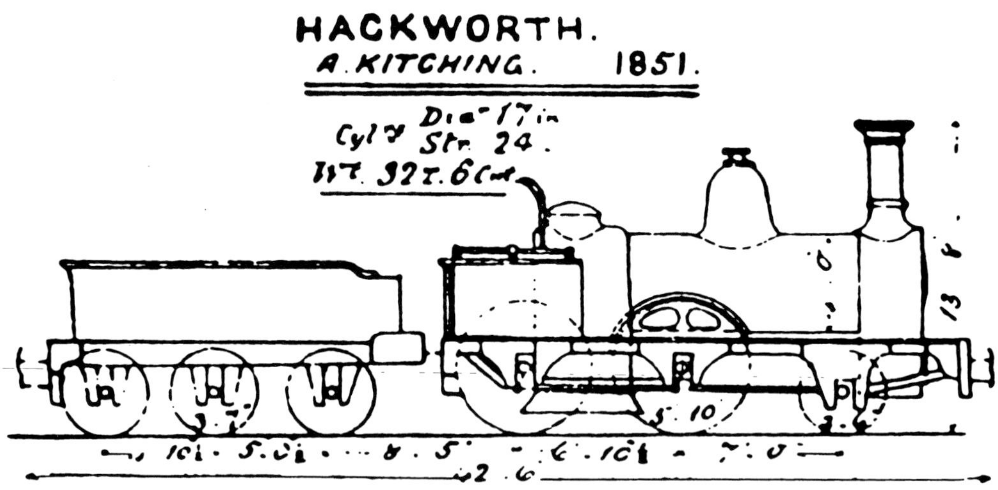

Fig.97 Theodore West's sketch of No.71 "Hackworth"

was overall, but Craven says that it was 10′-0″, which suits the later rebuild.

There are a number of puzzles about this engine. I have a drawing which came from the National Railway Museum and purports to show the engine as originally constructed. This is similar to the one by J. S. MacLean used in his book "The Locomotives of the North Eastern Railway", 1922 edition, and is almost certainly his work. This shows an engine as described by Inness, with a boiler shell of about 9′-0″ long, and a round topped firebox of about 4′-3″ long, but I have been told that it is inaccurate, and that it probably derives from the Theodore West sketch of the engine (Fig.97). This may be so, but the author was a very careful researcher, and must have believed that the drawing was a reasonable one. Also Theodore West must have seen this engine in its original state at the time; his sketches date to about 1885, and the engine was only rebuilt in 1873, at which time friend West was already working at the North Road Works.

The MacLean text agrees with Inness on the original engine in most respects, including a boiler diameter of 4′-0″. When rebuilt in 1873 it became a much bigger engine (see Chap.8).

However, I have recently received a poor photocopy of a drawing from Ken Hoole's files, which is a copy of an original Kitching works drawing of this engine (Fig.98), and this explains a lot. I have since obtained a better copy from the P.R.O. The main dimensions are similar, but there is no cab protection and the boiler fittings are different. It is obvious to me that there had been an earlier partial rebuild, possibly a reboilering, and that both West's and MacLean's drawings show it as then rebuilt with weatherboard, cleaded boiler and later steam dome and safety valves. The approximate dimensions scaled from this drawing are boiler barrel length 9′-3″, dia. 4′-0″, driving wheels 5′-8″ dia. on a wheelbase of 6′-0″ + 6′-6″, longer than that of the 'Woodlands' class and with the rear axle brought behind the firebox. It is difficult to believe that the wheelbase would have been changed so much then, and I suspect that West was less than accurate. Nevertheless, there is a note in the 1850-57 valuation lists that the position of "wheel" was altered in 1854,

Fig.99 Photograph of No.1071, possibly the rebuilt No.71 "Hackworth"

probably the leading axle only.

Inness gives no tube details, but says that the working pressure was 120 p.s.i and that the total weight of engine and tender was 42 tons. He makes a comment about the feed pumps; these as usual were eccentric driven, but fitted below the frame between the two sets of driving wheels, and they were driven from eccentrics keyed to the rear axle. He says that this engine was considerably rebuilt in the early 1860's, with larger cylinders of 17" dia., the wheelbase increased and a cab fitted. No photograph seems to be available of the engine in its original form, and the one that I give (Fig.99) is almost certainly of the later engine.

J. S. MacLean, however, in his book referred to above, says that No.71 was completely rebuilt in 1873 at the North Road Works and given the new works number of 84; it was then fitted with 6'-0" diameter coupled wheels and inside bearings only, and was renumbered again twice before being disposed of in 1901 (1717 in 1889, 1915 later). This conflicts with Inness's account; he says quite clearly that the first No.71 was scrapped in 1873 and a new engine built to replace it (see chapter 8). MacLean says also that the cylinders were 17" x 24", that the coupled wheel diameter was 5'-10", and that the boiler was reduced in length from the usual so as to locate the firebox between the two pairs of coupled wheels.

Craven gives the following details; boiler length 10'-0", with an oval section of 4'-1" x 3'-10", having 184 iron tubes 1$^{3}/_{4}$" dia. x 10'-4$^{1}/_{8}$" long. Since the other details in his account line up fairly well with what Inness and MacLean say about the original engine, he can hardly be talking about the later rebuild, and his dimensions agree well enough with the sketch above (Fig.98).

The 1868 list values the engine at £2100, its original price when new. It seems to me that Inness must have been right about the 1860 rebuilding, but without more details it is not possible to say which dimensions belong to which rebuild.

In March 1854 the Stockton & Darlington Railway bought two more second-hand engines for passenger work, at a cost of £1800 each. These were Stephenson locomotives of 2-4-0 wheel arrangement which had been built in 1849 for the Whitehaven & Furness Railway, who had paid £2800 each for them about five years previously, but only used them for a short time since they proved unsuitable for the line. They were apparently similar to four which had been supplied by the builders to the Newcastle & Berwick Railway in 1847, but Craven says that there is no evidence that they were actually built at Forth Street, and that they may have been subcontracted to T. Richardson of Hartlepool. (There is a gap of 25 engines in my Stephenson lists for 1849.)

They are described as being of 'long boiler' type, but although the firebox and ashpan were certainly outside the wheelbase, the leading pair of wheels was on the smokebox centre-line, which does not match the usual definition. They were given the designations No.78 "Lonsdale" and No.79 "Carlisle". The drive was on to the second pair of coupled wheels and the outside cylinders were mounted between the leading wheels and the first pair of drivers, on the frames. The valve motion was of link type. It would seem from the drawing (Fig.100) that the centre wheels were flangeless.

The driving wheels were 6'-0" in diameter and the leading pair 3'-8". The boilers were 13'-8" over the tubeplates x 3'-6" diameter, working at 100 p.s.i., but there are no details of the tubes or heating surface, except that John Dixon said that "Carlisle" had tubes of brass. The tenders were of the usual six wheeled type, carrying 1200 gallons of water and 3 tons of coal, and the total working weight was about 44 tons 5 cwt., quite big engines. They lasted until 1870 and 1875 respectively. Craven gives the cylinders as 15" x 22". "Carlisle" was overhauled and fitted with four new wheels in 1856.

Mr. David Baxter has informed me that in August 1854 the S. & D.R. bought a second-hand Fairbairn 0-4-2 locomotive from the Lancashire & Yorkshire Railway which had been their No.143 "Rhine". This had cylinders of 14" x 20", 4'-6" diameter driving wheels, weight 16 tons, and was built by William Fairbairn in 1842 for the Manchester & Leeds Railway, their No,43. This was never put into stock by the S. & D.R., and was probably sold again very soon, possibly to the West Hartlepool Harbour & Railway Co., who may have rebuilt it as an 0-4-0, their No.43 or 44, with a larger cylinder bore of 15".

Two more second-hand engines were bought in March 1855, from the Edinburgh & Glasgow Railway, their Nos.42 & 43. They had been built by Benjamin Hick & Son of the Soho Ironworks, Bolton, in 1847, and were 2-2-2 outside cylinder type with 5'-6" diameter driving wheels, and cylinders 15" x 22" (Fig.101). They were listed as No.93 "Uranus" and No.94 "Neptune". Their boilers were 10'-0" long x 3'-6" diameter, with 116 x 2" tubes and a working pressure of 100 p.s.i. They were apparently not very successful, and were disposed of in 1867, being replaced by two 0-6-0 mineral engines (see later).

Craven has the following to say about them: it appears that in 1854 the S. & D.R. had had to hire two engines from the L.N.W.R. southern division,

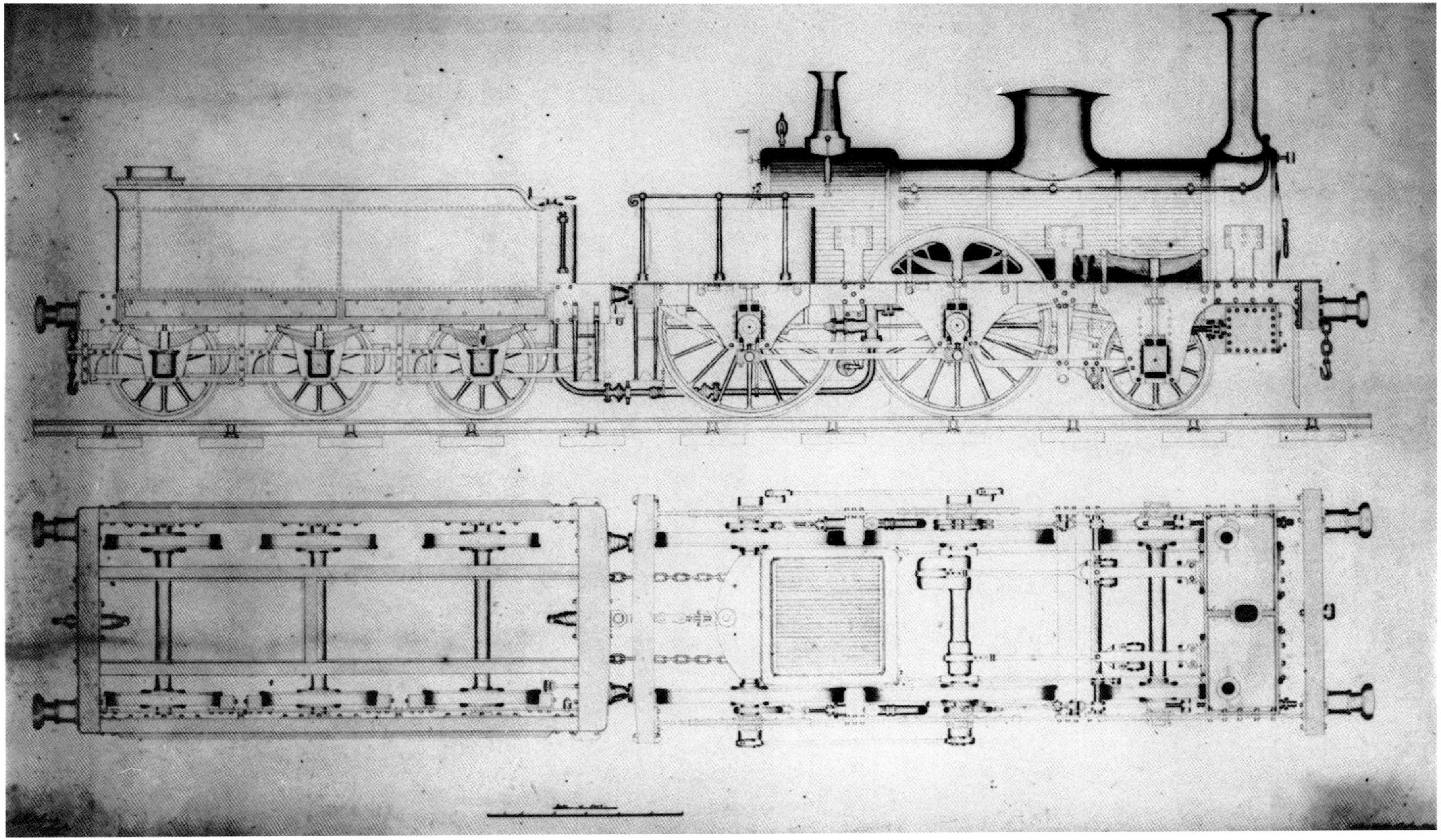

Fig.98 Kitching Works general arrangement drawing of No.71 "Hackworth"

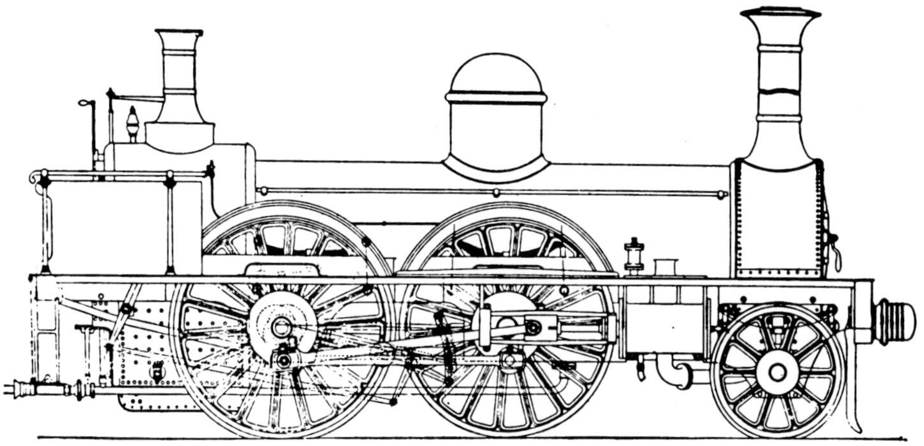

Fig.100 Drawing of No.78 "Lonsdale" & No.79 "Carlisle"

one of which was a Bury 0-4-0 with 4'-0" wheels similar to Nos.87-9. Cowans, Sheldon of Carlisle (in which firm Bouch had an interest) had (as agents) two engines for sale, in November, and he went to look at these at Edinburgh in February 1855. The asking price was £1200 each, half the original cost. They were Stephenson patent long boiler 2-2-2's, built under subcontract by Benjamin Hick in February 1847, of an unsuccessful type, and had been sold to the Edinburgh & Glasgow Railway in early 1848 at £2325 each. They had done little work on that line, who considered them "useless unless altered". They had 15" x 22" outside cylinders, wheels as above, and the Railway offered £2400 for them less £150 for necessary repairs. They were received in April after repairs in Glasgow, and kept their original names. They were sent to Shildon Works for disposal in August 1866. How useful they were is not recorded.

Two new passenger engines of double framed 2-4-0 type with outside bearings were bought from Kitchings in 1855, based on the 'Woodlands' type mentioned earlier, but slightly bigger. No.98 "Pierremont" (Fig.102) was delivered in May and No.101 "Marske" (Fig.103) in September, the price being £2400 each. The driving wheels were the same diameter as before, 5'-2", and the cylinders were also the same, 15" bore x 20" stroke, but the boiler was 12'-4" long x 3'-6" diameter, working at 100 p.s.i. The valve motion was of box link type, with the usual feed pumps behind the axle and driven by separate eccentrics. They were successful engines, and lasted No.98 until 1876 and No.101 until 1882. A very good general arrangement drawing of these engines dated 1855 is held among the Whessoe papers in the Durham County Records Office (Fig.104) but it is badly damaged.

Craven has no comments regarding these engines or any later passenger engines.

Another two, very similar, were built by the Shildon Works Co., No.99 "Ayton" (Fig.108) and

Fig.101 Drawing of No.93 "Uranus"

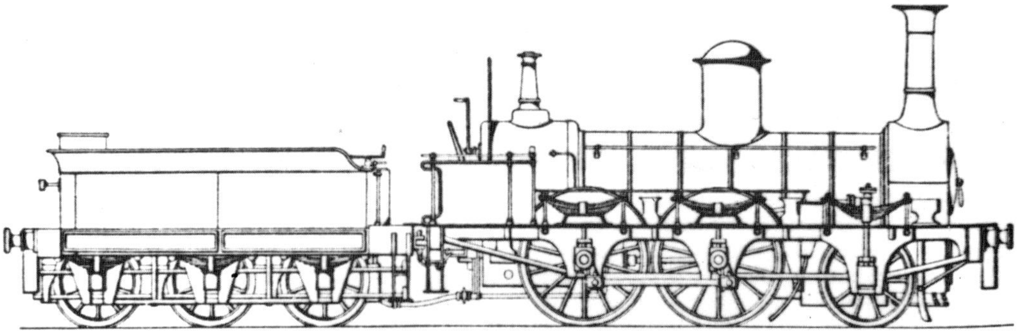

Fig.102 Drawing of No.98 "Pierremont"

No.100 "Stobart", at the same price as for the Kitching engines, and both delivered in July 1855. Apparently they did not compare well with the Kitching built engines, although in the 1868 stock list their values were very much the same. Nevertheless, they kept going until 1882 and 1877 respectively. The "Railway Engineer" says that they had smaller boilers and short connecting rods, and Lowes gives them 5′-4″ driving wheels and 16″ x 20″ cylinders.

According to the Hoole notes there were minor differences between these two pairs of engines, in the running boards, splashers and tender brakes.

The next passenger engines were from Stephensons, of inside framed 2-4-0 type, No.114 "Edward Pease" (Fig.105) and No.115 "Meynell" (Figs.106 & 107), both delivered in August 1856, at a price of £2100 each. They were very similar in size and details, with 5′-2″ driving wheels, 15″ x 22″ cylinders, and although the boiler length is not given, the diameter was 3′-7½″ with 109 tubes and a total heating surface of 822 sq. ft. at a pressure of 120 p.s.i. The weight of the engine in working order was 24 tons 14 cwt., tender not stated. Originally they were fitted with a weather board only, the cab being added later, and the original Stephenson chimney replaced by a plain tapered one. Inness says that their steaming capacity was poor, and they were both scrapped in 1877.

The two photographs (Figs.105 & 106) of the engines as built both contain an imposing figure in a tall silk hat, with a farm in the background and a dog in one picture. The site is just south of Shildon, the figure is believed to be George Graham, and the building is Tate's Farm, his home at the time, and still there today, I believe (see also Fig.195, Stephenson 4-4-0).

Gilkes, Wilson built the next two, at a price each of £2550. They were stated to be long boiler 2-4-0 engines, but were very similar to the others. These two were double framed, No.116 "Lartington", delivered in July 1856, and No.117 "Nunthorpe" (Figs.109 & 110) in August. The driving wheels were slightly smaller at 5′-1″ and the boilers 12′-0″ long x 3′-7″ diameter, working at 120 p.s.i. The tender carried 1300 gallons of water and 3 tons of coal, and the overall weight of engine and tender in working order was 41½ tons. Brakes were fitted to the tender only. They lasted until 1879 and 1880 respectively.

Alfred Kitching built a further double framed engine of this type, No.118 "Elm Field" (Fig.111) in

Fig.103 Photograph of No.1101, ex 101 "Marske", rebuilt

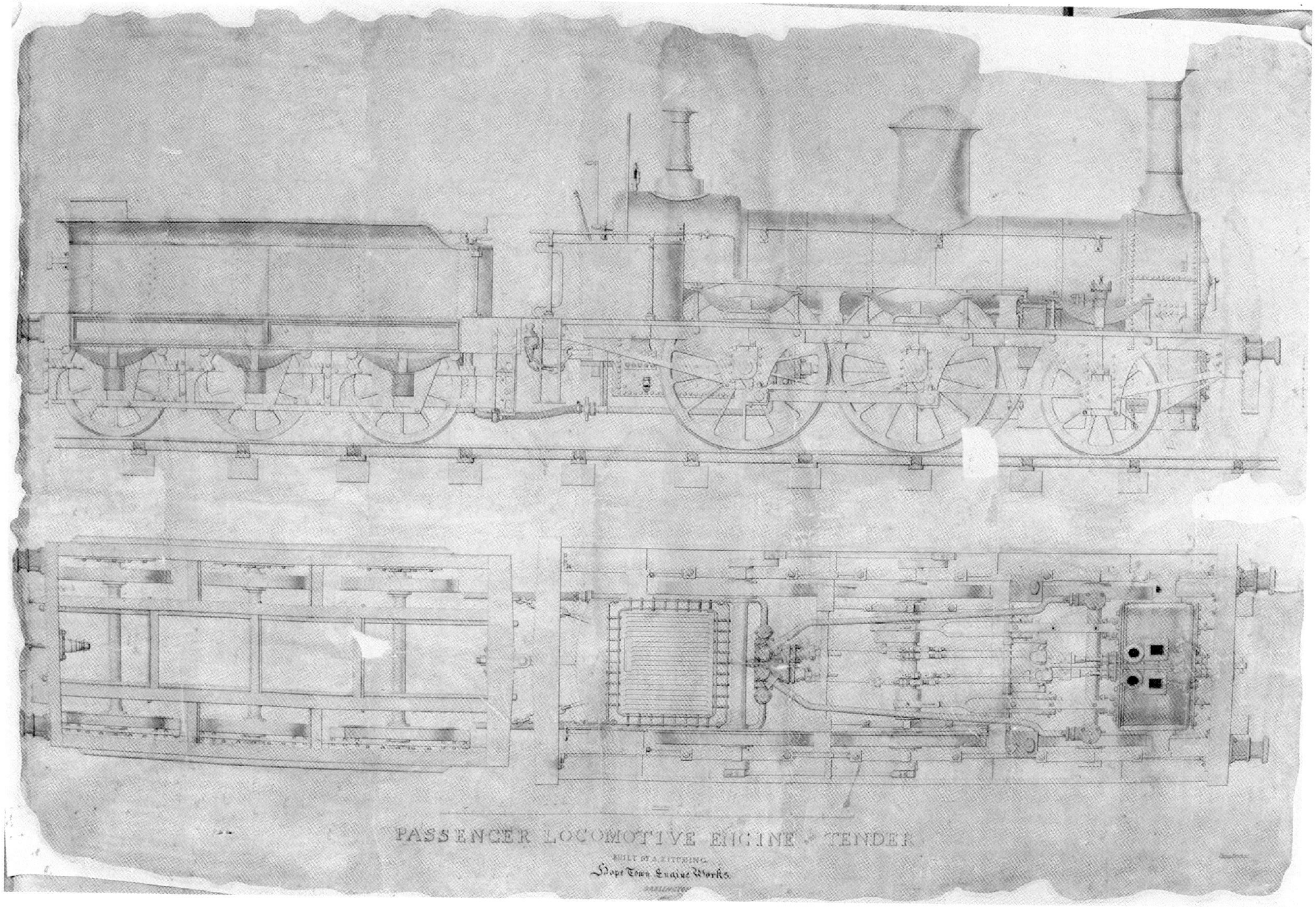

Fig.104 Kitching Works general arrangement drawing of No.98 "Pierremont" & No.101 "Marske"

Fig.105 Photograph of No.114 "Edward Pease" at Shildon

Fig.106 Photograph of No.115 "Meynell" at Shildon

Fig.107 Photograph of No.1115, ex 115 "Meynell", as rebuilt later

Fig.108 Photograph of No.1099, ex 99 "Ayton", as rebuilt later

Fig.109 Photograph of No.117 "Nunthorpe" in early condition

Fig.110 Photograph of No.117 "Nunthorpe", rebuilt and without nameplates

Fig.111 Photograph of No.118 "Elm Field"

May 1857, at a price of £2100, with a slightly shorter wheelbase; this engine lasted until 1880. He also built one more 'on spec.' according to Ken Hoole, No.166 "Oswald Gilkes" (Fig.112) delivered in November 1860, very similar to all the others above, which turned out to be the best and most economical that the Hopetown Foundry ever built, but sadly it was also the last main line locomotive that they built. It was rebuilt at the N.R.Works in 1875, scrapped in 1877 and replaced by a similar but larger Class 11.

By this time (1860) the Stockton & Darlington Railway was building 4-4-0 type passenger locomotives, which will be the subject of a separate chapter. No more 2-4-0 engines were built until 1873, well into N.E.R. days, and nearly outside the scope of this book.

J. S. MacLean groups most of these 2-4-0's as the 'Woodlands' class, and gives the following dimensions for No.99 "Ayton", No.116 "Lartington", No.117 "Nunthorpe", No.118 "Elm Field" and No.166 "Oswald Gilkes": cylinders 16" bore x 19" stroke, boilers 12'-6" long x 3'-8" diameter, tenders 1200 gallons water, 4 tons coal, and coupled wheels as 5'-0" diameter. Lowes often gives other dimensions again. Engines were rebuilt, cylinders bored out and replaced, and wheels wore; boilers were replaced less often, but dimensions did change over the years. Inness probably had access to the most accurate records, and I have used his figures where given.

Inness mentions two other second-hand locomotives received at Shildon in February 1854, from Wolverton on the L. & N.W.R. Apparently these two engines were so unsatisfactory that they were promptly returned with a letter of complaint to J. E. MacConnell, and no other details of them are known.

Fig.112 No.166 "Oswald Gilkes"

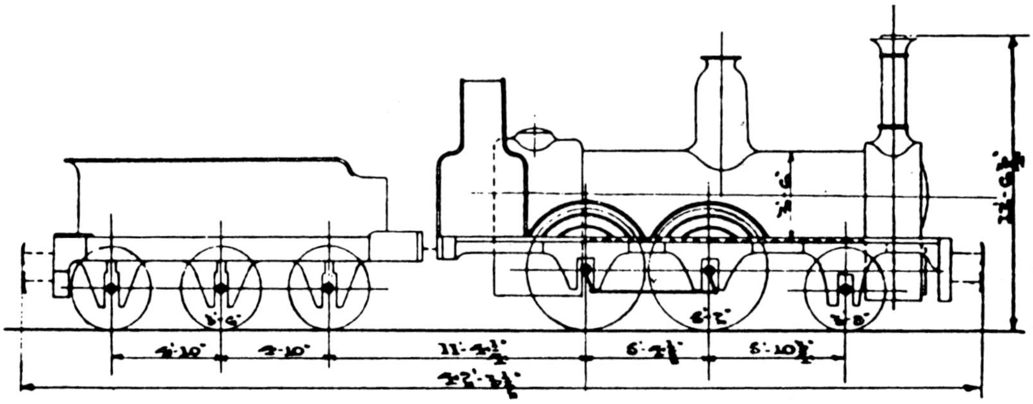

Data as early condition in general, and approximate. Nov.1840 list details not included;
n.k. = not known. n. = nominal. a = actual. refer to list, Chap.1, pp.

No.	Name	W.P. lb/in² (p.s.i.)	Boiler Length	Boiler Dia.	Surface Area F²	Cylinders Bore	Cylinders Stroke	Wheels Dia.	Wheelbase	Weight Tons Cwt.
26/ 51.	Arrow	n.k.	c.6'-0"	3'-4"	n.k.	1:17" 2:21"/22"	9"	L:3'-6" D:5'-0" T:3'-6"	5'-6" + 5'-5"	c.14 but no breakdown
		Tubes 135 x 1½". Tender 4 x 3'-6" wheels, no other details.								
27	Swift	n.k.	9'-8"	3'-4"	n.k.	10"	16"	4'-0"	6'-6"	c.8 but no breakdown
		Flue tube 4'-8" long + 102 tubes 5'-0" long x 1⅝" o.d. No details of tender.								
28/ 43.	Sunbeam	n.k.	n.k.	3'-2"	n.k.	12"	18"	L:4'-0" D:5'-0"	5'-0"	Inness: 10 14 in W.O.
29/ 40.	Queen	n.k.	n.k.	n.k.	n.k.	1:12" 2:14"	18" 19"	D:4'-6" T:3'-0"	n.k.	n.k.
		No other details.								
30.	Raby Castle	55	8'-0"	3'-4½"	c.700	1:12" 2:13" (but see text)	18" 18"	L:3'-6" D:5'-0" T:3'-6"	n.k.	n.k.
	No other details, but see Fig.76.									
42.	London									
	Inness:	60	7'-0"	2'-4"	n.k.	1:11" 2:11"	16" 18"	4'-6"	n.k.	n.k.
	Whishaw:					10"	18"			
		Firebox 2'-7½" long x 3'-3" high x 3'-7" wide, 2 x 3" firetubes, 72 x 1¾" smoke tubes.								
	Lowes:		7'-0"	2'-7"		11"	16"	4'-6"		
		71 x 1¾" x 7'-4" tubes.								
41.	Dart	2nd:70	8'-2"	3'-7⅛"	554	1:12" 2:13"	18" 16"	4'-5" 4'-6"	4'-10"	12 14 total in W.O.
		110 Low Moor iron tubes, 8'-5½" b.t.p. x 2", firebox 3'-11" long x 3'-7½" wide. Tender 4 x 2'-6" dia. wheels on 4'-10" base, 600 gals. water, 2 tons coal.								
52.	Comet	70	n.k.	n.k.	n.k.	12"	18"	L:n.k. D:5'-6" T:n.k.	11'-7¼" + 6'-3"	n.k.
		No other details.			Lowes:	14"	18"			
53	Manchester	60	n.k.	3'-3"	n,k.	12"	18"	L:3'-6" D:5'-0" T:3'-6"	6'-0" + 5'-0"	c.23 total.
		Tender 4 x 3'-0" dia. wheels on 5'-0" base, no other details.								
54	Tyneside	90	8'-11"	3'-5"	n.k.	1:13" 2:14"	18" 20"	L:4'-6" D:6'-0" T:4'-6"	5'-11" + 6'-4"	c.20 total in W.O.
		Firebox 3'-4½" long x 3'-10" wide, no tube details. Tender 4 x 3'-2" dia. wheels on 7'-2" base.								

Data as early condition in general, and approximate.
n.k. = not known. n. = nominal. a = actual.

No.	Name	W.P. lb/in² (p.s.i.)	Boiler Length	Boiler Dia.	Surface Area F²	Cylinders Bore	Cylinders Stroke	Wheels Dia.	Wheelbase	Weight Tons Cwt.
50.	Meteor	1:n.k.	8'-3½"	3'-6"	n.k.	12"	18"	L:3'-6" D:5'-0" T:3'-6"	7'-1" + 6'-8"	29 18 total in W.O.
		Firebox 3'-8" high x 3'-10" long x 2'-10" wide, no tube details. Tender 4 x 3'-1" dia. wheels on 7'-6" base, weight 10½ tons.								
		2:100	10'-3½"	n.k.	633	14"	19"	no change	no change	n.k.
		Firebox 3'-7" long x 4'-0" wide, 117 x 1¾" iron tubes.								

No.	Name	W.P. lb/in² (p.s.i.)	Boiler Length	Boiler Dia.	Surface Area F²	Cylinders Bore	Cylinders Stroke	Wheels Dia.	Wheelbase	Weight Tons Cwt.
44.	Sun	60	n.k.	n.k.	n.k.	13"	18"	L:n.k. D:5'-6"	5'-6"	13 total in W.O.
45.	Gannymede	"	n.k.	3'-0"	n.k.	13"	18"	5'-0"	5'-9½"	13 2 total in W.O.
46.	Antelope	"	n.k.	n.k.	n.k.	"	"	L:n.k. D:6'-0"	5'-6"	n.k.
47.	Unicorn	"	n.k.	n.k.	n.k.	"	"	"	"	n.k.
87.	Fryerage	90	11'-1"	3'-6"	782	15"	24"	5'-0" (Railway Eng'g; 4'-9")	7'-8"	20 12 eng. in W.O.
88.	Deanery	"	"	"	"	"	"	"	"	" "
89.	Huddersfield	"	"	"	"	"	"	"	"	" "
		Tenders: all four wheeled; No.45 wheelbase 7'-0"; Nos.87, 88, 89 wheels 2'-6" on 7'-5" base, 900 gals. water + 2 tons coal. Nos.87, 88, 89 firebox 3'-6" dia. x 4'-3" high, tubes 121 x 2"								

No.	Name	W.P. lb/in² (p.s.i.)	Boiler Length	Boiler Dia.	Surface Area F²	Cylinders Bore	Cylinders Stroke	Wheels Dia.	Wheelbase	Weight Tons Cwt.
48.	Active	80	n.k.	3'-6"	n.k.	14"	18"	D:5'-6" T:c.3'-0"	5'-0" + 6'-0"	20 17 see text.
55.	Wolsingham	1: 80 2:100	c.9'-0" n.k.	n.k. n.k.	n.k. n.k.	13" 14"	18" 18"	D:4'-6" T:3'-0"	5'-9¾" + 6'-3"	n.k.
38.	Rokeby	100	13'-9"	4'-2"	n.k.	14½"	24"	L:3'-0" D:6'-0"	4'-6" + 6'-3"	c.24¼ total in W.O.
39.	Ruby	"	"	"	n.k.	"	"	"	" "	"
		Both: 158 x 2" tubes; tenders 6 x 2'-11½" dia. wheels on 4'-6" base. 1500 gals. water, 3 tons coal.								

data as early condition in general, and approximate.
n.k. = not known. n. = nominal. a = actual.

No.	Name	W.P. lb/in² (p.s.i.)	Boiler Length	Boiler Dia.	Surface Area F²	Cylinders Bore	Cylinders Stroke	Wheels Dia.	Wheelbase	Weight Tons	Weight Cwt.
65.	Newmarket/ Stephenson (but see text)	95	14'-0"	3'-6"	1016	15"	22"	L:3'-8" D:5'-6"	5'-11" + 5'-10"	44 total in W.O.	7
66.	Priam	"	"	"	"	"	"	"	" "	"	"
67.	Orion	"	"	"	"	"	"	"	" "	"	"
68.	Brunswick	"	"	"	"	"	"	"	" "	"	"
69.	Clarendon	"	"	"	"	"	"	"	" "	"	2
70.	Alarm	"	"	"	"	"	"	"	" "	"	"
		Firebox 3'-6" long x 3'-8½" wide, casing 4'-2½" long x 4'-2" wide. 127 x 14'-3" long x 2" iron tubes. Tenders 6 x 3'-6" dia. wheels on 2 x 14'-3" base, 1200 gals. water + 3 tons coal, weight 17 tons.									
58.	Woodlands	120 (MacLean – 12'-6", 19")	12'-0"	3'-8"	n.k.	1:15" 2:16"	20" "	L:3'-8" D:5'-2"	5'-10½" + 5'-4"	40 total in W.O.	13
59.	Hallgarth	All as No.58 Woodlands. No firebox details. 129 x 1⅞" tubes. Tenders six wheeled, 1200 gals. water, 3 tons coal, weight 16 tons (MacLean says 4 tons coal).									
71.	Hackworth 1. Inness:	120	c.9'-0"	4'-0"	n.k.	1:15" 2:17"	20" "	L:3'-6" D:5'-9"	6'-2½" + 6'-10¾"	c.42 total in W.O.	
		Firebox 4'-3" long x 4'-1¾" wide, no tube details. Tender 6 x 3'-6" dia. wheels on 5'-0½" x 4'-10½" base, 1200 gals. water + 3 tons coal, weight 16 tons.									
	2. Works Drg.:		9'-3"	4'-0"	n.k.			L:3'-6" D:5'-8"	6'-0" + 6'-6"		
		Tender as 1.									
	3. MacLean:			4'-0"		17"	24"	L:3'-6" D:5'-10"			
	4. Craven:	120	10"	4'-1" x 3'-10"		15"	21"	L:3'-6" D:5'-9"	as Inness	26 eng. 16 tender	
		Copper firebox 3'-6" x 3'-6" x 5'-0", casing 4'-2" x 4'-1¾", 184 tubes 1¾" x 10'-4⅛"; Tender 6 wheels, 1104 gals. water, 3 tons coal.									
78.	Lonsdale	100	13'-8"	3'-6"	n.k.	1:15" 2:15¾"	22" "	L:3'-8" D:1:6'-0" 2:5'-3"	7'-9" + 6'-3"	44 total in W.O.	5
79.	Carlisle	"	"	"	n.k.	"	"	"	" "	"	"
		Firebox 3'-19" long x 4'-2" wide. No tube details. Tenders 6 x 3'-8" dia. wheels on 2 x 5'-0" base, 1200 gals water, 3 tons coal, weight 16 tons 4 cwt.									

Data as early condition in general, and approximate.
n.k. = not known. n. = nominal. a = actual.

No.	Name	W.P. lb/in^2 (p.s.i.)	Boiler Length Dia.	Surface Area F^2	Cylinders Bore Stroke	Wheels Dia.	Wheelbase	Weight Tons Cwt.
None	None	s/h Fairbairn 0-4-2 ex L. & Y.R. (their No.143 "Rhine").						
					14" 20"	D:4'-6"		16
93.	Uranus	100	10'-0" 3'-6"	n.k.	15" 22"	L:3'-6" D:5'-6"	6'-10" + 6'-0"	n.k.
94.	Neptune	"	" "	n.k.	" "	"	" "	n.k.
		Firebox casing 4'-10" long, 3'-10" wide; 116 x 2" tubes.						

Works Nos. are now added, but for previous locomotives they are included in Appendix IV where known.

No.	Name	Works No.	W.P. lb/in^2 (p.s.i.)	Boiler Length Dia.	Surface Area F^2	Cylinders Bore Stroke	Wheels Dia.	Wheelbase	Weight Tons Cwt.
98.	Pierremont	24.	100	12'-4" 3'-6"	n.k.	15" 20"	L:3'-8" D:5'-2"	5'-10¾" + 5'-4" total	42 10 in W.O.
101.	Marske	25.	"	" "	n.k.	" "	"	" "	" "
		Firebox casing 4'-0" long; Tenders 6 x 3'-6" dia. wheels on 4'-9" + 4'-8¼" base. 1200 gals. water, 3 tons coal, weight c.16 tons.							
99.	Ayton	n.k.	Generally similar to Nos.98 & 101.						
100.	Stobart	n.k.	"	" " "	" " "				
114.	Edward Pease	1071.	120	n.k. 3'-7½"	822	15" 22"	L:3'-8" D:5'-2"	6'-0" + 5'-4½"	24 14 eng. in W.O.
115.	Meynell	1072.	"	n.k. "	"	" "	"	" "	" "
		109 tubes. Tenders 4 x 3'-8" dia. wheels on 6'-6" base.							
116.	Lartington	74.	120	12'-0" 3'-7"	n.k.	1:16" 19" 2: " 22"	L:3'-8" D:5'-1"	6'-0" + 5'-4½"	41½ total in W.O.
117.	Nunthorpe	75.	"	" "	n.k.	1:16" 19" 2: " 20"	"	" "	" "
118.	Elm Field	Generally similar to Nos.116 & 117 26.				1:16" 19" 2: " 20"		Slightly shorter.	
166.	Oswald Gilkes	29.	120	12'-3" 3'-6"	n.k.	16" 19"	L:3'-8" D:5'-2"	5'-11" + 5'-4½"	25 16 2 eng. in W.O.
		No other boiler details; tender six wheeled, 1500 gals. water, 3½ tons coal.							

Chapter 7. Mineral Locomotives: 1845-75

In Chapter 3 I have mentioned the reason for the design of the long boiler locomotive, which was represented on the Stockton & Darlington Railway mainly by a very large number of 0-6-0 mineral locomotives. Apart from the long boiler principle, these derived directly from No.5 "Royal George" and its successors, including the later double tender classes.

However, before they really started to produce their own standard design of such an engine, and while both 'Tory' and 'Miner' classes were still being built, the railway bought in 1845 from the West Hartlepool Iron Co., then owned by Stephen Robinson and later by T. Richardson & Sons, a very much more up-to-date and even sophisticated looking 0-6-0 mineral engine, No.28 "Conside". Apparently no actual drawing of it exists, but Inness gave a drawing of what he said was a similar engine (Fig.113), apparently of slightly longer wheelbase. He says that the boiler had a diameter of 3'-10" and a length of 15'-3", but this length must have included the 2'-6" diameter cylindrical firebox and the smokebox unless the drawing is very much out of scale. It had 3'-9" wheels, Stephenson link gear and inside frames. The drawing shows also a Salter spring balance safety valve, a regulator valve lever on the backhead, lever reversing and cut-off gear, crosshead guides of a type that were still in use at the end of the century, cylinder drains with rod

Fig.113 Drawings of an 0-6-0 engine built by the Hartlepool Iron Co., 1845, and said to be similar to No.28 "Conside" built by them

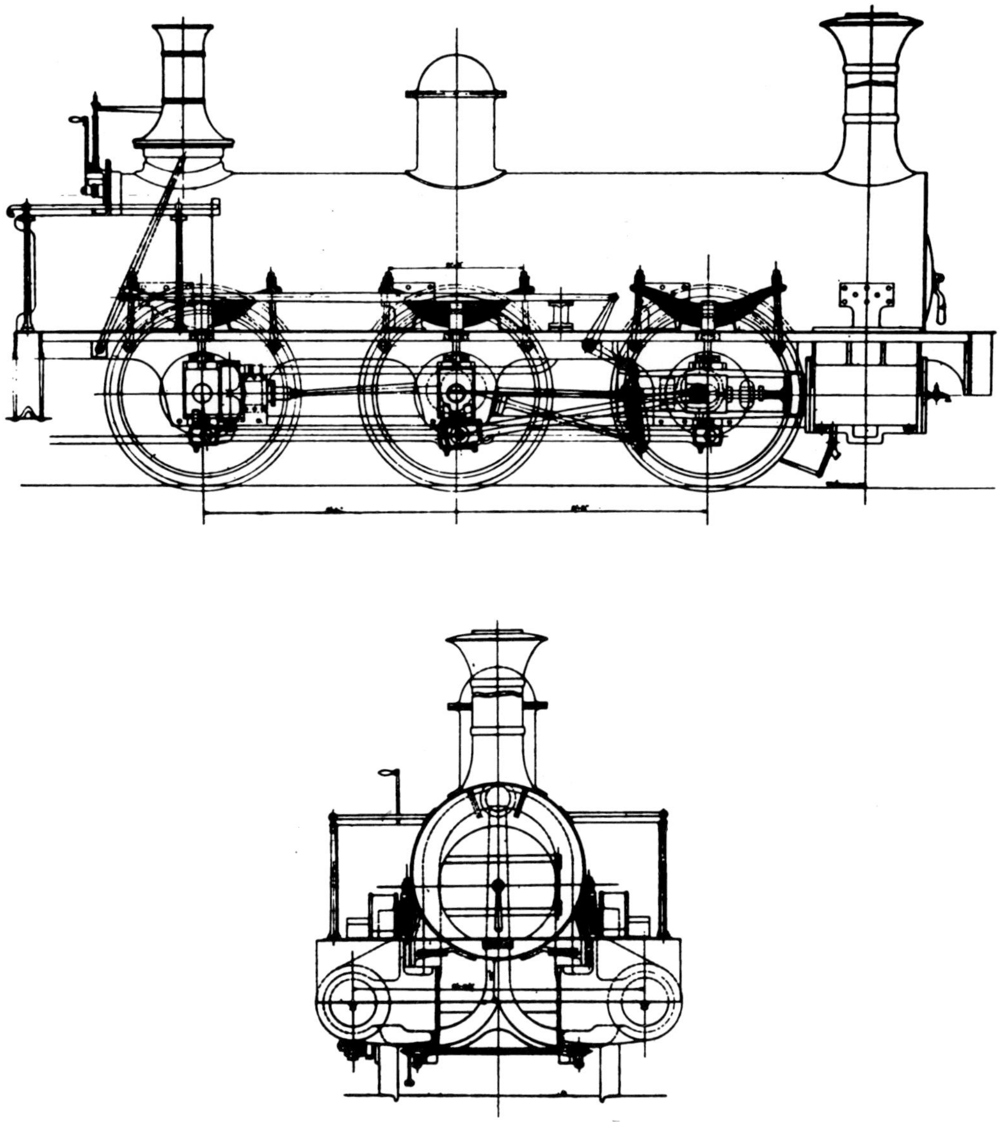

operation and a smokebox door with locking handle which would not have been out of place a century later.

In this drawing the cylinders are mounted horizontally outside the frames at the front and the connecting rods drive on to the centre pair of wheels. The bearings are in hornblocks with semi-elliptical springs above the footplate. This drawing bears the attribution 'Mécanicien', which I believe to have been R. H. Inness's pen name. A feed pump was driven from a separate eccentric on the centre axle.

No more such were bought, although apparently the builders did sell such engines elsewhere. Perhaps the true going price (for repeats) was a bit too high, and perhaps the performance did not match the appearance, or it was just not robust enough.

E. Craven's notes give a lot more information. He says that the engine was offered for sale in May, 1845, in Herapath's Railway Magazine, and was probably a speculative design. The prospective buyer was to apply to Mr. French at the works, Hartlepool. John Graham was instructed to inspect it, and it is minuted as bought on the 15th. May for £810 complete with tender. Craven describes it as having 3'-9" dia. wheels, outside cylinders 14" x 22", wheelbase 5'-3" + 5'-3", and says that it was sold to the Shildon Works Co. for use as materials on 15th. December 1858. This wheelbase matches the Inness drawing.

Craven also refers to an "Excellent photograph..." which apparently shows the engine appearing new, unpainted or in shop grey, with black smokebox. He says that the cylinders were placed obliquely above and between the first two pairs c.w. (carrying wheels). The drive was on to the rear pair. These wheels were of disc metal type, with eight spokes in relief and eight small holes pierced near the tyres. This sounds very much like two part cast iron wheels. Coupling and connecting rods were of circular section. There were no splashers to the wheels, the running-plate being raised in three areas to clear the wheels. Three piece chimney with flared top (a), small narrow dome, with short *outside* steampipes to the cylinders, which have flat valve chests on top (b) (Fig.114). One spring balance s.v. on dome, another (c) over the firebox, and a huge manhole between it and the dome. No cab or buffers, handbrake on tender only. The tender was four-wheeled, with outside bearings, and wheels of normal spoked pattern.

If the photograph concerned was accurately identified, this describes quite a different engine to that shown in Inness's drawing. There is no mention of it in the Hoole typed version of these notes, except a listed date of 2/66 for replacement by another engine. It is believed that Ken Hoole had seen this photograph, but it is not to be found now.

Inness gives no details of service, except to say that this engine was scrapped towards the latter end of the 1860's, which may mean that it was not scrapped immediately, but stored. Craven says that it was noted in an 1865 list as sold in 1858, and that it was handed over to the Shildon Works Co. for use as materials in December that year. It would have been interesting to have known its weight, and how it compared in tractive effort with the Shildon built engines of the time. It was not valued all that highly; £830 only in John Dixon's 1849 list, about two thirds of that of the surviving 'Tory' class engines, and somewhat higher than the value of the few remaining 'Director' class engines, but a little more than the purchase price. Lowes says that this firm built an 0-6-0 for the Stockton & Hartlepool Railway in the same year, 1845, but he does not list "Conside". Under Richardson's ownership the Hartlepool Iron Co. built much cruder engines of the Fossick & Hackworth type, and also others under sub-contract from other firms.

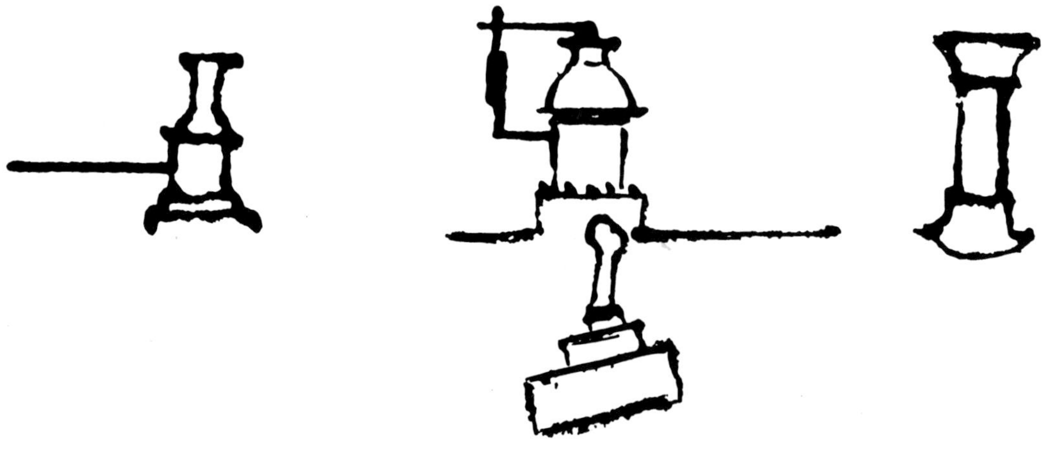

(c) (b) (a)
Fig.114 Sketches of cylinders and boiler mountings from Craven's notes (much enlarged)

Fig.115 Early photograph of No.35 "Commerce", taken at Shildon

Fig.116 Engraving made from Fig.115 photograph by J. Swain of the "Engineer"

Fig.117 Photograph of No.1035, ex 35 "Commerce", taken in connection with the 1875 celebrations

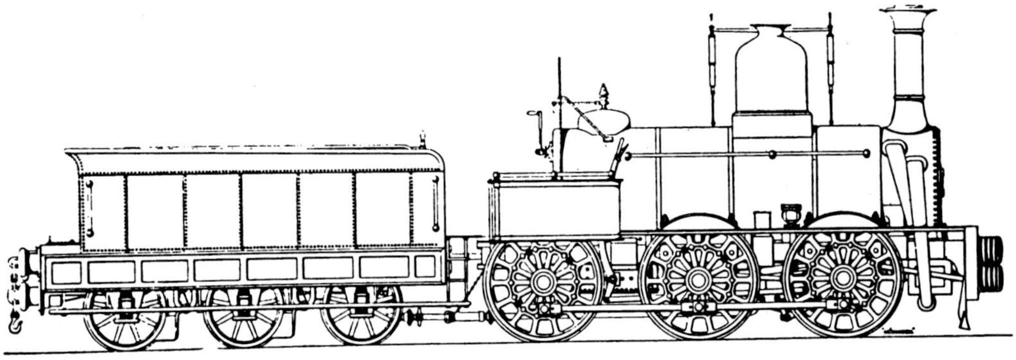

Fig.118 Drawing of No.63 "Birkbeck"

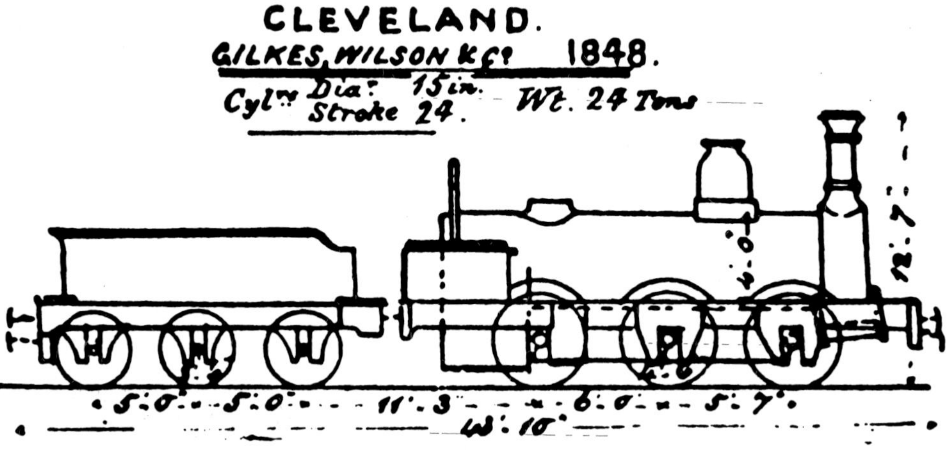

Fig.119 Theodore West's sketch of No.60 "Cleveland"

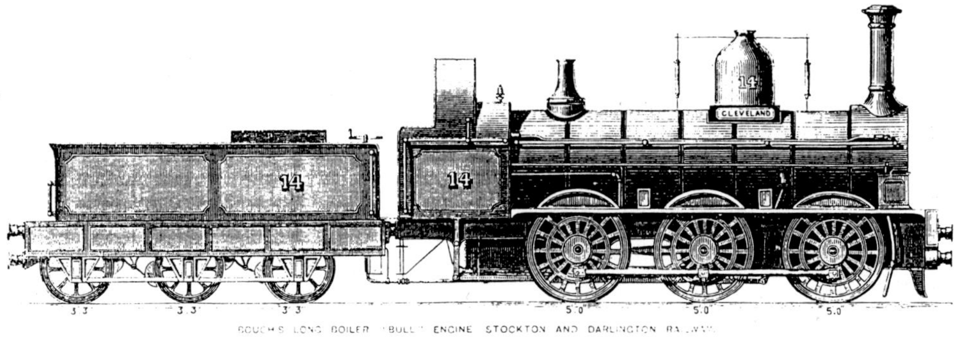

Fig.120 Drawing of No.60 "Cleveland", from the "Engineer" magazine and bearing the wrong number

William Bouch designed a small class of more advanced 0-6-0 mineral engines which had been ordered from the Shildon Works Co. in February 1846, and these became quite well known as the 'Commerce' class. The first of these, No.35 "Commerce", was delivered in January 1847, followed by No.36 "Guisbro'" in May and No.37 "Gem" in June. These engines still used the cast iron two part plug wheels, but otherwise were an improvement in design, with some features similar to the Inness drawing of the "Conside" and a few very original ones. It is not clear what the frames and springs were like, but unlike the previous classes there must have been such features. They were fitted with a high footplate and splashers. It is interesting to recall that when Churchward introduced similar high footplates on the G.W.R. some sixty years later there were outcries and protests! Between each pair of wheels was arranged a triangular wooden brake-block working upwards and operated by a steam brake cylinder mounted on the side of the boiler. The safety valves were on the steam dome and fitted with Salter spring balances. There is no mention of the valve gear, but it must have been of link type, probably of the Shildon design which came out that same year. Joseph Tomlinson associates the name of D. Dale with William Bouch in connection with these locomotives. When they were named, in November 1847, the minutes referred to them as merchandise engines - mixed traffic?

They were of long boiler type; the boiler was of 3/8" Low Moor iron plates, with a diameter of 4'-0" and a length of 13'-10" over the tubeplates, fitted with 107 2" iron tubes of 14 W.G., with a working pressure of 80 p.s.i. The heating surface was 826 sq.ft. A copper firebox was fitted. An article in the "Engineer" of December 26th. 1879 says that 93 brass tubes were fitted, but this may have been later. The tender had six wheels, possibly the first of this size on the line. The water capacity was 1400 gals. with 4 tons of coal, and the overall weight in working order was 25 tons 11 cwt. plus 15 tons for the loaded tender. It is obvious that the track had been upgraded!

The strangest feature was the design of the connecting rods. The outside cylinders (16" x 24") were kept as close together as possible, with the connecting rods next to the wheels and the coupling rods outside them; in order that they should clear the front crankpin the connecting rods had a large loop formed in each adjacent to the crosshead so that the crankpin could work within this loop. The crosshead was of massive construction so as to carry such a weight. This design feature was used again in the banking engine No.80 "The Duke" some seven years later, and No.135 "Eden" which followed. In "Commerce" and the rest of this class it was retained right up to the end, and never modified, although the engines were certainly rebuilt more than once. I have already mentioned a similar but differently constructed arrangement in Chapter 6, on the passenger engines No.38 "Rokeby" and No.39 "Ruby", which were built later the same year, and the possibility that restricted clearance in the Shildon Tunnel was the reason for keeping the cylinder centres close.

An engraving of No.35 "Commerce" appeared in the "Engineer" some years later (see above), based on the well known photograph of this engine at Shildon (Figs.115/6), but without any details to speak of in the accompanying text, except the tubes as mentioned above. It does say that the connecting rods were understood to have given no trouble. The engine is shown fitted then with a crude form of the standard S. & D.R. four buffer arrangement, but it is not known when the photograph was taken, or whether they were fitted originally (see previous comments). Similar buffers were still fitted in 1875.

Ahrons has rather more to say, and comments adversely on the very uneven distribution of weight over the three axles, an inevitable result of such a short wheelbase under a long engine; the leading axle carried far more weight than either of the others, no doubt due to the heavy crosshead assembly and the design of the front end of the connecting rods. The oscillation of this engine under load must have been very bad. Ahrons also says that the valve gear worked through rocking levers, and that a diagram of this gear was published in the "Engineer" of October 1875, but I have been unable to find this. Tomlinson made no comments except for what I have said above.

Inness comments on the hexagonal steam dome casing. The 1875 photograph (Fig.117) shows this clearly, whether a casting or built-up. My original photograph (Fig.115 again) also shows it, but not so clearly; later copies of this photograph with the background painted out do not show it, and although the "Engineer" engraving does show it, if one looks closely, the copy from this used by Tomlinson in his Presidential Address makes the dome appear to be round. The later photograph shows either only one safety valve, or more likely, two side by side, and the engine, of course, bears the N.E.R. No.1035. These locomotives worked for many years on mineral traffic, apparently successfully, before being relegated to shunting duties, and they were withdrawn and scrapped, No.36 in 1870 and the other two in 1876. There were very few outside cylindered mineral engines used on the S. & D.R.

apart from the double tender types; except for "Conside", the two bankers and a few strangers all those following had inside cylinders.

They were coal burning engines, and the question of fuel was under discussion at the time. Craven says that the Shildon Works Committee decided in February 1847 that no more coal burning engines were to be built, and Bouch was instructed to design "an engine that he would recommend, with a view to making several new ones for mineral or merchandise traffic". In October it was agreed to build two new coke burning mineral engines, and in November (on Guy Fawke's day!) the Committee reported , with regard to "altering the engines to a better type" that "the present class of engine may be converted from multitubular to firebox, or coke burning, engines at a cost of £200 each". Two were authorized to be so altered; these must have been double tender engines. In May 1848 they ordered experiments to be carried out as soon as possible "as to the relative economy of coal and coke burning engines, so that the Co. can decide on the propriety of adopting Fossick & Hackworths' patent tube in preference to coke-box engines". Finally, in June 1848, they reported "The Committee has now discussed the advantages and disadvantages of the present multitubular engines as compared with the remodelled fire-box one, No.30. Resolved, that in consideration of the savings to be made in the current repairs, No.31 and No.33 to be remodelled in like manner." (These would have been "Wear", "Redcar" and "Shildon"). This had apparently been completed by late November 1848, and it was agreed to do one more "of the six that were adapted for it", and Craven takes this to mean that nine engines were involved, Nos.29 to 37. However, we know that No.14 "Tees" was so modified in that year, among others, so he may not have been entirely correct.

The 'Commerce' class were long boiler engines, and as I have said already, it would be very interesting to know what the agreement on this was between the S. & D.R. and Stephensons, who held a patent on the type (1841, and including inside cylinders and frames). Possibly a licence fee was paid; I have never seen it mentioned, apart from a few inconclusive remarks in John Dixon's notes. Anyhow, the next class of 0-6-0 mineral engines were quite short locomotives; Inness attributes their ancestry to the Sharp Bros. 'Sphynx' type, but the 'Sphynx' was a long boiler locomotive. These two were obviously the two mentioned above; they were of Bouch design and built at the Shildon Works, No.63 "Birkbeck" (Fig.118) and No.64 "Larchfield", both delivered in February 1849. These were the first of the long series of inside cylindered 0-6-0's on this line. John Dixon says that No.64 was fitted with iron tubes made by the Patent Lap Welded Tube Co., who were still going in my younger days, but I think that their later products were produced by electric resistance welding.

The cylinders were 15" x 24", their boilers were similar to but shorter than those of the 'Commerce' class; the length was down to 8'-8 ", but the number of tubes was increased to 168 x 2" and the heating surface was 893.3 sq.ft., a quite dramatic increase, while for the first time boiler pressure was above 100 p.s.i. (110). They had link gear of the Shildon type, with the eccentrics outside the cranks. The connecting rods were forked, presumably to enable them to work round the first axle and to drive on to the centre axle (later 0-6-0 mineral engines, certainly from No.56 "Shotley" on, had the cylinders mounted higher and at an angle so as to overcome this problem). Inness gives some details of the coupling rods which I interpret as meaning that they were each made in two separate halves, with a double bearing on the centre crankpin rather than the more usual method of a joint in the coupling rod itself next to the centre crankpin (see also Chapter 9, the final 0-6-0's). The weight of engine and tender in working order was about 41 tons, little different from the much longer 'Commerce' class. Hackworth plug wheels were still being used, but of 4'-6" dia. Inness points out the peculiar arrangement of steam and exhaust pipes on the outside of the smokebox, without explanation.

These two engines lasted until 1875 for "Birkbeck" and 1873 for "Larchfield". "Birkbeck" was overhauled in 1857 to the extent that the engine was virtually renewed.

The next types of 0-6-0 mineral engines were the last to be fitted with the two part cast iron 'Hackworth' plug wheels, and the first to be ordered from (but not the first supplied by; see previous Chapter) the Tees Engine Works, Messrs. Gilkes, Wilson & Co. Two engines were ordered to begin with, at £1800 for the first and £1740 for the second; the learning curve was coming into its own! Two similar were ordered from Kitchings at the same time and same prices. Eventually three engines, No.60 "Cleveland" (Fig.119), No.61 "Star" and No.62 "Southend", were supplied by the Tees Engine Works in November 1848, January 1849 and February 1849 respectively, and the price paid for the third was the same as for the second. They are said to have been similar to Nos. 63 and 64 but of long boiler type, which makes me suspect that Nos. 63 and 64 were very soon rebuilt with longer boilers.

Fig.121 Drawing of No.72 "Peel" in early condition.

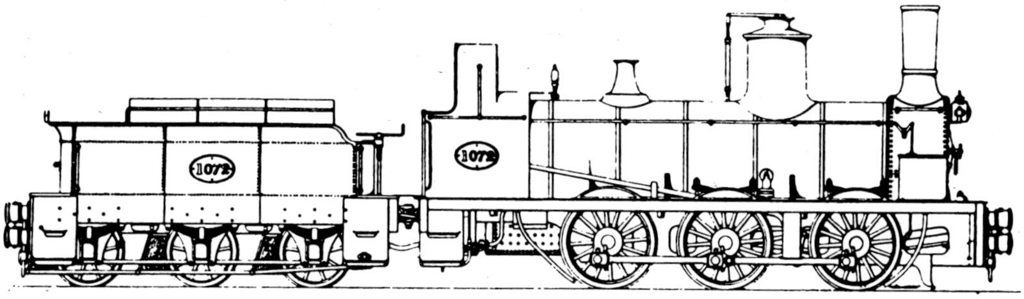

Fig.122 Drawing of No.1072, ex No.72 "Peel", as rebuilt later

Fig.123 Photograph of No.1072 as rebuilt

Fig.124 Photograph of No.1072 with railway personnel

Fig.125 Photograph of No.2226, ex 73 "Aberdeen"

Fig.126 Photograph of No.1708, ex 84 "Albert"

Fig.127 Photograph of No.82 "Hawthorn" outside Shildon Roundhouse.

John Dixon says that they were to J. Graham's design, which surprises me. In his 1849 on list he makes other comments which conflict with previous information about these early 0-6-0 engines. He says that No.61 "Star" and No.62 "Southend", which were supposed to be similar to No.60, were exactly the same as No.2 "Times" and No.3 "Graham", built later the same year by Kitchings, and that No.64 "Larchfield" was also the same except for the tubes mentioned above. Dixon was the civil engineer for the line, not a locomotive engineer, and may have been mistaken, but he was a good engineer, and what he says cannot be ignored. So we have another mystery. Inness is quite clear about which engines were the same as which, and I have followed him. I am inclined to think that Dixon was generalizing somewhat, and that although there may have been a general similarity it was not an exact one in all respects. After all, these engines were quite different from those preceding them, and to that extent were much alike.

I said above that the driving wheel diameter for these three engines had now been increased to 4'-6", whereas Nos.2 and 3 had 4'-0" diameter wheels. The boilers for Nos. 60, 61 and 62 were 13'-0" long, with 162 x 2" tubes, heating area not stated but obviously much greater, by something like 50%. The first two lasted until 1870 and the third until 1876. Gilkes, Wilson supplied four other engines of similar type to the Leeds & Thirsk Railway, and since they were basically to Bouch design (John Dixon says clearly "to plans and specifications supplied by the Shildon Works Co.), it is difficult to believe that they would have been allowed to do so independently. Craven says that these four engines had very heavy fuel consumption, and were sold in 1855 to the South Yorkshire Railway through E. B. Wilson of Leeds (confused earlier with Robert Wilson of Forth Street and "Chittapratt" fame), who probably altered them first.

Ken Hoole's notes say that No.61 "Star" was apparently named "Perseverance" at first, and described as a "very short engine".

Fig.120 is taken from an article in the "Engineer" for February 1885. The artist is unknown; it seems strange that the "Engineer" should make such a mistake regarding the engine No., shown as 14 instead of 60. but a drawing of "Rokeby" on the same page also bears a wrong number, 194 instead of the correct No.38. The works number of "Cleveland" was 16, and Gilkes, Wilson's Works No.14 was not supplied to the S. & D.R., so it cannot be that.

The article gives few details, but does say that the class as designed had short boilers, which did not steam well, and so were lengthened (see John Dixon's remarks above). It may also account for the remarks a a little later regarding the boilers of the 'Peel' class. Another interesting point in the article is that it claims that the earlier class of 0-6-0 mineral engines was known as the 'Driver' class (the last of the 'Miner' class was No.34 "Driver") and that the 'Cleveland' class were known as the "Bulls".

Alfred Kitching also supplied two similar engines, No.3 "Times" in March 1849 and No.2 "Graham" in July, to Shildon design. They were different in some details, as mentioned above and as listed, and returned to 4'-0" diameter wheels, according to Inness, presumably of plug type. He also stated 15" x 20" cylinders and a few other details which I have listed (see also the last paragraph of Chap.5 regarding names).

Craven, however, has a lot more to say. He says that the wheels were 4'-6" dia. as the three from Gilkes, Wilson, and that the Railway charged Kitchings £120 for the use of the drawings, so the engines should have been similar. No names were allocated to start with, and "York" and "Utility" were suggested in March 1852, but rejected in favour of those above. He gives cylinders 15" x 24", boilers 8'-8$^{1}/_{8}$" x 4'-0", 166 x 2" x 8'-11$^{1}/_{2}$" copper tubes, six wheeled 1400 gal. tender carrying 3 tons of coke, and other dimensions. It is a great pity that no Kitching drawing has survived as far as is known. Both were replaced in 1865, and little more seems to be known about them.

The 'Hackworth' two part cast iron plug wheel was used no more, but the long boiler type lived on, and continued to develop. Inness has some interesting remarks about these wheels in service (Fig.25). This drawing shows them as finally developed. The two halves were secured together by four 1$^{1}/_{4}$" iron bolts with heavy washers, and in each of the plug holes, one to each spoke (14 on this drawing) a square section tapered plug was driven, so that the two halves had a slight gap between them. This is said to have had a cushioning effect, and I have seen it said also that some adjustment of the outer radius was possible by means of the plugs, but I would not have thought this of any great significance considering the amount of wear accepted. When the plugs dried out they shrank, sometimes to the extent that the wheels rattled, and the enginemen had to throw buckets of water over them to swell the wood again. A similar drawing appears in one of the Whessoe histories. It is noticeable that the tyres were not only shrunk on, but secured further by seven tapered pins held in place by nuts inside the rim (see also earlier remarks, p.41-42).

William Bouch next designed a new series of larger engines to be built by Gilkes, Wilson over the next few years, fitted with so-called welded wrought iron spoked wheels with cast iron centres, similar to the wheels used on the current passenger engines and mentioned in the last chapter. Eleven engines altogether were built in this series, but nearly all of them had slight differences to start with. Nevertheless, they established a standard which held for many years. They were all rebuilt later except for the first two, renumbered more than once, and lasted well into the last years of the century, while a few of them survived into the twentieth century.

The first two were No.56 "Shotley", in January 1852, and No.57 "Tow Law" in March, (Craven says March & June) these being the numbers said to have been given at first to the first two 'Priam' class engines, and later changed. They had 4'-0" driving wheels and inside frames, the boilers were 13'-0" long x 4'-0" in diameter, working at 100 p.s.i. The cylinders were 17" bore x 18" stroke, and set slightly above the wheel centre line at an inclination of 1 in 12, so that the connecting rods should clear the front axle. Their combined weight was 44 tons for engine and tender in working order. The six-wheeled tenders were fitted with screw brakes, with blocks, probably of wood, working on both sides of the wheels, which is surprising. No brakes appear to have been fitted on the engines. Although they gave good service, these two were replaced in 1866; it is not known whether they were scrapped or sold.

The next nine had wheels of 4'-2½" diameter and their wheelbase differed slightly, as listed. They were No.72 "Peel" (Figs.121-124) in October 1852, No.73 "Aberdeen" (Fig.125) in the same month, No.74 "Emperor" in July 1853, No.75 "Baring" in August, No.76 "Prince of Wales" in February 1854, No.77 "Alexander" the same month, and No.83 "Victoria", No.84 "Albert" (Fig.126) and No.85 "Hardinge" all in July 1854. These are the first engines shown by Inness in his line drawings with sand boxes fitted, and in the drawing of "Peel" as later rebuilt (Fig.122) he shows the tender with large wooden brake blocks working on the rear side of each wheel. Nos.74 and 75 had a longitudinal division in the firebox (a mid-feather). John Dixon says that they were to Robert Stephenson & Co. patent (see earlier comment on long boiler design).

Their boilers were 13'-6" long x 4'-0" diameter, of ⅜" Low Moor iron plates, fitted with 158 x 2" tubes, again working at 100 p.s.i. I have been told that these nine engines were not built originally as long boiler locomotives, but were rebuilt in this form later, in which case John Dixon's note does not seem to make much sense. It is my belief that only a few of those built in 1849, with short boilers, were so rebuilt, but I could be wrong.

No.72 "Peel" lost its chimney at one time, going under a bridge. Long boiler engine oscillation?

No.76 "Prince of Wales" became the first engine to be rebuilt at the new North Road Works in 1864. All the others were rebuilt eventually, as listed where known. Some of them were fitted with injectors instead of feed pumps, additional safety valves of later pattern and other improvements, plus cabs, of course. They were all renumbered later, some more than once, and not always in the N.E.R. 1000 class.

A banking engine, No.80 "The Duke" (Fig.128), was designed by William Bouch, Craven says in April 1851, and built by the Shildon Works, entering service in May 1854. This engine had the same type of driving gear as the 'Commerce' class, and was of 0-6-0 long boiler type, with wheels of 3'-11½" diameter; the boiler was similar to the usual mineral locomotive type, 13'-0" long x 4'-0" diameter, with a huge steam dome that made the engine look squat and powerful. The boiler was fitted with 158 x 2" tubes and the steam pressure was 120 p.s.i., with cylinders of 18" bore x 18" stroke. The wheelbase was fairly short at 10'-0" equally divided, and the suspension was unusual, the first two pairs of wheels sharing an inverted ellipsoid leaf spring on each side, while the trailing pair had a transverse spring, a feature used again later. The total weight of engine and six wheeled tender was about 47 tons 5 cwt. The cost was £1800 less tender. The engine was intended for banking duties on the severe gradients in the South Durham coalfields (John Dixon says the Howden incline at Crook), and presumably was successful, since a second similar locomotive, No.135 "Eden" (Fig.129), was built, and put into service in November 1858. "The Duke" was supplied with a new tender in 1856.

"Eden" was rebuilt as a mineral engine under the N.E.R., and so possibly was "The Duke". Both engines lasted until 1882.

Also in 1854, due to delay in delivery of mineral engines on order, the railway bought two second-hand 0-6-0 locomotives, long wheel base engines with outside frames, that had been built by R. & W. Hawthorn for the Edinburgh & Glasgow Railway in February 1846, but had not been suitable for the duties, being too heavy for the track over which they had to work. They were acquired in May 1854, and became No.81 "Miller" and No.82 "Hawthorn" (Fig.127). They had large cylinders of 18" bore x 24" stroke, but a working pressure of only 80 p.s.i., so were not as powerful as they looked. The wheels

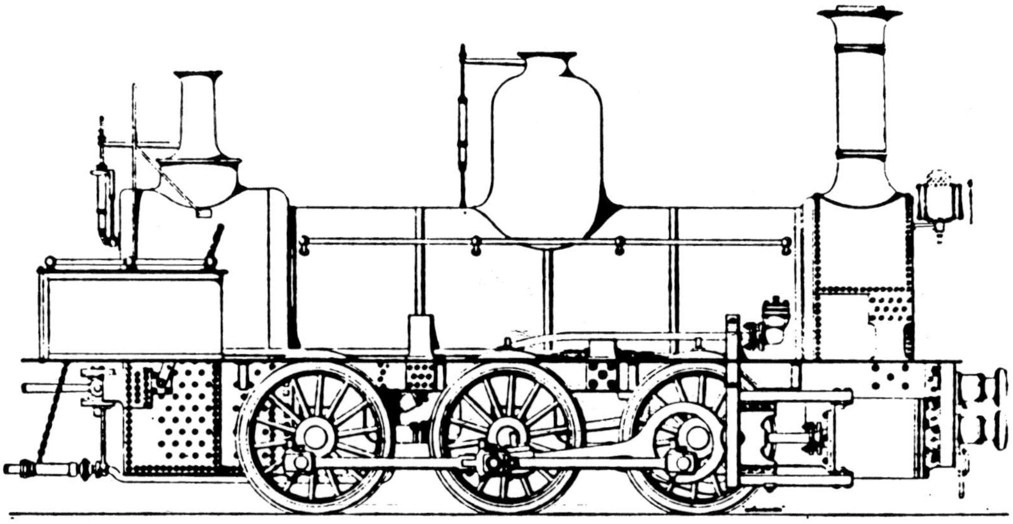

Fig.128 Drawing of No.80 "The Duke"

Fig.129 Photograph of No.1135, ex 135 "Eden", rebuilt as a mineral engine

Fig.130 Photograph of No.107 "Barnard Castle"

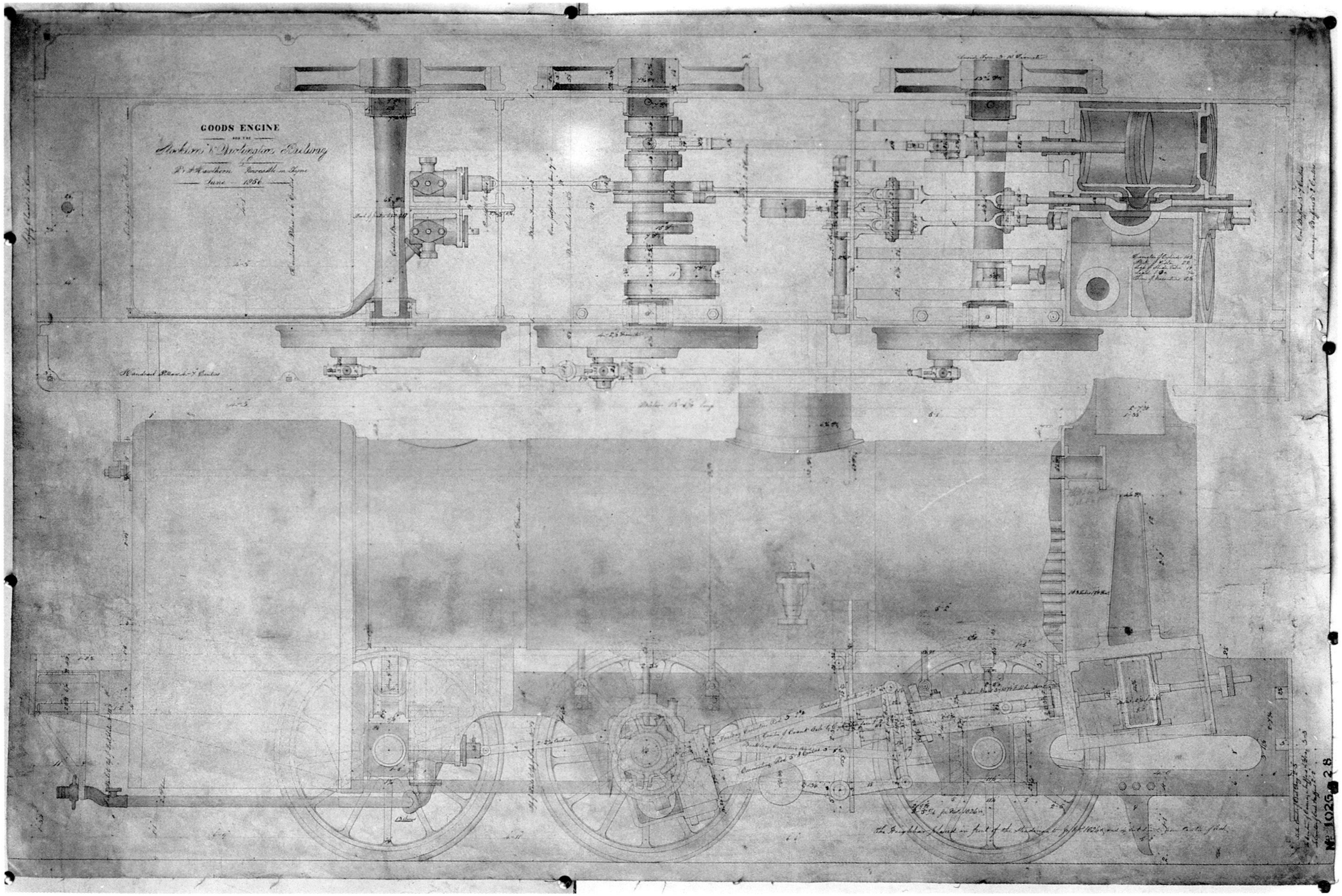

Fig.131 R. & W. Hawthorn Works drawing of later "Peel" class, Nos.108-113, works Nos. 959-964, of 1856

Fig.132 Photograph of No.2259, ex 104 "Durham"

Fig.133 Photograph of No.1836, ex 104 "Durham", as rebuilt again later

Fig.134 Photograph of No.1112, ex 112 "Lion"

Fig.135 Photgraph of No.1916, ex 122 "Kestrel"

Fig.136 Photograph of No.125 "Gazelle"

Fig.137 Photograph of No.1129, ex 129 "Stanley"

were of 4'-3" diameter (Baxter says that when built the wheels were 5'-0" diameter on a wheelbase 6'-7" + 7'-2"), and the boiler smaller than usual; 10'-0" long x 3'-10" diameter, fitted with 135 x 2" tubes. Total weight of engine and tender was 42 tons in working order. No.81 only lasted until 1867, but No.82, which was later reboilered with an increased pressure of 100 p.s.i. with 17" cylinders, kept going until 1881, although part of its final duty was as a winding engine!

Craven has a few other details; he says that they were built in about May 1847, works Nos.532/3, E. & G.R. Nos.81/2, and that the names were those originally held. He says also that they had been in use as banking engines on Cowlain Bank.

A somewhat similar engine with a shorter wheelbase, also Hawthorn built, was bought in August 1854, again second-hand, after an inspection by Bouch. It had been Blackburn, Darwen & Bolton Railway No.4, and then Lancashire & Yorkshire Railway No.219. The price was £1550. For some reason Inness says that it was the only one of its type on the S. & D.R., but apart from size it seems to have been very like the previous two. It was, according to Inness, of long boiler type, while Nos.81 and 82 were not, so this is probably what he meant; it was the only outside framed long boiler engine that they had. The original date of building was probably 1847, Craven says July. It was given the No.86 and the name "Zetland". The wheels were of 4'-8" diameter; the boiler dimensions are not known, but the steam pressure was 85 p.s.i. and the cylinders 16" bore (Baxter and Craven give cylinders 16" x 24", Baxter gives boiler 10'-0" x 3'-11", total heating surface 605 sq,ft., W.P. 110 p.s.i, wheelbase 6'-9" + 5'-6", weight 23 tons 11 cwt., also Lancashire & Yorkshire Railway No.219 after its B.D. & B.R. days.). It was rebuilt in about 1869 with a larger boiler uprated to 120 p.s.i., 12'-10" x 4'-0" diameter, 170 x 1⅞" tubes, and lasted until 1881, its final years being spent on the Normanby branch hauling ironstone trains.

Craven, of course, differs in one respect; he says that it was *not* a long boiler locomotive. If Baxter is right about the boiler length he would seem to be correct. However, Craven's notes cease at this point, so from now on Inness rules more or less alone.

It would seem that the Railway was so short of locomotives at this time that they had to hire some from the L.N.W.R., at a cost of £417-1-6d. (April 1855), one being a Bury 0-4-0. (see earlier).

A further twelve engines, similar to the 'Peel' class, were built by Gilkes, Wilson, starting in February 1855, the first one being No.90 "Merchant", followed by No.91 "Australia" and No.92 "Yarm" in March, No.95 "Elephant", No.96 "Middleton" and No.97 "Gainford" all in June, No.102 "Cotherstone" in September, No.103 "Darlington" and No.104 "Durham" (Figs.132 & 133) in October, No.105 "Teesdale" in December, No.106 "Staindrop" in February 1856 and finally No.107 "Barnard Castle" (Fig.130) in March 1856.

Six more such engines were delivered by R. & W. Hawthorn during 1856, with slightly different dimensions. The boilers were 13'-6½" long x 4'-0" diameter, with 153 x 1⅞" tubes, working at 120 p.s.i. and the driving wheels were of 4'-3" diameter (Fig.131). They had six wheeled tenders capable of carrying 1300 gals. of water and 3 tons of coal. They were delivered as follows; No.108 "Eagle", No.109 "Falcon" and No.110 "Hawk" in June, No.111 "Stag" and No.112 "Lion" (Fig.134) in August, and finally No.113 "Camel" in September 1856.

No early photographs of these engines seem to exist, and we will have to assume that they looked very much like Inness's drawing of No.72 "Peel", as the next Gilkes, Wilson built batch certainly did. The works drawing (Fig.131) does not show their external appearance. These next engines had driving wheels of 4'-2½" diameter to begin with, possibly still of the built-up type, difficult to tell from the photographs. The boiler dimensions were similar to those of the 'Peel' class, and the overall weight of engine and tender in working order was about 48 tons. The names, Nos. and delivery dates follow: No.119 "Greta" and No.120 "Brough", December 1856; No.121 "Tiger", March 1857; No.122 "Kestrel" (Fig.135) May; No.123 "Belle Vue" June; No.124 "Brookside" October; No.125 "Gazelle" (Fig.136) February 1858; No.126 "John Wakefield" May; No.127 "Polam" June; and finally No.128 "East Mount" in August 1858.

Inness does not give the steam pressure, but it may be assumed to be 120 p.s.i. as the previous engines, since some of this class were later reboilered with a higher pressure of 140 p.s.i. No.120 "Brough" suffered a boiler explosion in June 1865, near the north end of Shildon Tunnel, with four serious casualties.

No.125 "Gazelle" was involved in a contest soon after being built with the N.E.R. 0-6-0 locomotive No.134, a Fletcher designed and Gateshead built engine known as "Big Bella", although, of course, N.E.R. locomotives at that time had no official names. This appears to have been an official series of trials and took place on the newly completed east coast main line. "Gazelle" was rewheeled for the purpose for some of these trials, and the 4'-2½" drivers were replaced by wheels of 4'-11½" diameter, which may have increased the speed under light load, but at the same time would certainly have decreased the tractive effort, so I assume that with the larger wheels speed trials only were involved. It would appear that although the S. & D.R. and the makers were not dissatisfied with "Gazelle's" performance, the honours very clearly went to the N.E.R. engine, but there was a belief that the latter had the safety valves screwed down to increase the working pressure by 20 p.s.i.! No.125 "Gazelle" was also used in some trials on the Sunniside incline at Crook (see historical notes in Chapter 1.). All these trials took place within 1858. Craven wrote some notes about them, and these are appended at the end of the chapter.

Nos.126 and 128 were reboilered to a higher pressure of 140 p.s.i. and fitted with steam brakes later. No.126 was used on the Rosedale branch during the later part of its life.

All these engines had a fairly long life, but the next three did very well. two of them lasting for 47 years and one for 51 years. These were ordered from R. & W. Hawthorn in 1857 and all delivered in 1858, No.129 "Stanley" (Fig.137) in April; No.130 "Hutton" (Fig.138) and No.131 "Long Hull" in May. They had slightly smaller cylinders but the main change in external appearance was the steam dome cover, which was rounded at the top instead of the 'lid' type used for most of the preceding engines in this chapter.

Another two were supplied by Kitchings in the same year; No.132 "Appleby" in August and No.133 "Kirkby Stephen" in September, with a reversion to 17" cylinders. These two had 4'-3" wheels. Stephensons supplied a further three in the same year; No.136 "Tebay" in November: No.137 "Deepdale" and No.138 "Beelah" (Figs.139 & 141) in December, all very similar to the previous batch.

Nos.112, 132 and 138 found their way to the coal staiths at Gateshead later, under the N.E.R.

The next seven were from Gilkes, Wilson, all except one being similar to the others above, and delivered as follows; No.134 "Kendal", December 1858; No.139 "Crossfell" (Fig.140) in February 1859; No.140 "Penrith" in April; No.141 "Excelsior" (the odd one) and No.142 "Wycliffe" in July; No.143 "Windermere (Fig.142) in November and No.144 "Ulleswater" (Fig.143) in March 1860. No.141 "Excelsior" was built with cylinders of 18" bore x 24" stroke, a steam pressure of 130 p.s.i. and 4'-11" driving wheels. The boiler dimensions were 14'-10" long x 4'-1½" diameter, with a heating surface of 1728 sq.ft. Unfortunately this engine lived up to its name only too well, and blew up at Simpasture Junction on May 10th. 1867, killing the driver, whose name was Thomas Parker, and badly injuring another man, but the engine was rebuilt and put back into service by October that year.

No.139 "Crossfell" was another to go to the Rosedale branch later. The engines in use on this high and exposed branch were in general fitted with cabs on the tender in addition to the usual type (Fig.140).

The next batch was ordered in April 1859, especially for the new line from Barnard Castle to Tebay, soon to be opened, and were somewhat more powerful. Six were supplied by R. & W. Hawthorn, being No.145 "Panther" (Figs.144-146) in January 1860; No.146 "Ostrich" and No.147 "Leopard" in February; No.148 "Zebra" (Fig.148) in March; No.149 "Fox" in April, and No.150 "Mastiff" (Fig.147) in May. They had 5'-0" diameter driving wheels and cylinders of 17" bore x 26" stroke. The boilers were 13'-7" long x 4'-1" diameter, with 191 x 2" o.d. tubes and a heating surface of 1483 sq.ft. total, at 120 p.s.i. They were fitted with Stephenson link motion and eccentric driven feed pumps as previously described, and also with the Bouch feed water heating jacket round the chimney, with a bypass valve for direct feed if required. In order to protect the enginemen from the Stainmore winter weather these engines were fitted with what was a very large cab for the time (Fig.145). The total weight of engine and tender in working order was 49½ tons.

No.149 "Fox" was used to haul the first excursion train to the Lake District in 1861, and was derailed 2½ miles above Bowes on the return journey, running down the embankment with the train (six coaches and two vans). The driver was killed and many injured, one passenger, a William Fagget, getting £800 compensation for a broken collar-bone.

These engines were fitted with transverse rear axle springs similar to those on the two banking engines, "The Duke" and "Eden".

A further nine engines with 24" stroke cylinders but otherwise very similar to the last class was built by Gilkes, Wilson, whose deliveries seem to have been slower than the other suppliers. These were No.151 "Mercury", in June 1860; No.152 "Venus" in July; No.153 "Mars" in October, and No.154 "Jupiter in November, followed by No.155 "Saturn" (Fig.149) in January 1861; No.156 "Herschel" in February; No.157 "Planet" in August; No.158 "Lune" in October and No.159 "York" in February 1862. They all had 5'-0" wheels, but No.157 "Planet" had them changed to 4'-2" later. They were a little lighter than the Hawthorn engines; 48 tons 16 cwt. total working. The boilers seem to have been the same as the

Fig.138 Photograph of No.130 "Hutton"

Fig.139 Photograph of No.1138, ex 138 "Beelah", shunting at Greencroft Quarry

Fig.140 Photograph of No.1139, ex 139 "Crossfell"

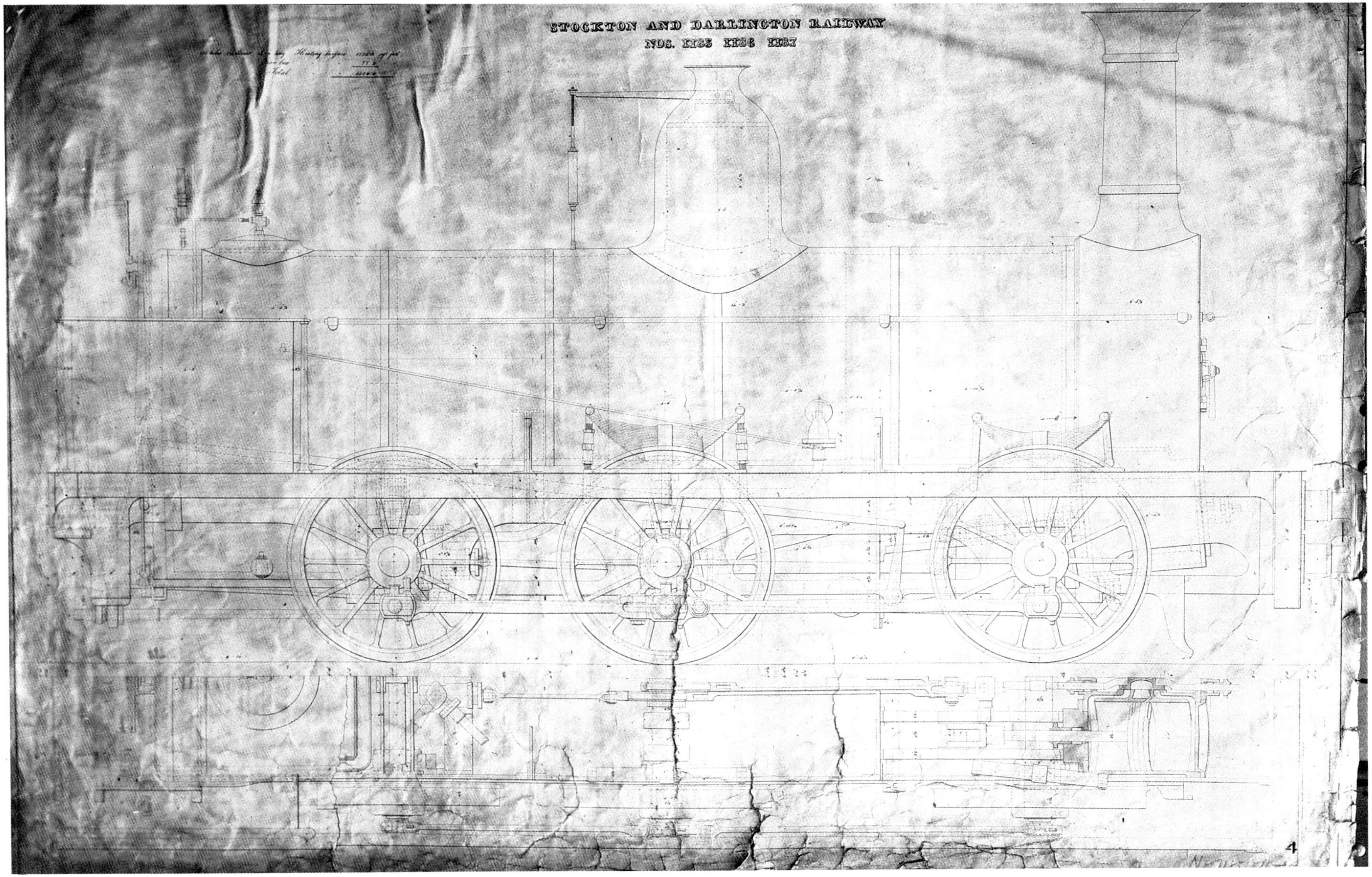

Fig.141 R. Stephenson & Co. Works drawing of Nos.136/7/8, "Tebay" class, works Nos. 1185/6/7, of 1858

Fig.142 Photograph of No.1725, ex 143 "Windermere"

Fig.143 Photograph of No.1715, ex 144 "Ulleswater"

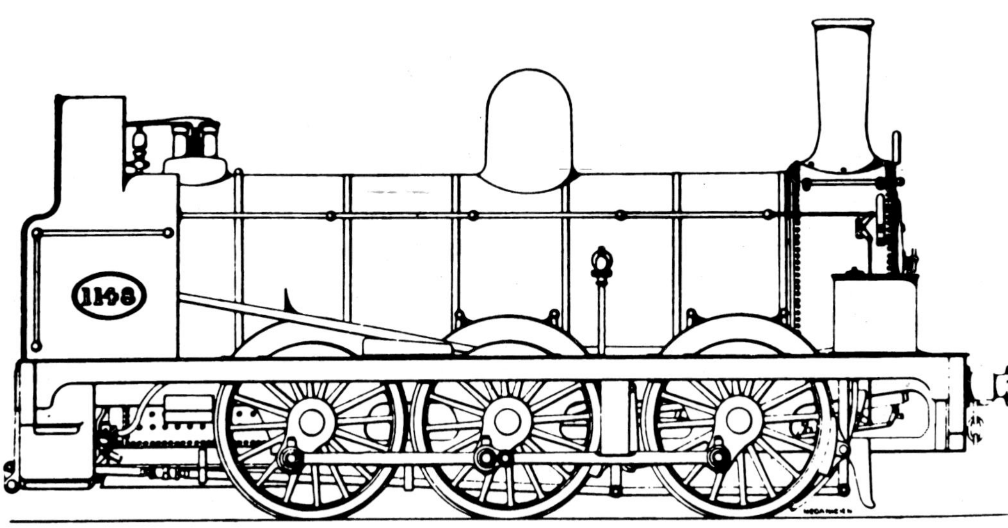

Fig.148 Drawing of No.1148, ex 148 "Zebra", as rebuilt

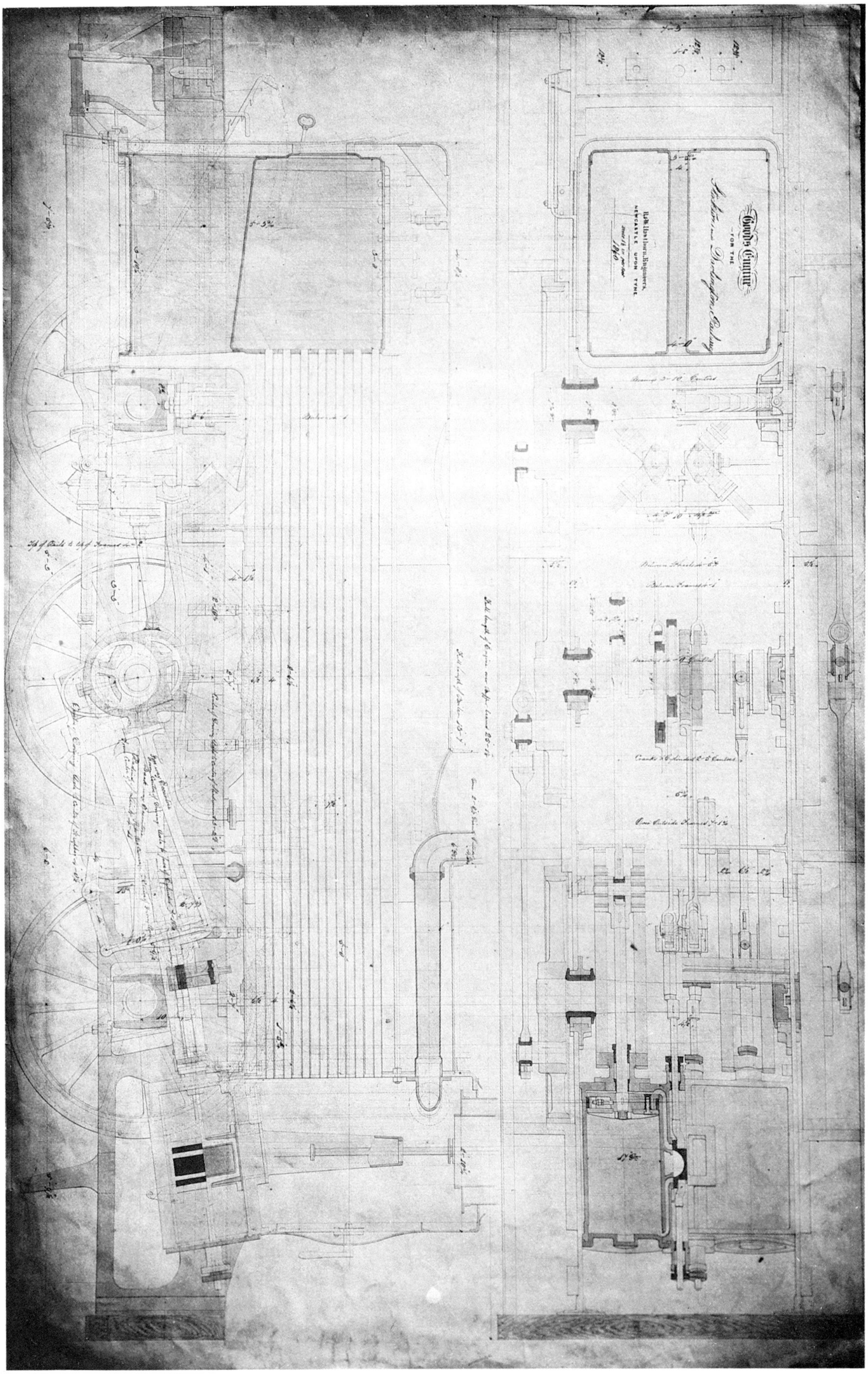

Fig.144 R. & W. Hawthorn Works drawing of 1860, probably of "Panther" class, Nos.145-150, works Nos.1074-1079 (for tender see Fig.146)

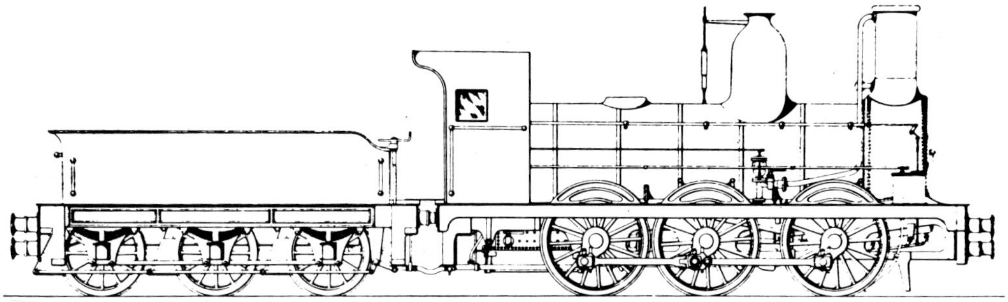

Fig.145 Drawing of No.145 "Panther"

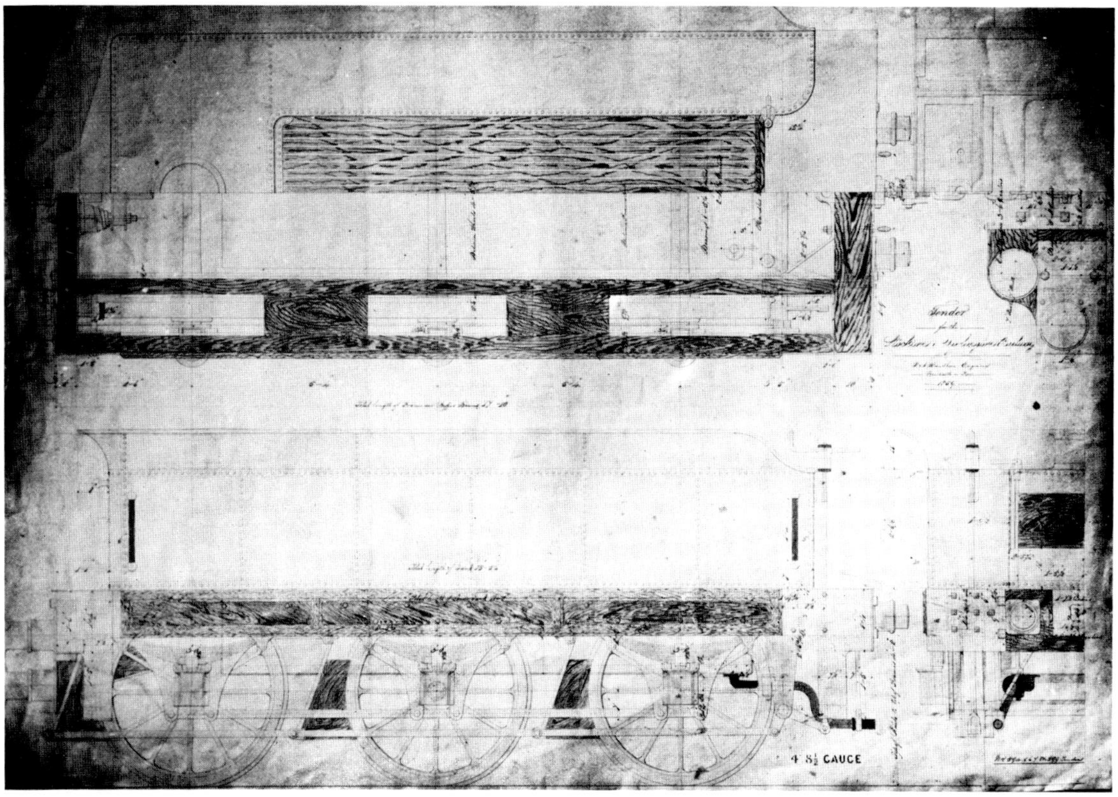

Fig.146 R. & W. Hawthorn works drawing of wood framed tender for Nos.145-150, works Nos.594-599

Fig.147 Photograph of No.1150, ex 150 "Mastiff"

Fig.149 Photograph of No.1728, ex 155 "Saturn"

Fig.150 Photograph of No.1169, ex 169 "Tufton", at Middlesbrough Shed.

Fig.151 Photograph of No.170 "Reliance"

Fig.152 Photograph of No.1188, ex 188 "Lily"

Hawthorn batch. No.159 "York" may have been the engine sold as N.E.R. No.1674 to Pelaw Main Collieries in 1905 and converted to an 0-6-0ST. No.153 badly damaged the motion by striking a large stone between the rails sometime later.

Gilkes, Wilson built a further four similar engines a little later; No.167 "Newlands" in November and No.168 "Clifton" in December 1862; No.169 "Tufton" (Fig.150) in May 1863, and No.170 "Reliance" (Fig.151) in January 1864. The total weight is given as 51 tons 8 cwt. working, most of this increase being in the tender. Apparently No.170 was sent north under the N.E.R. to Percy Main shed on coal staiths work, and stayed there for many years.

A further two were built by the same firm, No.187 "Iron Age" delivered in December 1864, and No.188 "Lily" (Fig.152) in August 1865. These two were similar to Nos.167-170 above. The independence of the Darlington Committee may be seen here; after the official merger with the N.E.R. in July 1863 they were still giving their new engines names; strictly against N.E.R. policy.

This merger between the Stockton & Darlington Railway and the North Eastern Railway had been in preparation for some time. It benefited the latter line considerably and helped it financially for many years, so they were in no position to impose their own conditions on the S. & D.R., whose main problem was not financial. On the contrary, at the time they were in fine fettle, making good profits, paying high dividends and sound in every sense, which is a lot more than could be said for their competitors. The problem was that there was nowhere further to expand, and they were surrounded by other lines, which brought continual commercial conflict. The Great North of England Railway had been sold to Hudson much earlier, since the Pease plans to expand their empire north had been frustrated, and it was already part of the N.E.R. Had that plan succeeded in 1842 it might well have been a different story. Maybe the merger was a wise and far-seeing move commercially, the Quakers, as usual, being very astute in their business dealings, but it seems a shame. They could quite possibly have continued successfully as an independent line until 1923.

Four engines were built at the Shildon Works in 1863/4, very similar to the previous 0-6-0's but with larger wheels, and a reversion to the steam dome covers with 'lids' on (Fig.153). These were No.171 "Gladstone" and No.172 "Barrow", in December 1863; No. 173 "London" and No.174 "John Dixon" both in June 1864. They were fitted with driving wheels of 4'-11" diameter, possibly as a result of the trials referred to earlier. The cylinders were of 17" bore x 24" stroke, the proportions changing in line with the higher pressures now being used and the larger wheels with a bigger crank throw. The boilers had a length of 13'-0" with a diameter of 4'-0", and were fitted with 158 x 2" tubes, working at 120 p.s.i.

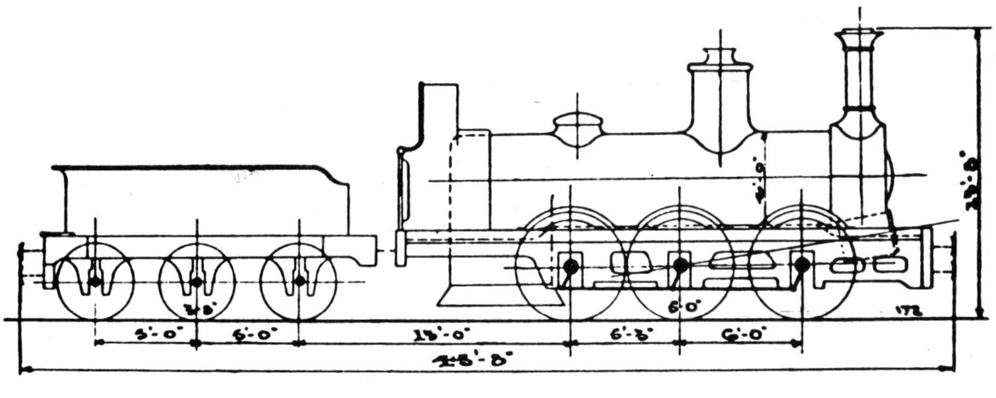

Fig.153 Drawing of Nos.171-174

The total working weight of engine and tender was 49 tons 10 cwt. "Gladstone" lasted until 1889, but the other three kept going into the early 1900's.

The Shildon Works Co. moved into the new North Road Works, Darlington, later in 1864, and the first locomotive built there was another 0-6-0 mineral engine, No.175 "Contractor" (Fig.154), delivered to service in October. Wheels and cylinders were as the previous four, but the wheelbase was increased from 11'-3" to 11'-8" and the boiler lengthened to 14'-0", with 158 x 2" tubes, working at a pressure of 130 p.s.i. This engine seems to have been the first to be fitted with Bouch's improved valve gear (Fig.10) which had fairly minor differences from the Stephenson type, mainly in details of construction and with improved and case-hardened bearing surfaces. Inness says that No.175 "Contractor" ran with the original valve motion until scrapped in 1908, having in 44 years service covered 1,070,488 miles. Strangely enough the stated weight of engine and tender in working order was down to 48 tons 14 cwt.

Fig.154 Photograph of No.175 "Contractor" at North Road, Darlington

The next ten engines were ordered from Stephensons, and were of distinctly unusual appearance. They were long boilered 0-6-0's still, but the wheelbase was increased to 13'-0", bringing the leading pair well forward. The boiler had a length of 14'-0", similar to "Contractor", but the smokebox was longer and of large diameter, and the round-topped firebox was also somewhat longer, giving them an even 'longer' look than usual. The large round topped steam dome was on the first ring of the boiler, giving them an ungainly 'lurching forward' look, and it was fitted with the usual Salter spring balance safety valve, supplemented later by an additional 'Adams' type safety valve in a 'Beyer' bonnet just in front of the firebox. As built they were fitted with Bouch's chimney jacket feed water heater, which increased the bulky forward appearance, and with his 'steam retarder' brakes (see later comments). I assume that the link motion was Stephensons' own (Figs.155 & 156). It is of interest that the drawing of the engine is dated August 4th. 1864, but the drawing is still headed Stockton & Darlington Railway.

The boiler diameter was 4'-0" minimum over four rings; number of tubes not stated by Inness and cannot be read on the drawing (158?) but were 2" dia. The heating surface totalled 1308 sq.ft. and the working pressure was 130 p.s.i. The water capacity of the tenders was 1800 gals. with $3^1/_2$ tons coal, and the overall weight of engine and tender totalled 52 tons 6cwt. in working order. These tenders had full screw operated brakes fitted in pairs on each wheel, with metal blocks.

They were delivered as follows; No.176 "Windsor" in April 1865; No.177 "Osborne" (Figs.157 & 158) in May; No.178 "Balmoral" in June; No.179 "Edinburgh" and No.180 "Dublin" in September; No.181 "Malton" and No.182 "Elton" (Fig.159) both in December 1865; No.183 "Acklam" and No.184 "Lark" in March 1866, and finally No.185 "Swallow" in April that year. The royal flavour seems to have run out half way through! They were rebuilt more than once, and lasted well into the first decade of the twentieth century, two of them reaching 1910.

Two similar engines were built at the North Road Works, and delivered No.186 "Union" in February 1865, and No.191 "Autumn" (Fig.160) in April 1866. They were somewhat heavier, totalling 56 tons for engine and tender, but otherwise apparently much the same: possibly the valve motion would have been of Bouch's type. It would seem that the original Shildon Works were still in production, as two engines of the 'Gladstone' class were built there and delivered in September 1865; No.189 "Spring" and No.190 "Summer" (Figs.161 & 162). One of these is a well known photograph of "Summer" outside the Shildon paint shop at the Soho Works with many of the workmen, including 'Putty' Bowran, the foreman painter. Large sand-boxes were beginning to appear on all the locomotives, and apparently earned the soubriquet of 'Shildon Tea Caddies' or 'Quakers' Pickle Jars'! The paint shop still exists as part of the Hackworth Railway Museum.

These last two engines lasted in rebuilt form until 1906 and 1908 respectively. The 'Windsor' class was described in some detail, with drawings of the side elevation, end views and cross section, and of the tender, in the "Engineer" magazine of August 30th., 1872. Among other things this article gives the number of boiler tubes as 168.

Before I move on to the next batch of standard engines, I will describe a peculiar trio of absorbed locomotives which came into stock in 1865, when the West Hartlepool Harbour & Railway Co. was amalgamated with the N.E.R., and these three were

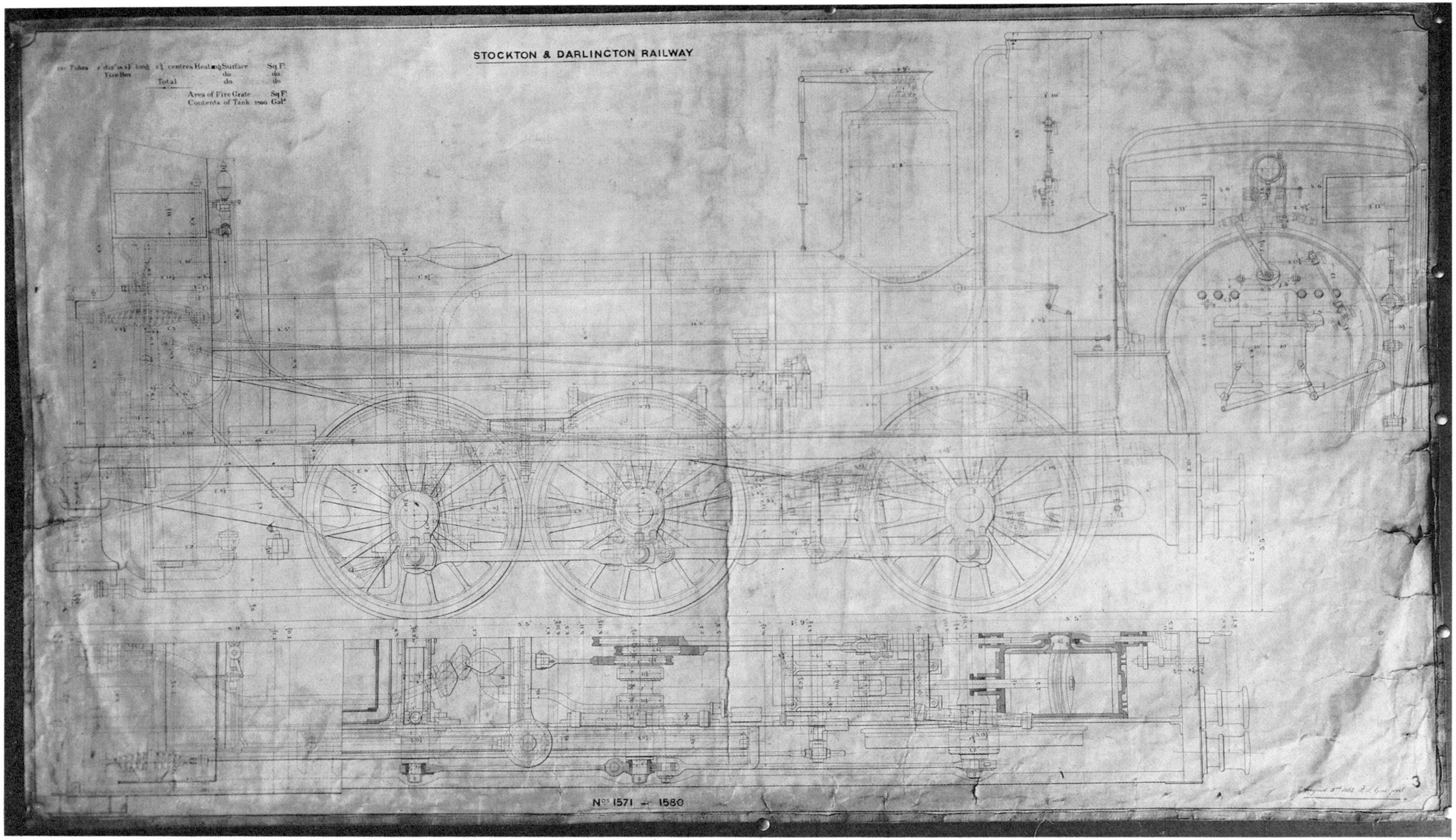

Fig.155 R. Stephenson & Co. Works drawing of 1864 for Nos.176-185, "Windsor" class, works Nos.1571-1580 (for tender see Fig.156)

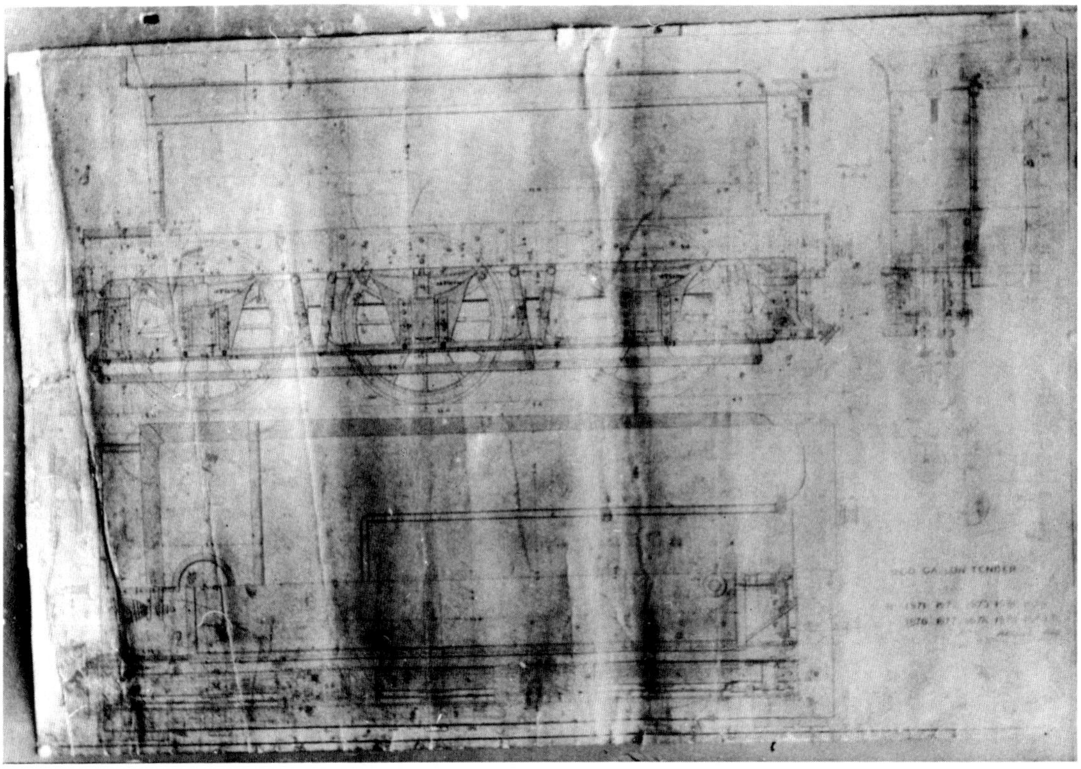

Fig.156 R. Stephenson & Co. Works drawing of 1800 gal. tender for Nos.176-185

Fig.157 Photograph of No.1177, ex 177 "Osborne"

Fig.158 Photograph of No.1177 as rebuilt later, at Kirkby Stephen in 1893

Fig.159 Photograph of No.182 "Elton", with Bouch feed heater chimney, outside Shildon Roundhouse

Fig.160 Photograph of No.191 "Autumn"

Fig.161 Photograph of No.190 "Summer" outside Shildon (Soho Works) paint shop, 1870

Fig.162 Photograph of No.1190, ex 190 "Summer" as rebuilt later

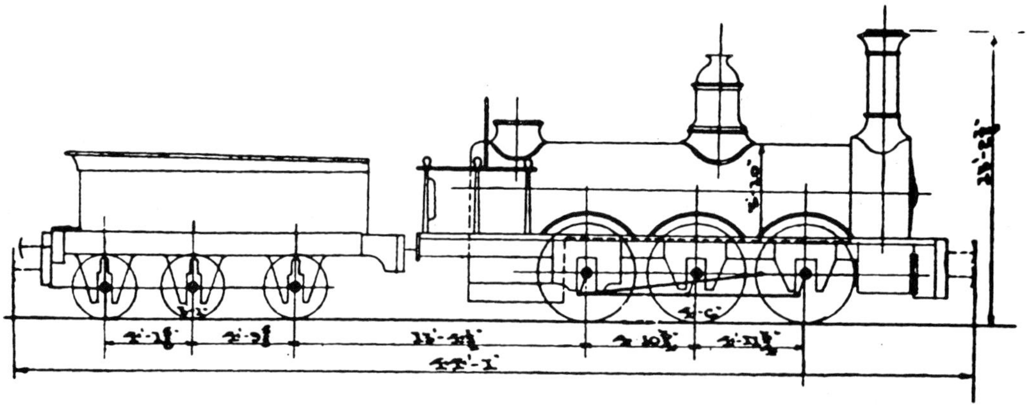

Fig.163 Drawing of No.19 "Tweed"

allocated to the Darlington Section, probably because they were there already. They had been built by that railway (W.H.H.R.) at their works in Stockton, near the junction, in 1861, designed by Mr. James Irving Carson, their Locomotive Superintendent, and to a very archaic form by then. Their lineage was early Stockton & Darlington Hackworth, right back to the 'Tory' and 'Miner' classes. In Chapter 5 I mentioned that one engine had been built by Timothy Hackworth's brother Thomas and his partner Fossick, at Stockton. Fossick & Hackworth improved on this design and supplied quite a lot of engines of the type to neighbouring lines such as the Clarence, the West Hartlepool, the Llanelly Railway and others, but their design was by now very out-of-date. These three were very similar, and derived from them. They were short wheelbase long boiler 0-6-0 engines, one of them, which was designated No.19 "Tweed" (Fig.163) with outside horizontal cylinders, and the other two, No.20 "Normanby" and No.21 "Cleveland" (Fig.164) had outside cylinders mounted diagonally on the smokebox centre line in a very similar position to those on "Stockton" in West's sketch, on the 'Miner' class and on the rebuild of No.14 "Tees". Both types drove on to the trailing pair of wheels, which meant very long connecting rods.

Almost certainly they had inside frames, and were fitted with Stephenson link motion working the valves through rocking levers. The wheels were of cast iron, presumably spoked, and the engines were bigger than appears at first sight. No.19 "Tweed" had 4'-6" wheels and $15^3/_4$ bore x 22" stroke cylinders. The boiler was 13'-0" long x 3'-10" diameter, working at 120 p.s.i. The other two had 4'-2" wheels, cylinders $15^3/_4$" bore x 21" stroke, and boilers 11'-6" long x 3'-10" diameter, again working at 120 p.s.i. The tenders carried 1200 gals. of water and $3^1/_2$ tons of coal. These engines were used on the old Cleveland Railway Co's mineral line at first, two of them working from Slapewath to the Normanby incline, and the other from the foot of this incline to Normanby Jetty. This brings memories of reading in W. W. Tomlinson about the 'Battle of the Tees' in 1860, when this jetty was first built. These engines were not rebuilt, and finished their lives shunting in the Middlesbrough area, being scrapped in 1877.

Photographs exist of several engines built by Fossick & Hackworth, Fossick & Blair, T. Richardson and other local firms, all much like these three, and the family resemblance to the second No.4, "Stockton" is very strong.

The next batch of standard engines reverted to 4'-$2^1/_2$" diameter wheels, four of them from the North Road Works and the last two from Shildon

Fig.164 Drawing of No.20 "Normanby" & No.21 "Cleveland"

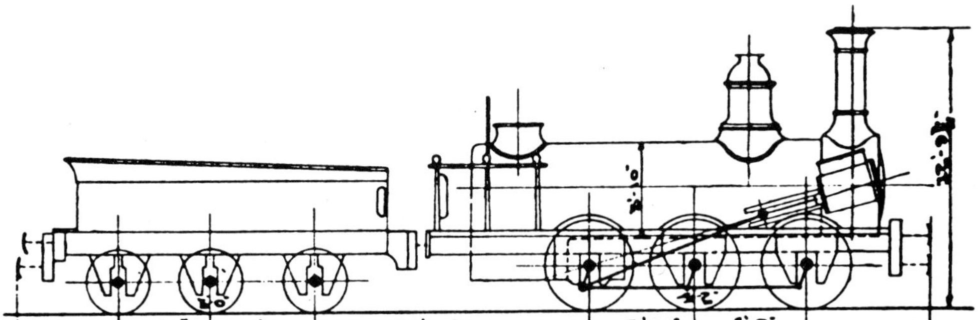

Fig.165 Photograph of No.1192, ex 192 "Winter", as rebuilt later

Works. They were No.2 and No.3, both unnamed, in December 1865, No.192 "Winter" (Fig.165) in July 1866; No.193 "Princess" in August; No.194 "Alice" (Fig.166) and No.195 "Helena" both in October 1866. They were very similar in appearance to No.175 "Contractor", and the Bouch feed water chimney was not fitted. Cylinders were 17" bore x 24" stroke and the boilers 13'-8" long x 4'-$0^{3}/_{4}$" diameter, with a surface area of 1246.8 sq.ft., working at 130 p.s.i. The total weight of engine and tender, working, was 53 tons 6cwt. After many rebuilds and renumberings they were all scrapped between 1901 amd 1910.

Gilkes, Wilson had now become Hopkins, Gilkes & Co., and amalgamated with the Teesside Iron Works, although I suspect that the Pease family still held a stake in them. They now built a further fifteen engines over about two and a half years, only two of which received names (Fig.167). Deliveries were as follows; No.9, November 1865; No.23 December; No.28 February 1866; No.40 March; No.42 May; No.43 June; No.46 August; No.49 October; No.51 December; No.81 in January 1867; No.200 "Eskdale" in February; No.201 "Carlton" April; No.204 and No.205 in July, and the last one, No.206, in April 1867. They were the usual long boiler 0-6-0's but with 5'-0" diameter wheels again, and looked much like the 'Panther' class. The cylinders were 17" bore x 24" stroke, the boilers 14'-0" long x 4'-0" diameter, with 158 x 2" tubes, working at a pressure of 130 p.s.i. The tender carried 1500 gals of water and the usual $3^{1}/_{2}$ tons of coal, while the overall weight of engine and tender in working order was 52 tons 5 cwt.

Of these engines, No.40 worked in the Scarborough and Whitby area for many years, and worked goods trains all over that section, Forge

Fig.166 Photograph of No.1194, ex 194 "Alice"

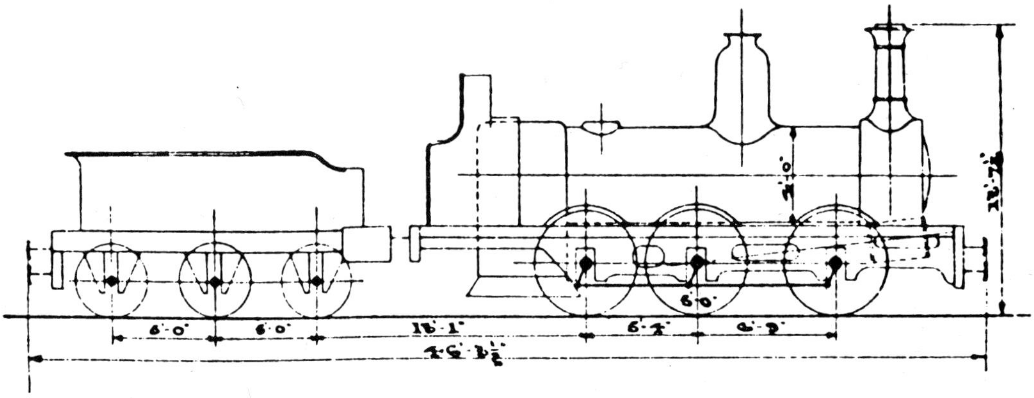

Fig.167 Drawing of 0-6-0 mineral engines built by Hopkins, Gilkes from November 1865-April 1867

Valley, Pickering, Helmsley and Malton. No.51 as 1051 got up to Gateshead, Greensfield Shed, on Dunston coal staiths, and No.201 ended up at Hull, shunting in the docks. No.204 was sent to the Northern Division. None of them lasted later than 1909.

Two small wheeled engines similar to No.192 "Winter" etc., were built at Darlington in 1866; No.56 in April and No.57 in June, replacing the earlier engines bearing those numbers ("Shotley" and "Tow Law"), and lasted until 1909 and 1903 respectively.

R. & W. Hawthorn received the next orders, for twelve and five engines respectively. These were large engines similar to the 'Windsor' class, but none of them named. They had wheels of 5'-0½" diameter, cylinders of 17" bore x 26" (first batch) or 28" (second batch) stroke, with boilers 14'-6" long x 4'-0" diameter, 168 x 2" tubes, working at 130 p.s.i. and with a heating surface totalling 1432 sq.ft. Stephenson link motion was fitted. The fireboxes were flat topped, a forerunner of the later Belpaire type, and they were fitted with the Bouch chimney jacket feed water heater, steam retarder brakes and eccentric driven feed pumps. The reversing gear was of Bouch type, fitted earlier to most recent classes, with a cigar shaped screw and a sector lever; still to be seen, when I wrote this, on No.1275 in the North Road Station Museum (see also Chapter 8). The tenders were increased in capacity, carrying 1800 gals. of water, but only 3 tons of coal. The overall weight of engines and tender in working order was 64 tons 1 cwt. They were delivered between December 1867 and November 1870; the individual dates listed by Inness are in the list later. The numbers ran from 207-218 and 221-225 (Figs.170-173).

Several of these engines, Nos. 207, 208, 210, 211 and 223, were rebuilt as long wheelbase locomotives in the early 1880's (Figs.168 & 169) with longer round topped fireboxes, but the remainder stayed out their careers largely unaltered except for necessary reboilering etc. Where later type boilers were fitted the type and arrangement of the fittings and mountings were also changed, of course, which altered the appearance. No.213 lasted until 1910, and the others did almost as well.

The North Road Works delivered three engines in early 1867 similar to the 5'-0" wheeled Hopkins, Gilkes 1865 series. These were No.53 in March, No.93 "Uranus" (Fig.174) in April and No.94 "Neptune" in May. All three lasted until 1909. They replaced the earlier engines of the same numbers, No. 53 "Manchester" of 1840, and the two second-hand Hicks 2-2-2 engines of the same names bought in 1835. "Uranus", as N.E.R. No.1093, went to Rosedale, and was fitted with the double cab used there (Fig.175).

Two similar engines were built at Shildon and delivered in June 1867; No.202 "Ireland" and No.203 "England". These were the last two locomotives to carry names on the Darlington Section, and also the last two to be built at Shildon Works. They lasted until 1905 and 1909 respectively, after reboilering.

The next series were built at the North Road Works, and comprised twentynine locomotives between April 1868 and December 1870, with numbers quite unrelated to the position in the series, no doubt as the previous engine bearing that number was withdrawn so that number would be reallocated, but this was not always the case (see remarks under General Note, Chap.1). They all had 5'-0½" wheels, but the cylinders, which were all of 17" bore, varied between 24" and 26" stroke as listed. The boilers were very much as the Hopkins, Gilkes 1865 series, and the weight only a little more, at 53 tons 8 cwt. total working. None of them lasted beyond 1910 but they began the N.E.R. 1001 class. One of them, No.219, had a spectacular derailment together with a Fletcher engine (No.1135), when working double headed near Winston, in the Primrose Cutting between Darlington and Barnard Castle, on October 24th. 1905. By the look of the photograph*, this part of the cutting was an embankment! No.220 ran away

Fig.168 Photograph of No.1207, ex 207, rebuilt as a long wheelbase 0-6-0

Fig.169 Photograph of No.1210, ex 210, rebuilt as a long wheelbase 0-6-0

Fig.170 Photograph of No.214 in early condition and with Bouch feed heater chimney

Fig.171 Photograph of No.217 in early condition and with Bouch feed heater chimney

Fig.172 Photograph of No.1221, ex 221, as rebuilt

Fig.173 Photograph of No.1222, ex 222, in early condition but without Bouch feed heater chimney

Fig.174 Photograph of No.93 "Uranus" (2nd.) in early condition, probably at Darlington North Road Works

Fig.175 Photograph of No.1093, ex 93 "Uranus", as rebuilt later with double cab, and at Rosedale

Fig.176 Photograph of No.1012, ex 12, as rebuilt later, and at Shildon

on the Skinningrove zig-zag in 1880, with tragic results. The young son of a locomotive inspector riding on the footplate with his father was trapped by the tender and scalded to death. (Figs. 176-185)

(*"The Locomotive" magazine, August 15, 1929; not reproduced here)

These engines were fitted with transverse springs to support the trailing axle, made up of 6" wide plates and 3'-10" between centres. Probably volute springs at each axle box were also fitted.

A further five of this type were built by Hopkins, Gilkes from September 1870, to April 1871, Nos.226-230 (Figs.186-189). Only the wheelbase was at at all different from the previous ones, being a little longer at 12'-1". They were all scrapped by 1909. No.229, after reboilering in 1882, ran between Sunderland and Durham until 1901, when it was again rebuilt, and ended up on mineral traffic on the old Londonderry Railway between Seaham Harbour and South Docks at Sunderland.

At the same time the North Road Works was turning out a further series of the 11'-10" wheelbase type, similar to the 1868-70 batch, together with Hopkins, Gilkes from the end of 1872 (Figs.190-194). Between them they built 36 of these engines, 24 at North Road and 12 at Middlesbrough (see list). The first one was delivered by North Road in August 1870, and the last one by Hopkins, Gilkes in May 1875. They were all 26" stroke engines, and the boiler diameter was given as 4'-0 ", with 1268 sq.ft. heating surface, but these are minor differences. In addition to the two feed pumps they also had a No.8 Robinson & Gresham injector. They were fitted with the Bouch Steam Retarder, now described as a means of using the main cylinders as brakes by letting steam in against the pistons, short of actually putting the reverse lever over. The suspension with its transverse spring over the trailing pair of wheels was as described previously, and the boilers and tubes were still of Low Moor iron. Bouch's valve motion was used, of course. Many of them were later reboilered, and one, No.259, lasted until 1915. No.1245 was derailed at Lingdale Incline in June 1900. No.1263 had a spell on the Easingwold Railway in 1911, hired to them while the Easingwold engine No.2 was under overhaul.

These 12 were the last main line engines built by Hopkins, Gilkes at the Tees Engine Works.

It will be seen that in the sixteen years between 1847 and 1863 the Stockton & Darlington Railway had received 83 new mineral locomotives of their own design, and most of the engines of the 'Tory and 'Miner' classes were still then in service. This does not include No.28 "Conside" or the two 0-6-0 banking engines. John Proud has listed two more, Nos. 64 & 65 of 1873.

Not counting the ex West Hartlepool H. & R. engines, or the second-hand purchases, you will see also that a further 136 new 0-6-0 mineral engines were built between 1863 and 1875 under the Darlington Committee.

There we will leave the mineral engines for the time being, as the next series were somewhat different.

Low Moor iron, from the famous iron works just south of Bradford, remained the most highly rated material for boilers and tubes for most of this long period.

J. S. MacLean, in his "The Locomotives of the North Eastern Railway", 1922 edition, part II, has some interesting remarks about these and later 0-6-0's as formed into the N.E.R. 1001 class.

E. Craven Notes on "Gazelle" Trials.

Engine 125 Gazelle built. by GW in February 1858, was used for various sets of trials. In March & April 1858, fitted with 4'0" wheels, it worked over the S & D from Shildon to Darlington, & forward over the NER to MNN(?), hauling up to 50 loaded wagons & returning with up to 55 empties. 7 such return trips were made. In July of the same year the same engine made 9 similar return trips, but this time fitted with 5'0" wheels, and further tests with 5'0" wheels took place on 4 days in August 1858 over the same route.

Starting from Darlington the trains were observed to be reaching the foot of Dalton Bank at 30 MPH with 5'0" wheels, and at 22-23 MPH with 4'0" wheels. With 5'0" wheels the summit of the bank ($2^1/_4$ M at 1 in 377) was passed at 10-11 MPH, 3-4 MPH less than with 4'0" wheels.

In Sept. & Oct. 1858 No.125 was tested up Sunniside Inc., north of Crook, with $1^3/_4$ miles @ 1 in 32/19/30/19/16/13.4/16. Trials were made running light & with four loaded chaldron wagons, both up the full length of the incline, & up the last 616 yds. @ 1 in 13·4 & 1 in 16. During the latter tests the summit was topped at 24 MPH light & 12 MPH loaded.

Trials were also carried out using 125 on trains between the foot of Sunniside Inc. & Upleatham ironstone mines.

(Craven's abbreviations have in general been expanded)

Fig.177 Photograph of No.1001, the second and only other No.1, as rebuilt

Fig.178 Photograph of No.1006, ex No.6

Fig.179 Photograph of No.7 in original condition

Fig.180 Photograph of No.1011, ex 11, as rebuilt later, at Darlington

Fig.181 Photograph of No.18 in original condition

Fig.182 Photograph of No.1004, ex 4, as rebuilt later

Fig.183 Photograph of No.1078, ex 78, as rebuilt later

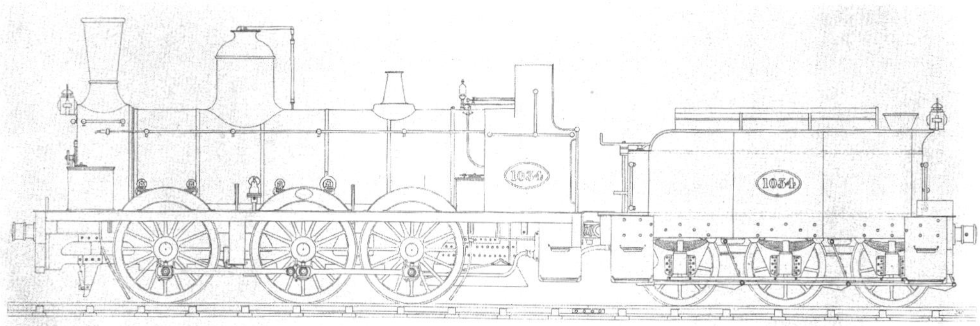

Fig.184 Drawing of No.1034 (Draughtsman unknown)

Fig.185 Photograph of No.1030, ex 30, as rebuilt later

Fig.186 Photograph of No.227 at Eden Colliery, in original condition

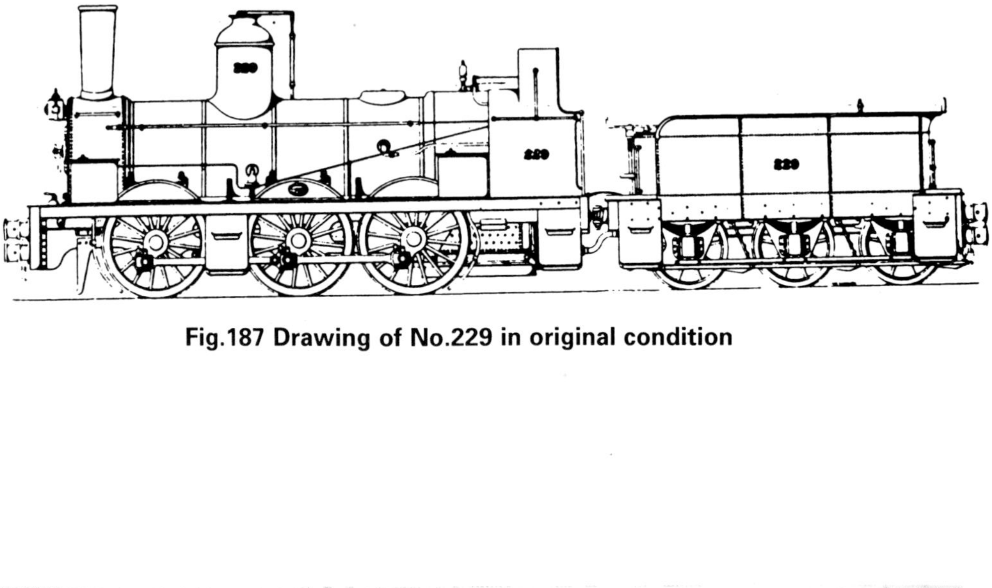

Fig.187 Drawing of No.229 in original condition

Fig.188 Photograph of No.1229, ex 229, as rebuilt later

Fig.189 Photograph of No.1230, ex 230, as rebuilt later

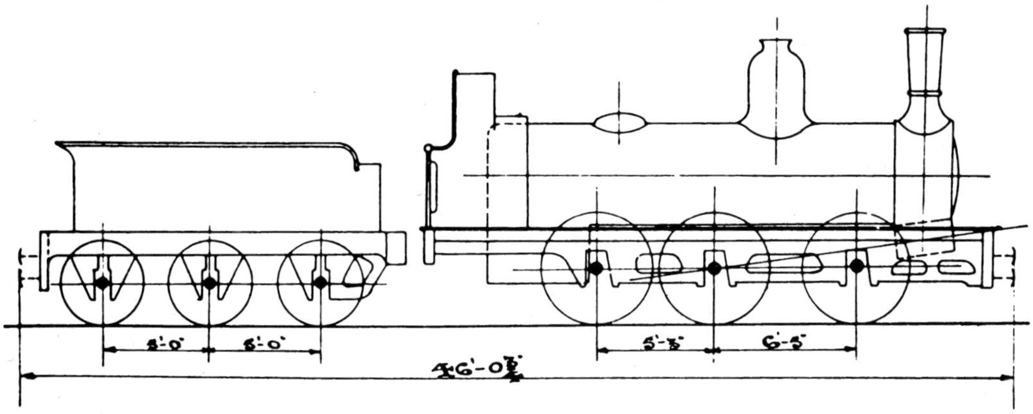

Fig.190 Drawing of 0-6-0 mineral locomotives as built by both Hopkins, Gilkes and by the North Road Works between September 1870 & May 1875

Fig.191 Photograph of No.1088, ex 88, as rebuilt later

Fig.192 Photograph of No.1249, ex 249, as rebuilt later

Fig.193 Photograph of No.1253, ex 253, as rebuilt later

Fig.194 Photograph of No.1256, ex 256, as rebuilt later

Data as early condition in general, and approximate.
n.k. = not known. n. = nominal. a = actual.

No.	Name	Works No.	W.P. lb/in² (p.s.i.)	Boiler length Dia.	Surface Area F²	Cylinders Bore Stroke	Wheels Dia.	Wheelbase	Weight Tons Cwt.
28.	Conside		70	15'-3" 3'-10"	n.k.	14" 22"	3'-9"	2 x 5'-3"	n.k.
	Firebox cylindrical, 2'-6" dia.								
35.	Commerce		801	3'-10" 4'-0"	826	16" 24"	4'-0"	2 x 4'-4"	25 11 eng. in W.O. 15 tender
36.	Guisbro'		"	" "	"	" "	"	" " "	" "
37.	Gem		"	" "	"	" "	"	" " "	" "
	Copper firebox 3'-9" high, 3'-5" long, 3'-0" wide. 107 x 14'-0" x 2" tubes, grate area 10 sq.ft. Tender 6 x 3'-0" dia. wheels on 2 x 4'-6" base, 1400 gals. water, 4 tons coal.								
63.	Birkbeck		110	8'-8⅛" 4'-0"	893.3	15" 24"	4'-6"	5'-7" + 6'-0"	c.41 total in W.O.
64.	Larchfield	"	"	" "	"	" "	"	" "	" "
	Copper firebox, 168 x 2" tubes, tenders 6 x 3'-0" dia. wheels on 2 x 4'-6" base, 1500 gals. water, 3 tons coal.								
60.	Cleveland	16.	110	13'-0" 4'-0"	n.k.	15" 24"	4'-6"	5'-7" + 6'-0"	c.43 total in W.O.
61.	Star	17.	"	" "	n.k.	" "	"	" "	" "
62.	Southend	18.	"	" "	n.k.	" "	"	" "	" "
3.	Times	21. 1.	Inness:	Similar to Nos.	60, 61, 62.	15" 20"	4'-0"	5'-7" + 6'-0"	n.k.
		2.	Craven:	8'-8⅛" 4'-0"	827.10				Eng. 27 Tend. 20
	166 x 8'-11½" x 2" copper tubes, grate area 12 sq.ft.								
2.	Graham	22.	all as No.3 Times.						" "
	Tenders: 1. as Nos.63 & 64. 2. 6 wheeled, 1400 gal. water, 3 tons coke.								
56.	Shotley	30.	100	13'-0" 4'-0"	n.k.	17" 18"	4'-0"	6'-7½" + 6'-0¼"	44 total in W.O.
57.	Tow Law	31.	"	" "	n.k.	" "	"	" "	" "
	No firebox or tube details. Tenders 6 x 3'-6" dia. wheels, 16 tons.								
72.	Peel	33.	100	13'-6" 4'-0"	n.k.	1:17" 18" 2: " 24"	4'-2½"	6'-0" + 5'-3"	44 total in W.O.
73.	Aberdeen	34.	"	" "	n.k.	" "	"	" "	" "
74.	Emperor	39.	"	" "	n.k.	" "	"	6'-7½" + 6'-0"	" "
75.	Baring	40.	"	" "	n.k.	" "	"	as 72 & 73	" "

Data as early condition in general, and approximate.
n.k. = not known. n. = nominal. a = actual.
Renumbering given separately in Appendix II.

No.	Name	Works No.	W.P. lb/in² (p.s.i.)	Boiler length Dia.	Surface Area F²	Cylinders Bore Stroke	Wheels Dia.	Wheelbase	Weight Tons Cwt.
76.	Prince of Wales	46.	100	13'-6" 4'-0"	n.k.	1:17" 18"	4'-2½"	6'-5" + 5'-3"	44 total in W.O.
80.	The Duke		120	13'-0" 4'-0"	n.k.	18" 18"	3'-11¼"	2 x 5'-0"	47 5 total in W.O.
135.	Eden		"	" "	n.k.	" "	"	" "	" "
							Eden only as rebuilt		48-10-1
	Copper firebox 3'-4½" long, 4'-8" wide max., 5'-8" high, 4" mid-feather; 158 x 2" Low Moor iron tubes. Tenders 6 x 3'-6" dia. wheels on 2 x 5'-0" base, weight 16 tons 4 cwt.								
81.	Miller	532.	80	10'-0" 3'-10"	n.k.	18" 24"	4'-3*	6'-7" + 7'-2"	42 total in W.O.
82.	Hawthorn	533.	1: 80 2:100	" "	n.k.	1:18" 24" 2:17" "	"	" "	" "
	135 x 2" Low Moor iron tubes, six wheeled tenders, weight c.16 tons.								
86.	Zetland		1: 85 2:120	n.k. n.k. 12'-10" 4'-0"	n.k.	16" 18"	4'-8"	6'-9" + 5'-6"	n.k.
	2nd. boiler; copper firebox 3'-8" long, 3'-7½" wide, 5'-8½" high; 170 x 1⅞" tubes. Usual six wheeled tender, 1200 gals. water, 3 tons coal.								
77.	Alexander	47.	100	13'-6" 4'-0"	n.k.	1:17" 18" 2: " 24"	4'-2½"	6'-5" + 5'-3"	44 total in W.O.
83.	Victoria	51.	"	" "	n.k.	" "	"	" "	" "
84.	Albert*	52.	"	" "	n.k.	1:17" 18" 2: " 20"	"	" "	" "
85.	Hardinge	54.	"	" "	n.k.	1:17" 18" 2: " 24"	"	" "	" "
	No firebox details, mid-feathers in Nos.74 & 75, 158 x 2" tubes; tenders as Nos.56 & 57.								
*84	Albert, as rebuilt at North Road Works, February 1891:		140	13'-6" 4'-3"	1278.5	17" 20"	no change	no change	50 2
	Firebox casing 4'-8¾" long, 4'-0" wide, 165 x 2" tubes, tender 16 tons.								
90.	Merchant	55.	Similar in general to No.72.			17" 18"	as 72.	n.k.	as 72.
91.	Australia	59.	"	" " "	"	1:17" 18" 2: " 20"	"	n.k.	"
92.	Yarm	60	"	" " "	"	1:17" 18" 2: " 24"	"	n.k.	"

Data as early condition in general, and approximate.
n.k. = not known. n. = nominal. a = actual.
Renumbering given separately in Appendix II.

No.	Name	Works No.	W.P. lb/in² (p.s.i.)	Boiler length Dia.	Surface Area F²	Cylinders Bore Stroke	Wheels Dia.	Wheelbase	Weight Tons Cwt.
95.	Elephant	61.	Similar in general to No.72.			1:17" 18" 2: " "	as 72.	n.k.	as 72.
96.	Middleton	62.	"	" " "	"	1: " 18" 2: " 20"	" "	n.k. n.k.	" "
97.	Gainford	63.	"	" " "	"	" "	"	n.k.	"
102.	Cotherstone	64.	"	" " "	"	1:17" 18" 2: " 22"	"	n.k.	"
103.	Darlington	68.	"	" " "	"	17" 18"	"	n.k.	"
104.	Durham	69.	"	" " "	"	" "	"	n.k.	"
105.	Teesdale	70.	"	" " "	"	" "	"	n.k.	"
106.	Staindrop	71.	"	" " "	"	" "	"	n.k.	"
107.	Barnard Castle	72.	"	" " "	"	" "	"	n.k.	"
	No other details.								
108.	Eagle	959.	120	13'-6½" 4'-0"	N.K.	1:16" 22" 2:17" 24"	4'-3"	6'-6" + 4'-11"	46 14 total in W.O.
109.	Falcon	960.	"	" "	n.k.	" "	"	" "	" "
110.	Hawk	961.	"	" "	n.k.	" "	"	" "	" "
111.	Stag	962.	"	"	n.k.	" "	"	" "	" "
112.	Lion	963.	"	" "	n.k.	" "	"	" "	" "
113.	Camel	964.	"	" "	n.k.	" "	"	" "	" "
	153 x 1⅞" tubes; Tenders six wheeled, 1300 gals. water, 3 tons coal, weight 15 tons 12 cwt. as rebuilt; engine weight 33 tons 4 cwt. against original 31 tons 2 cwt.								
119.	Greta	82	Similar in general to No.72.			1:17" 20" 2: " 22"	4'-2½"	6'-7½" + 5'-4"	c.48 total in W.O.
120.	Brough	83.	"	" " "	"	1: " 20" 2: " 24"	"	" "	" "
121.	Tiger	84.	"	" " "	"	" 20"	"	" "	" "

Data as early condition in general, and approximate.
n.k. = not known. n. = nominal. a = actual.
Renumbering given separately in Appendix II.

No.	Name	Works No.	W.P. lb/in² (p.s.i.)	Boiler length Dia.	Surface Area F²	Cylinders Bore Stroke	Wheels Dia.	Wheelbase	Weight Tons Cwt.
122.	Kestrel	85.	Similar in general to No.72.			1:17" 20" 2: " 22"	4'-2½"	6'-7½" + 5'-4"	c.48 total in W.O.
123.	Belle Vue	88.	"	" " "	"	" "	"	" "	" "
124.	Brookside	89.	"	" " "	"	1: " 22" 2: " 24"	"	" "	" "
125.	Gazelle	90.	"	" " "	"	" 22"	"	" "	" "
126.	John Wakefield	91.	"	" " "	"	" "	"	" "	" "
127.	Polam	92.	"	" " "	"	" "	"	" "	" "
128.	East Mount	93.	"	" " "	"	" "	"	" "	" "
	Standard six wheeled 16 ton tenders as No.72 etc. See text regarding change of wheels to 4'-11½" dia. on No.125 "Gazelle".								
129.	Stanley	1026.	Similar to above.			1:16½" 22" 2:17" 24"	4'-2½"	6'-6" + 4'-11"	n.k.
130.	Hutton	1027.	"	" "		" "	"	" "	n.k.
131.	Long Hull	1028	"	" "		" "	"	" "	n.k.
	Weight of engine after first rebuild 48 tons 0 cwt. 3 qr. in working order. Standard six wheeled 16 ton tender as No.72 etc.								
132.	Appleby	27	Similar to above.			1:17" 20" 2: " 24"	4'-3"	6'-7" + 5'-4"	n.k.
133.	Kirkby Stephen	28	"	" "		" "	"	" "	n.k.
136.	Tebay	1185.	120	14'-2" 4'-1"	1307	17" 22"	4'-2½"	7'-5½" + 5'-6"	46 3 in W.O.
137.	Deepdale	1186.	"	" "	"	" "	"	" "	" "
138.	Beelah	1187.	"	" "	"	" "	"	" "	" "
	No details of original boilers, but see Fig.139. 185 tubes x 14'-6" x 1¾" Assume standard 16 ton tenders.								
134.	Kendal	95.	120	Similar in general to No.72.		1:17" 22" 2: " 24"	4'-2½"	6'-7½" + 5'-4"	46 5 total in W.O.

Data as early condition in general, and approximate.
n.k. = not known. n. = nominal. a = actual.
Renumbering given separately in Appendix II.

No.	Name	Works No.	W.P. lb/in² (p.s.i.)	Boiler length Dia.	Surface Area F²	Cylinders Bore Stroke	Wheels Dia.	Wheelbase	Weight Tons Cwt.
139.	Crossfell	96.	120	Similar in general to No.72.		1:17" 22" 2: " 24"	4'-2½"	6'-7½" + 5'-4"	46 5 total in W.O.
140.	Penrith	97.	"	" " "	" "	" "	"	" "	" "
141.	Excelsior	98.	130	14'-10" 4'-1½"	1728	1:18" 24"	4'-11"	6'-9½" +	n.k.
			As rebuilt after explosion:			2:17" "		5'-4½"	51-18-2
142.	Wycliffe	99.	120	As 134 above.		17" 22"	4'-2½"	6'-7½" + 5'-4"	46 5 total in W.O.
143.	Windermere	100.	"	" " "		1: " " 2: " 24"	"	" "	" "
144.	Ulleswater	104.	"	" " "		" "	"	" "	" "
	Tenders six wheeled, 1400 gals. water, 3½ tons coal, weight n.k.								
145.	Panther	1074.	120	13'-7" 4'-1"	1483	17" 26"	5'-0"	6'-2" + 5'-3"	49 10 total in W.O.
146.	Ostrich	1075.	"	" "	"	" "	"	" "	" "
147.	Leopard	1076.	"	" "	"	" "	"	" "	" "
148.	Zebra	1077.	"	" "	"	" "	"	" "	" "
149.	Fox	1078.	"	" "	"	" "	"	" "	" "
150.	Mastiff	1079.	"	" "	"	" "	"	" "	" "
	Firebox 5'-3¼" high, 3'-8" long at top, 3'-10¼" long at bottom, 3'-5¾" wide, casing 4'-4½" long, 4'-0" wide. Copper side, back & tubeplates, iron crown plate, mid-feather 4" wide. 191 Low Moor iron tubes 13'-10" long b.t.p. x 2" o.d., 14 w.g.								
	Tenders 6 x 4'-0" dia. wheels, 1500 gals. water, 3½ tons coal, c.17 tons weight.								
151.	Mercury	110.	Similar in general to No.145.			17" 24"	5'-0"	12'-0¾" div. n.k.	48 16 total in W.O.
152.	Venus	111.	"	" " "	"	" "	"	"	" "
153.	Mars	112.	"	" " "	"	" "	"	"	" "
154.	Jupiter	113.	"	" " "	"	" "	"	"	" "
155.	Saturn	114.	"	" " "	"	" "	"	"	" "
156.	Herschel	115.	"	" "	" "	" "	"	"	" "

Data as early condition in general, and approximate.
n.k. = not known. n. = nominal. a = actual.
Renumbering given separately in Appendix II.

No.	Name	Works No.	W.P. lb/in² (p.s.i.)	Boiler length Dia.	Surface Area F²	Cylinders Bore Stroke	Wheels Dia.	Wheelbase	Weight Tons Cwt.
157.	Planet	119.	Similar in general to No.145.			17" 24"	1:5'-0" 2:4'-2"	12'-0¾" Div. n.k.	48 16 total in W.O.
158.	Lune	120.	"	" "	" "	" "	5'-0"	"	" "
159.	York	125.	"	" "	" "	" "	"	"	" "
	Other details as No.145.								
167.	Newlands	139.	Similar in general to No.145.			" "	"	"	51 8 total in W.O.
168.	Clifton	140.	"	" "	" "	" "	"	"	" "
169.	Tufton	147.	"	" "	" "	" "	"	"	" "
170.	Reliance	158.	"	" "	" "	" "	"	"	" "
	Larger tender, 18 tons 4 cwt., used with these engines.								
187.	Iron Age	186.	Similar in general to No.145.			" "	"	"	" "
188.	Lily	187.	"	" "	" "	" "	"	"	" "
171.	Gladstone		120	13'-6" 4'-0"	n.k.	17" 24"	4'-11"	6'-0" + 5'-0"	49 10 total in W.O.
172.	Barrow		"	" "	n.k.	" "	"	" "	" "
173.	London		"	" "	n.k.	" "	"	" "	" "
174.	John Dixon		"	" "	n.k.	" "	"	" "	" "
	Firebox casing 4'-8" long, copper firebox with mid-feather, 158 x 2" tubes. Tender weight 17 tons, no other details.								
175.	Contractor		130	14'-0" 4'-0"	n.k.	17" 24"	4'-11"	6'-5" + 5'-3*"	48 14 total in W.O.
	Copper firebox with mid-feather, 158 x 2" tubes, no other details. Tender six wheeled, 1500 gals. water, 3% tons coal, weight 15 tons 16 cwt.								
176.	Windsor	1571.	130	14'-0" 4'-0" min.	1308	17" 26"	5'-0½"	7'-9" + 5'-3"	52 6 total in W.O.
177.	Osborne	1572.	"	" "	"	" "	"	" "	" "
178.	Balmoral	1573.	"	" "	"	" "	"	" "	" "
179.	Edinburgh	1574.	"	" "	"	" "	"	" "	" "

Data as early condition in general, and approximate.
n.k. = not known. n. = nominal. a = actual.
Renumbering given separately in Appendix II.

No.	Name	Works No.	W.P. lb/in² (p.s.i.)	Boiler length Dia.	Surface Area F²	Cylinders Bore Stroke	Wheels Dia.	Wheelbase	Weight Tons Cwt.
180.	Dublin	1575.	130	14'-0" 4'-0" min.	1308	17" 26"	5'-0½"	7'-9" + 5'-3"	52 6 total in W.O.
181.	Malton	1576.	"	" "	"	" "	"	" "	" "
182.	Elton	1577.	"	" "	"	" "	"	" "	" "
183.	Acklam	1578.	"	" "	"	" "	"	" "	" "
184.	Lark	1579.	"	" "	"	" "	"	" "	" "
185.	Swallow	1580.	"	" "	"	" "	"	" "	" "
	Firebox casing 5'-3" long, mid-feather fitted, tubes 158 x 2". Tenders 6 x 4'-0" dia. wheels on 2 x 5'-6" base, 1800 gals. water, 3½ tons coal, weight 17 tons 14 cwt.								
186.	Union	Generally similar to Nos.176-185.							56 total in W.O.
191.	Autumn	"	"	" "					" "
	Tender weight 20 tons 14 cwt.								
189.	Spring	Generally similar to Nos.171-174.							
190.	Summer	"	"	" "					
19.	Tweed		120	13'-0" 3'-10"	n.k.	15¾" 22"	4'-6"	4'-11¾" + 4'-10¾"	n.k.
	Tender 6 x 3'-1" dia. wheels on 4'-9⅝" + 4'-1⅝" base, 1200 gals. water, 3½ tons coal.								
20.	Normanby		120	11'-6" 3'-10"	n.k.	15¾" 21"	4'-2"	5'-2" + 4'-10"	n.k.
21.	Cleveland		"	" "	n.k.	" "	"	" "	n.k.
	Tenders 6 x 3'-0" dia, wheels on 4'-10" + 4'-1½" base, 1200 gals. water, 3½ tons coal.								
2.	unnamed		130	13'-8" 4'-0¾"	1246.8	17" 24"	4'-2½"	6'-5" + 5'-3"	53 6 total in W.O.
3.	"		"	" "	"	" "	"	" "	" "
192.	Winter		"	" "	"	" "	"	" "	" "
193.	Princess		"	" "	"	" "	"	" "	" "
194.	Alice		"	" "	"	" "	"	" "	" "

Data as early condition in general, and approximate.
n.k. = not known. n. = nominal. a = actual.
Renumbering given separately in Appendix II.

No.	Name	Works No.	W.P. lb/in² (p.s.i.)	Boiler length Dia.	Surface Area F²	Cylinders Bore Stroke	Wheels Dia.	Wheelbase	Weight Tons Cwt.
195.	Helena		130	13'-8" 4'-0¾" o.d.	1246.8	17" 24"	4'-2½"	6'-5" + 5'-3"	53 6 total in W.O.
	Standard six wheeled tenders, no details. No other boiler details.								
56.	unnamed	Generally similar to Nos.192-195.							
57.	"	"	"	" "					
9.	"	219.	130	14'-0" 4'-0" i.d.	n.k.	17" 24"	5'-0"	6'-9" + 5'-4"	52 5 total in W.O.
23.	"	220.	"	" "	n.k.	" "	"	" "	" "
28.	"	221.	"	" "	n.k.	" "	"	" "	" "
40.	"	222.	"	" "	n.k.	" "	"	" "	" "
42.	"	223.	"	" "	n.k.	" "	"	" "	" "
43.	"	224.	"	" "	n.k.	" "	"	" "	" "
46,	"	225.	"	" "	n.k.	" "	"	" "	" "
49.	"	226.	"	" "	n.k.	" "	"	" "	" "
51.	"	227.	"	" "	n.k.	" "	"	" "	" "
81.	"	228.	"	" "	n.k.	" "	"	" "	" "
200.	Eskdale	229.	"	" "	n.k.	" "	"	" "	" "
201.	Carlton	230.	"	" "	n.k.	" "	"	" "	" "
204.	unnamed	248.	"	" "	n.k.	" "	"	" "	" "
205.	"	249.	"	" "	n.k.	" "	"	" "	" "
206.	"	250.	"	" "	n.k.	" "	"	" "	" "
	Firebox casing 4'-4½" long, 158 x 2" o.d. iron tubes.								
	Tenders 6 x 3'-6" dia. wheels on 2 x 5'-0" base, 1500 gals. water, 3½ tons coal.								

The following class were similar to but larger than No.176 "Windsor" - No.185 "Swallow". Where dimensions are the same they have been omitted in the following list. None of them were named.

No.	Works No.	Delivery date	Boiler length	Surface Area F^2	Cylinders Bore	Cylinders Stroke	Weight Tons	Weight Cwt.
207.	1410.	December 1867	14'-6"	1432	17"	26"	64	1
208.	1411.	January 1868	"	"	"	"	"	"
209.	1412.	" "	"	"	"	"	"	"
210.	1413.	" "	"	"	"	"	"	"
211.	1414.	February 1868	"	"	"	"	"	"
212.	1415.	" "	"	"	"	"	"	"
213.	1416.	March 1868	"	"	"	"	"	"
214.	1417.	" "	"	"	"	"	"	"
215.	1418.	April 1868	"	"	"	"	"	"
216.	1419.	" "	"	"	"	"	"	"
217.	1420.	May 1868	"	"	"	"	"	"
218.	1421.	" "	"	"	"	"	"	"
221.	1486.	August 1870	"	"	"	28"	"	"
222.	1487.	" "	"	"	"	"	"	"
223.	1488.	September 1870	"	"	"	"	"	"
224.	1489.	October 1870	"	"	"	"	"	"
225.	1490.	November 1870	"	"	"	"	"	"

Flat topped copper firebox, 5'-9½" high, 4'-8⅛" long, 3'-6⅛" wide, casing 5'-3" long, 4'-0" wide at bottom, 5'-0" wide at top, mid-feather 2½" wide inside, 2'-4" deep. 168 x 2" o.d. iron tubes, 14'-9½" long b.t.p. Tenders 6 x 3'-7" dia. wheels on 2 x 5'-6" base, 1800 gals. water, 3 tons coal, weight 26 tons 10 cwt.

Long wheel base rebuilds of Nos.207, 208, 210 & 223 as follows:

Wheels as before, 5'-0½" dia., on 8'-0" + 8'-6" base. Boilers fitted with longer round- topped firebox giving larger surface area. Chimney feed heaters removed; no other details.

No.	Name	Works No.	W.P. lb/in^2 (p.s.i.)	Boiler length	Boiler Dia.	Surface Area F^2	Cylinders Bore	Cylinders Stroke	Wheels Dia.	Wheelbase	Weight Tons	Weight Cwt.
53.	unnamed	Generally as No.206 etc. above.					17"	24"	5'-0"	6'-9" + 5'-4"	52	5 total in W.O.
93.	Uranus	"	"	"	"	"	"	"	"	" "	"	"

Data as early condition in general, and approximate.
n.k. = not known. n. = nominal. a = actual.
Renumbering given separately in Appendix II.

No.	Name	Works No.	W.P. lb/in² (p.s.i.)	Boiler length	Dia.	Surface Area F²	Cylinders Bore	Stroke	Wheels Dia.	Wheelbase	Weight Tons	Cwt.
94.	Neptune	Generally as No.206 etc. above					17"	26"	5'-0"	6'-9" + 5'-4"	52	5 total in W.O.
202.	Ireland	"	"	"	"		"	"	"	" "	"	"
203.	England	"	"	"	"		"	"	"	" "	"	"

No.	Works No.	Delivery date	W.P. lb./in² (p.s.i.)	Boiler length	Dia.	Surface Area F²	Cylinders Bore	Stroke	Wheels Dia.	Wheelbase	Weight Tons Cwt. in W.O.
12.		April 1868	130	14'-0"	4'-0" i.d.	n.k.	17"	24"	5'-0½"	6'-7" + 5'-3"	53 8 total
44.		" "	"	"	"	n.k.	"	"	"	" "	" "
1.		May 1868	"	"	"	n.k.	"	"	"	" "	" "
26.		June 1868	"	"	"	n.k.	1:17" 2: "	24" 26"	"	" "	" "
8.		April 1869	"	"	"	n.k.	17"	24"	"	" "	" "
25.		" "	"	"	"	n.k.	"	"	"	" "	" "
6.		May 1869	"	"	"	n.k.	"	"	"	" "	" "
31.		" "	"	"	"	n.k.	"	"	"	" "	" "
7.		June 1869	"	"	"	n.k.	"	"	"	" "	" "
15.		" "	"	"	"	n.k.	"	"	"	" "	" "
219.		September 1869	"	"	"	n.k.	"	"	"	" "	" "
220.		" "	"	"	"	n.k.	"	"	"	" "	" "
5.		December 1869	"	"	"	n.k.	"	"	"	" "	" "
11.		" "	"	"	"	n.k.	"	"	"	" "	" "
18.		" "	"	"	"	n.k.	"	26"	"	" "	" "
22.		" "	"	"	"	n.k.	"	24"	"	" "	" "
10.		April 1870	"	"	"	n.k.	"	"	"	" "	" "

Data as early condition in general, and approximate.
n.k. = not known. n. = nominal. a = actual.
Renumbering given separately in Appendix II.

No.	Works No.	Delivery date	W.P. lb./in² (p.s.i.)	Boiler length	Dia.	Surface Area F²	Cylinders Bore	Stroke	Wheels Dia.	Wheelbase	Weight Tons Cwt. in W.O.
13.		April 1870	130	14'-0"	4'-0" i.d.	n.k.	17"	24"	5'-0½"	6'-7" + 5-3"	53 8 total
24.		May 1870	"	"	"	n.k.	"	"	"	" "	" "
4.		June 1870	"	"	"	n.k.	"	24"	"	" "	" "
16.		" "	"	"	"	n.k.	"	"	"	" "	" "
17.		" "	"	"	"	n.k.	"	"	"	" "	" "
36.		" "	"	"	"	n.k.	"	"	"	" "	" "
61.		" "	"	"	"	n.k.	"	26"	"	" "	" "
78.		September 1870	"	"	"	n.k.	"	"	"	" "	" "
54.		October 1870	"	"	"	n.k.	"	"	"	" "	" "
60.		" "	"	"	"	n.k.	"	"	"	" "	" "
47.		November 1870	"	"	"	n.k.	"	"	"	" "	" "
34.		December 1870	"	"	"	n.k.	"	"	"	" "	" "
		Copper firebox with mid-feather, casing 4'-8¾" long, 4'-0" wide, 158 x 2" o.d. iron tubes. Tenders 6 x 3'-9½" dia. wheels, 1600 gals. water, 3 tons coal, weight n.k.									

No.	Works No.	Delivery date		Wheelbase	Weight
226.	271.	September 1870	Generally similar to above series except:	6'-9" + 5'-4"	as above
227.	272.	October 1870	" " " " " "	" "	" "
228.	273.	January 1871	" " " " " "	" "	" "
229.	274.	March 1871	" " " " " "	" "	" "
230.	275.	April 1871	" " " " " "	" "	" "

No.	Works No.	Delivery date	W.P. lb./in² (p.s.i.)	Boiler length	Dia.	Surface Area F²	Cylinders Bore	Stroke	Wheels Dia.	Wheelbase	Weight Tons Cwt. in W.O.
231.		August 1870	140	14'-0"	4'-0¾" o.d.	1268	17"	26"	5'-0½"	6'-7" + 5'-3"	52 12 total
232.		September 1870	"	"	"	"	"	"	"	" "	" "
233.		October 1870	"	"	"	"	"	"	"	" "	" "
29.		December 1870	"	"	"	"	"	"	"	" "	" "

Data as early condition in general, and approximate.
n.k. = not known. n. = nominal. a = actual.
Renumbering given separately in Appendix II.

No.	Works No.	Delivery date	W.P. lb./in² (p.s.i.)	Boiler length	Dia.	Surface Area F²	Cylinders Bore	Stroke	Wheels Dia.	Wheelbase		Weight Tons	Cwt. in W.O.
30.		February 1871	140	14'-0"	4'-0¾" o.d.	1268	17"	26"	5'-0½"	6'-7" +	5'-3"	52	12 total
14.		March 1871	"	"	"	"	"	"	"	"	"	"	"
48.		April 1871	"	"	"	"	"	"	"	"	"	"	"
88.		May 1871	"	"	"	"	"	"	"	"	"	"	"
39.		June 1871	"	"	"	"	"	"	"	"	"	"	"
234.		February 1872	"	"	"	"	"	"	"	"	"	"	"
235.		" "	"	"	"	"	"	"	"	"	"	"	"
236.		April 1872	"	"	"	"	"	"	"	"	"	"	"
237.		" "	"	"	"	"	"	"	"	"	"	"	"
242.		December 1872	"	"	"	"	"	"	"	"	"	"	"
246.	293.	" "	"	"	"	"	"	"	"	"	"	"	"
243.		January 1873	"	"	"	"	"	"	"	"	"	"	"
247.	294.	" "	"	"	"	"	"	"	"	"	"	"	"
244.		February 1873	"	"	"	"	"	"	"	"	"	"	"
245.		" "	"	"	"	"	"	"	"	"	"	"	"
248.	295.	March 1873	"	"	"	"	"	"	"	"	"	"	"
249.	296.	April 1873	"	"	"	"	"	"	"	"	"	"	"
250.		June 1873	"	"	"	"	"	"	"	"	"	"	"
251.		July 1873	"	"	"	"	"	"	"	"	"	"	"
252.		" "	"	"	"	"	"	"	"	"	"	"	"
257.	327.	September 1873	"	"	"	"	"	"	"	"	"	"	"
258.	328.	October 1873	"	"	"	"	"	"	"	"	"	"	"
253.		" "	"	"	"	"	"	"	"	"	"	"	"
254.		November 1873	"	"	"	"	"	"	"	"	"	"	"

Data as early condition in general, and approximate.
n.k. = not known. n. = nominal. a = actual.
Renumbering given separately in Appendix II.

No.	Works No.	Delivery date	W.P. lb./in^2 (p.s.i.)	Boiler length	Boiler Dia.	Surface Area F^2	Cylinders Bore	Cylinders Stroke	Wheels Dia.	Wheelbase	Wheelbase	Weight Tons	Weight Cwt. in W.O.
255.		December 1873	140	14'-0"	4'-0¾" o.d.	1268	17"	26"	5'-0½"	6'-7" +	5'-3"	52	12 total
256.		" "	"	"	"	"	"	"	"	"	"	"	"
259.	329.	January 1874	"	"	"	"	"	"	"	"	"	"	"
260.	330.	February 1874	"	"	"	"	"	"	"	"	"	"	"
261.		December 1874	"	"	"	"	"	"	"	"	"	"	"
262.	349.	January 1875	"	"	"	"	"	"	"	"	"	"	"
263.	350.	March 1875	"	"	"	"	"	"	"	"	"	"	"
264.	351.	May 1875.	"	"	"	"	"	"	"	"	"	"	"

Copper firebox 5'-7" high, 4'-0⅝" long, 3'-6" wide, casing 4'-8⅝" long, 4'-0" wide, mid-feather fitted, 158 x 2" o.d. tubes, 14'-3½" long b.t.p. Tenders 6 x 3'-9" dia. wheels on 2 x 5'-0" base, 1500 gals. water, 3½ tons coal. weight 18 tons 2 cwt.

Fig.195 Photograph of Stephenson 4-4-0 passenger locomotive, loaned to S. & D.R. for trial, early 1860, shown near Shildon

Fig.196 Photograph of No.160 "Brougham", at Alpha Place, Saltburn.

Fig.197 Photograph of No.1160, ex 160 "Brougham", with rebuilt cab, at York.

Chapter 8. The 4-4-0's: 1860-74, and The "Gamecocks": 1873-76

The 4-4-0's

In 1859 William Bouch was interested in larger passenger engines. The Stainmore line would soon be complete, the Saltburn line was under construction, and the Eden Valley line would be coming along about a year later. Apart from the need to reach the Peases' iron stone mines at Upleatham more easily, it was hoped to make the little fishing village of Saltburn into a popular sea-side resort, which would generate more passenger traffic. The Stainmore and Saltburn lines were eventually opened in mid 1861, and the Eden Valley line in mid 1862.

Robert Stephenson & Co. built in the same year (1859) some four-coupled engines for the Ottoman Line (Smyrna to Aidin), the first tender engines to which they had fitted four wheeled front bogies, although they had designed one in 1844, and built similar tank engines in 1854. Knowing of Bouch's interest, and eager to make a sale, they offered one of these early 4-4-0's to him on trial before delivering it to the purchaser, and there is a good photograph of it standing at Tate's Farm, Shildon (Fig.195), probably in early 1860. Warren gives another photograph of the type. It was a peculiar looking engine with a huge steam dome on top of the firebox, and a scallop edged canopy instead of a cab roof. However, presumably it was a successful trial, and orders were promptly placed for two 4-4-0 locomotives; good salesmanship on Stephensons' part. Whether they were entirely Stephenson design, or whether Bouch, as usual, had a hand in the basic specification, I do not know.

Ken Hoole says that Bouch obtained approval for six such engines, four for the Stainmore line and two for the Eden Valley line. He says also that of the original order for fifteen engines for the Ottoman line, only ten were confirmed, and of these only six were actually exported, but that the others were sold to the London, Chatham and Dover Railway.

The first two for the S. & D.R. were fitted with a large American type cab, which C. Hamilton Ellis described as "Strange and exotic". I had thought that these two might have been part of a frustrated export order, which would explain the speed with which they were delivered, but this seems not to have been so. As has been more logically suggested, they were designed for severe Stainmore winters, like the 'Panther' class 0-6-0's. Anyhow, it would seem that the enginemen did not like being so enclosed, and they were rebuilt to the usual pattern in due course.

No.160 "Brougham" (Figs.196 & 197) was delivered in August 1860, and No.161 "Lowther" (Figs.198-202) in October, which seems very smart work by Stephensons. The price was £2500 each. Ken Hoole says that the first one was ready before it was required, and was loaned in the meantime to the Londonderry Railway, but quotes a works No. of 1206; this does not agree with the Inness list, but is one of the batch for the Smyrna & Aidin Ry. My list of Stephenson works Nos. also gives two sets for the two S. & D.R. engines, the first ones being 1212/3, within the Smyrna & Aidin batch. They were handsome engines, with the dome on the second ring of the boiler, the usual Salter spring balance safety valves, a wrapper plate round the smokebox and cylinders in a nice double ogee curve, and the bogie centred under the smokebox and chimney; 6'-0" diameter driving wheels with the drive on to the first pair, and neat double splashers over the front pair of driving wheels and crankpins. The bogie wheels were 3'-6" in diameter with inside bearings, and they were mounted on large inverted leaf springs, a change in both ways from the makers' previous practice. The bogie pivot was in the form of a large ball joint with no sideways motion, shown clearly in the Colburn drawing.

The cylinders were of 16" bore by 24" stroke, and the boilers were about 10'-6" long x 4'-0" diameter with a heating surface of 1048 sq.ft., at a pressure of 120 p.s.i. Stephenson link motion was fitted, of course. The six wheeled tender shown in a well known photograph of "Brougham" outside Alpha Place at Saltburn (Fig.196) has a distinctly M.R. or later N.E.R. look about it, with outside springs and the typical outside framing pierced with semi-circular openings; this tender carried 1500 gals. of water and 3½ tons of coal. The total weight of engine and tender in working order was more than 48 tons 14 cwt. (see also Fig.197). It is believed by many people that No.160 "Brougham" initiated the Saltburn service in 1861, but it is unlikely that it would have been used in regular service on this line.

They were very modern looking locomotives, especially when rebuilt with the more usual S. & D.R. cabs and a shorter MacDonnell chimney (Fig.197). These first two lasted until 1888.

Figs.199 & 200 show No.161 "Lowther" with S. & D.R. breakdown crane No.7, the leading pair of

Fig.198 Photograph of No.161 "Lowther"

Fig.199 Photograph of No.161 "Lowther" partly derailed, with No.7 breakdown crane

Fig.200 Photograph of No.161 "Lowther" after rerailing

wheels obviously having left the track. It is possible that this happened more often than reported, due to the design of the bogie.

The second batch, of four engines, was delivered by Stephensons in 1862; No.162 "Saltburn", No.163 "Morecambe" (Fig.206), No.164 "Belfast" (Fig.207) and No.165 "Keswick" (Fig.208). They were stated to be bigger than the first two, but had a lesser heating surface area of 1001 sq.ft.(2 less tubes). The driving wheels were bigger, having a diameter of 7'-0$^{1}/_{2}$"; the bogie wheels were the same as before and so were the cylinders. The boiler length was 10'-2", x 4'-0" diameter, fitted with 174 tubes, at the same pressure of 120 p.s.i. The tenders were similar to but heavier than the first batch, and the total weight was up to 52 tons 3 cwt. Instead of the ample cabs they had only a meagre weatherboard.

These engines were fitted with what Inness called Bouch's Steam Retarders, described as a form of steam brake operating band brakes fitted with wooden blocks over the top of the driving wheels. (n.b. Inness quite clearly calls this "Bouch's Steam Retarder", but I understand this name to apply only to the later device for admitting steam to the exhaust side of the pistons, mentioned in the previous chapter.) The addition of these brakes and other modifications brought the price up to £2593 each.

Zerah Colburn illustrated the 'Saltburn' class in two excellent and very well detailed plates, and Warren used the side elevation from these (Figs.203/4; Fig.205 shows the sections) and also provided a good works general arrangement of this locomotive (Fig.203). "Keswick" was never reboilered, and was scrapped in 1886. The other three were rebuilt as listed; "Saltburn" lasted until 1880 and the coupled wheels were then used for a new 2-4-0 locomotive (No.1162, Fig.209) built at the North Road Works. "Morecambe" did better, was renumbered 1736 and lasted until 1888, while "Belfast" was scrapped in 1882.

According to MacLean, No.1162 had cylinders of 17" bore x 26" stroke and ran until 1914. "Morecambe" was first renumbered 1163, and was used as spare engine at York until disposed of, while "Keswick" became No.1165 and was used in a similar role at Scarborough.

The next series of 4-4-0's were those known as 'Ginx's Babies', in late 1871, but although actually built under the N.E.R. regime they carried S. & D. numbers to start with, another measure of the Darlington Section's independence. The nickname came from a political novel by Edward Jenkins, the M.P. for Dundee, written as a comment on Poor Law Reform, a leading issue of the time, and was a result

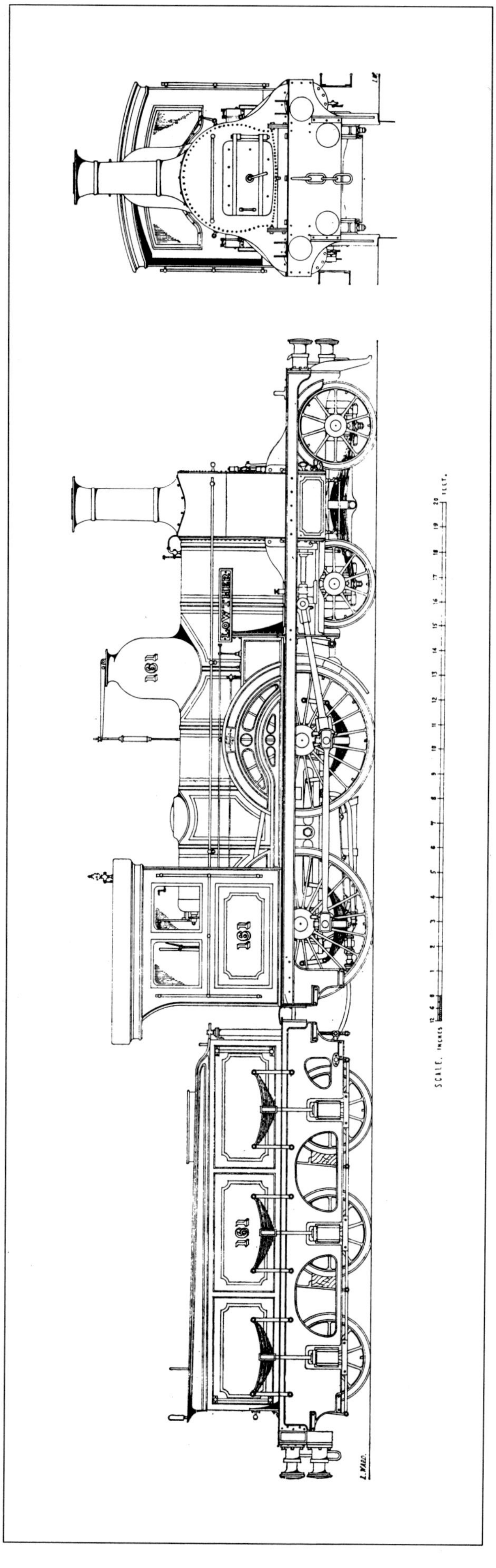

Fig.201 Drawing of No.161 "Lowther" by Laurie Ward

of the trouble these engines had with their valves, which were of piston type, not a new development (see Chapter 3) but unusual. Mr. Ginx couldn't get rid of his babies, and Bouch couldn't get rid of his valve problems, probably because he was stubbornly reluctant to admit a basic design fault. They were all built at the North Road Works, and looked very like the earlier engines as rebuilt, although the chimney had a distinctively N.E.R. appearance.

The use of names had now been dropped. No.238 was ready in December 1871, followed by No.239 in August 1872, No.240 in October and No.241 in November. Then there was a gap of eighteen months, and N.E.R. practice came into the picture, with No.1265 in May 1874; No.1266 in July; No.1267 in August; No.1268 in September; No.1269 in November and No.1270 in December 1874. There were also some differences of external appearance between the two batches, mostly in the relative sizes of the chimneys and steam dome covers, and the type and number of openings in the splashers (Figs.211-222).

The bogie was pivoted a little behind the chimney and smokebox centre-line, and the pivot was of ball joint type, but the other way up to that used by Stephensons on the earlier classes, the ball being accommodated in a socket in the frame of the bogie (Fig.210). There have been a number of statements that these bogies were not mechanically secured to the engine except by weight, and Ahrons is said to have commented that "it seems a somewhat perilous/ferocious bogie to let loose on an unoffending railway!". However, it is my belief that the drawing is misleading in this respect, and that there is a securing plate round the neck of the ball column above the ball itself. The bogie wheels had a diameter of 3'-7" and the driving wheels were 7'-1" diameter, on a wheelbase of 7'-3$^{1}/_{2}$" which allowed very little clearance between the flanges when new. The continuous splashers over the crankpins were extended right up to the front of the smokebox, which partly concealed the cylinders and broke the smooth curve of the wrapper plate. The cylinders were bigger, of 17" bore x 30" stroke, and the boilers were 11'-0" long x 4'-0" diameter, fitted with 167 x 1$^{3}/_{4}$" tubes, giving a heating surface of about 1000 sq.ft. for the first four; Inness says that this figure was increased to 1217 sq.ft. for the second batch without saying why. The boilers appear to be the same, at least externally.

The first four of these engines were fitted with transverse springs over the trailing pair of driving wheels. This consisted of an inverted leaf spring 8$^{1}/_{2}$" wide and comprising nine plates plus packing plates, with a distance between centres of 3'-9". In addition volute springs were fitted at each axlebox.

The trouble with the valves seems to have arisen from the fact that the valves themselves were of solid brass, 13" in diameter, the inference being that no rings were fitted, however unlikely this may seem. Since the coefficient of expansion of brass is much higher than that of cast iron, and since also the cylinders were outside and even if well lagged would have radiated and so lost some heat, the valves expanded excessively and partial seizure took place, with heavy wear. All types of lubricant were tried, including graphite, but without improvement. The graphite may have been in colloidal form but was more usually flaked, and from my experience this, combined with the cylinder oil, would probably have resulted in the formation of hard balls in the valve chests and cylinders, instead of providing continuous lubrication. When rebuilt later by Fletcher inside cylinders were fitted, and, no doubt, conventional slide valves, since these engines continued in service for many years.

Another small difference was the use of bushed ends to the connecting rods instead of the more usual split bearings with loose brasses, gib (wedge) and cotter. Bouch's steam brakes were fitted, and also screw brakes on all tender wheels. The tenders were similar to those used with the earlier classes, but the springs were underhung. The tender used with the first of the class, No.238, was differently shaped, as the drawing and photograph show (Figs.211 & 212), but this was soon altered to the same form as the others. This may have applied also to a few of the others; the tender on No.240 in Fig.213 is certainly of the same form, and appears to bear the right number for the engine. The water capacity was increased to 1600 gals., with 3$^{1}/_{2}$ tons of coal. The overall weight of engine and tender in working order was 63 tons 16 cwt.

When new they were fitted with reversing gear of the Bouch type, with a cigar shaped screw and a sector lever, as was being fitted to the mineral engines at this time. A similar screw arrangement developed by Stephensons is illustrated by Warren on his page No.371.

After Bouch's death, Fletcher rebuilt all these engines, turning them into 2-4-0's, with a variety of boilers and differences in external appearance at various rebuilds. No.1270 was exhibited in original form at the 1875 celebrations (Figs.211-222).

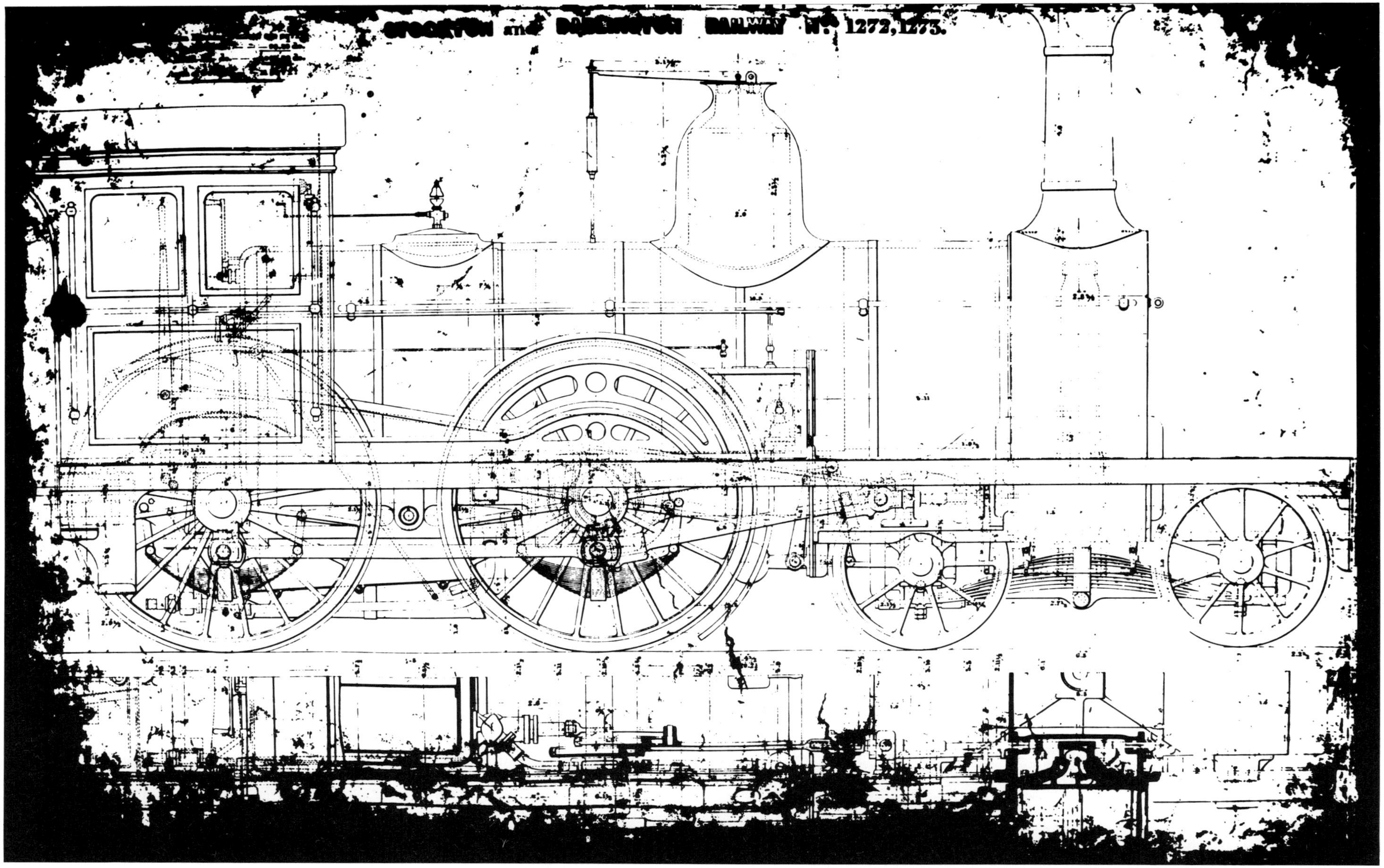

Fig.202 R. Stephenson & Co. works drawing of Nos.160 & 161, works Nos.1272/3

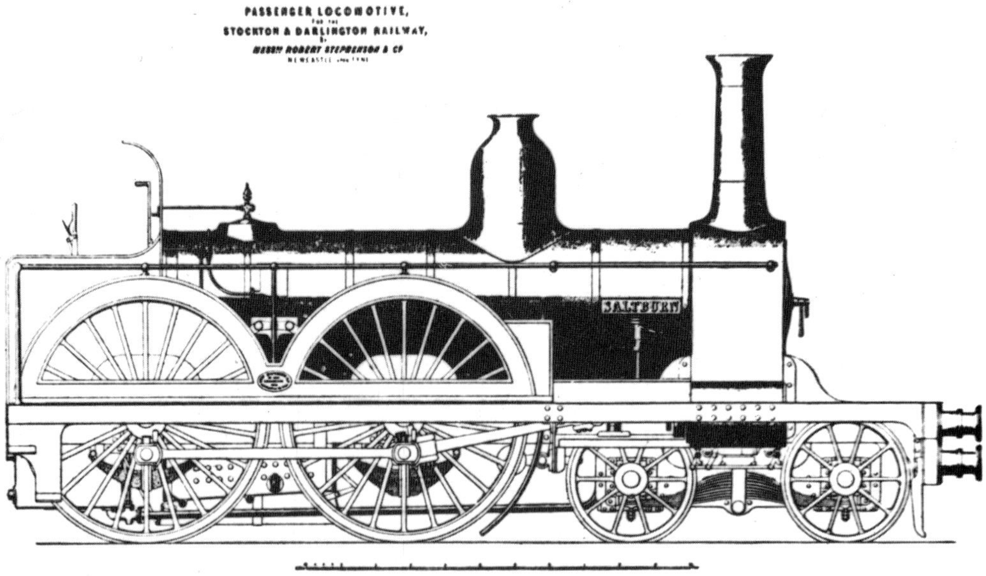

Fig.203 General Arrangement drawing of No.162 "Saltburn"

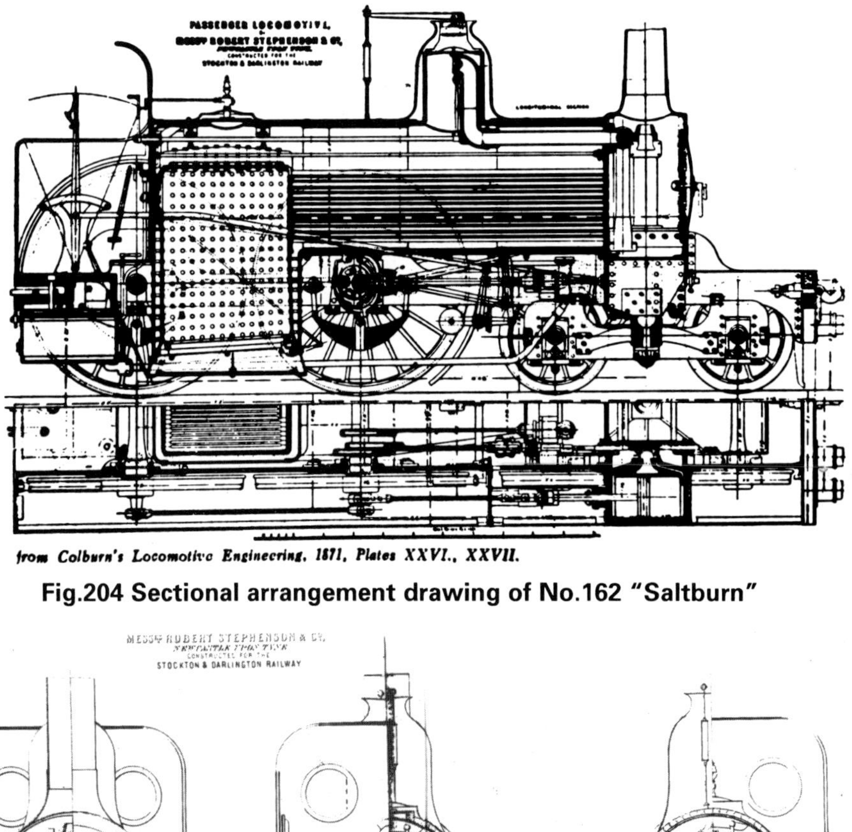

from Colburn's Locomotive Engineering, 1871, Plates XXVI., XXVII.

Fig.204 Sectional arrangement drawing of No.162 "Saltburn"

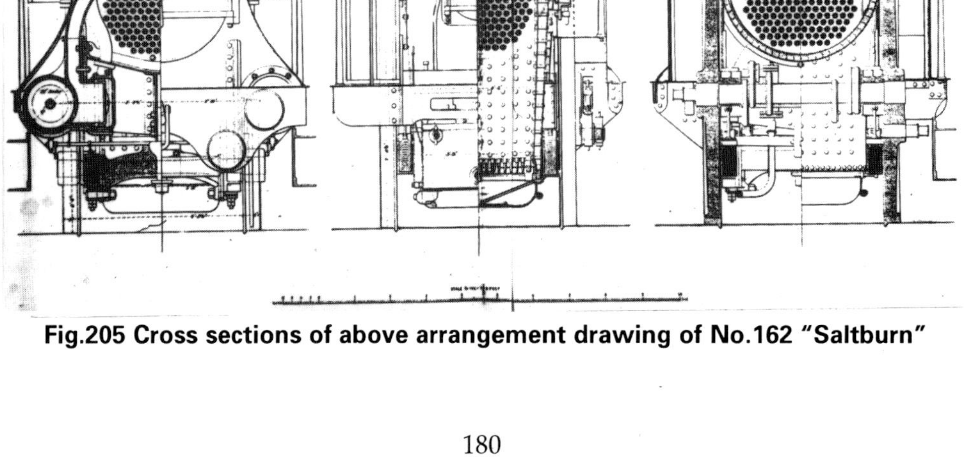

Fig.205 Cross sections of above arrangement drawing of No.162 "Saltburn"

Fig.206 Photograph of No.163 "Morecambe" in original condition.

Fig.207 Photograph of No.164 "Belfast" with rebuilt cab.

Fig.208 Photograph of No.165 "Keswick" at Alpha Place, Saltburn.

Fig.209 Photograph of No.1162, ex 162 "Saltburn", rebuilt as a 2-4-0

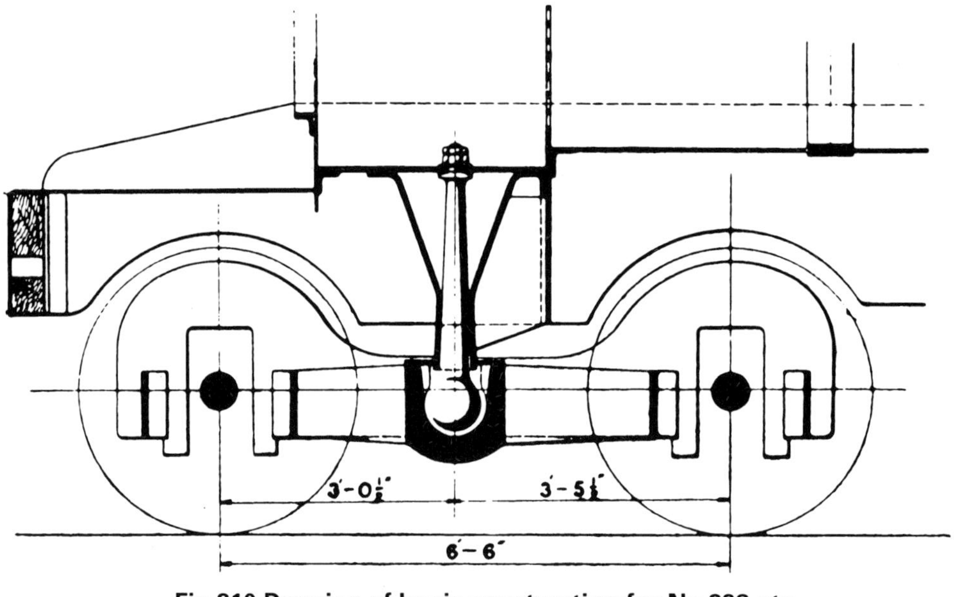

Fig.210 Drawing of bogie construction for No.238 etc.

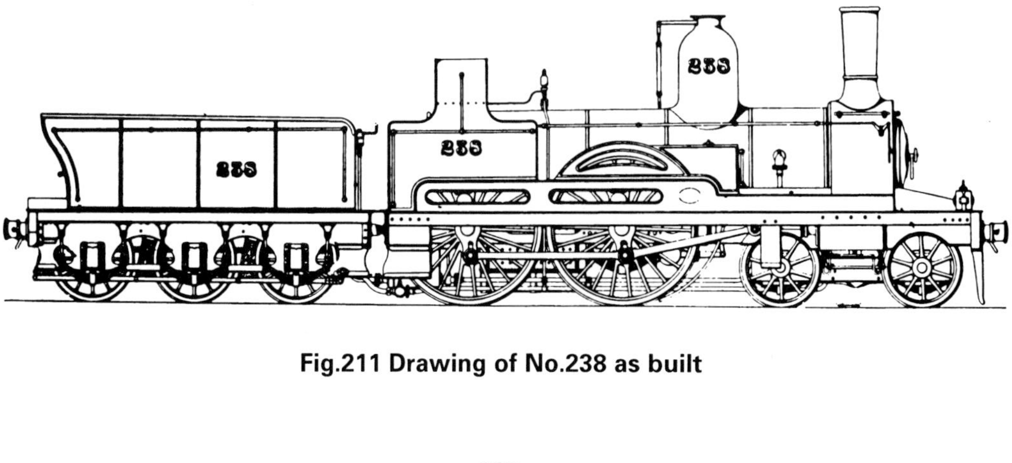

Fig.211 Drawing of No.238 as built

Fig.212 Photograph of No.238 in original condition and with original tender

Fig.213 Photograph of No.240 in original condition, at Saltburn

Fig.214 Photograph of No.1240, ex 240, rebuilt as a 2-4-0

Fig.215 Photograph of No.1240 as rebuilt again later

Fig.216 Photograph of No.1241, ex 241, rebuilt as a 2-4-0, at Shildon

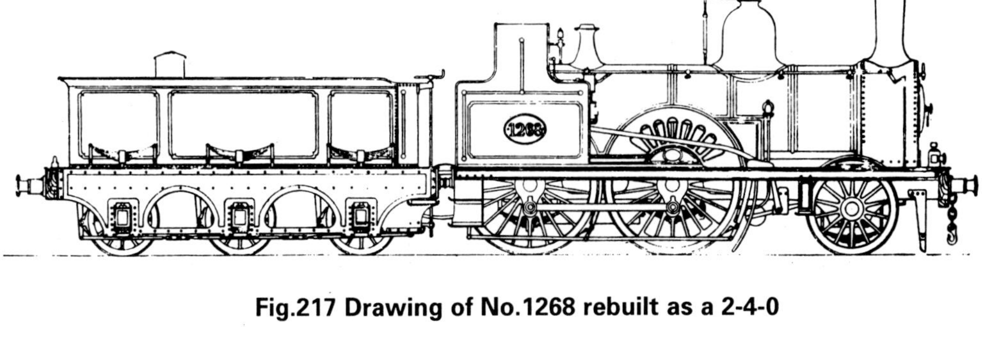

Fig.217 Drawing of No.1268 rebuilt as a 2-4-0

Fig.218 Photograph of No.1268 rebuilt as a 2-4-0 as shown in Fig.217

Fig.219 Photograph of No.1269 in early condition

Fig.220 Photograph of No.1269 rebuilt as a 2-4-0

Fig.221 Photograph of No. 1270 in early condition

Fig.222 Photograph of No.1270 rebuilt as a 2-4-0, with the old Saltburn Brine baths behind.

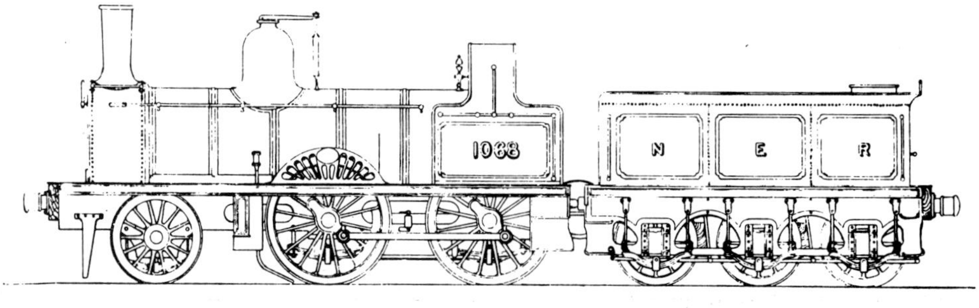

Fig.223 Drawing of No.1068 in original condition.

The "Gamecocks"

In June 1873, the North Road Works delivered a new inside cylindered 2-4-0 passenger engine, No.1071, to replace the earlier No.71 "Hackworth", built in 1851 by Kitchings. Apparently the designer was John Kitching. This engine had 6′-0″ diameter coupled wheels, and 17″ bore x 24″ stroke cylinders. No details of the boiler are given, but since this engine is regarded as the prototype of the 1068 class which followed, and which became known as the "Gamecocks", I assume that the boiler was much the same as was fitted to that class (see also comments in Chapter 6). The photograph, Fig.99, referred to there is certainly of the 1873 engine, whether a rebuild of the earlier engine or a complete replacement. No.1071 was a successful engine, was reboilered in 1886 and renumbered twice before its final years at the Hull Paragon shed before being scrapped at York in 1901. The overall weight of engine and tender in working order was 58 tons 15 cwt.

No.1068 (Figs.223-225), the first of the following class, was delivered from the North Road Works in August 1875, in time to take part in the S. & D.R. Jubilee celebrations. It was followed by No.1050 in November; No.1035 in March 1876; No.1066 in April; No.1062 in May and finally No. 1098 in July 1876. They were popular and successful passenger engines, so good at coping with the severe gradients on the system that they were given the soubriquet of "Gamecocks" for their pluckiness. The driving wheels were of 6′-0½″ diameter, with leading wheels of 4′-8″, and at first all of them had cylinders of 17″ bore x 26″ stroke. The boilers were 11′-2″ long x 4′-0¾″ diameter, fitted with 158 x 2″ o.d. tubes, with a total heating surface of 1091.2 sq.ft. and working at 140 p.s.i.

Two of them, Nos.1068 and 1050, were later fitted with shorter stroke cylinders as No.71, i.e 24″, but since none of the others were so altered this change would appear to have produced very little improvement. They were fitted with Bouch's valve motion; no mention is made of feed pumps but two No.8 Friedmann injectors were fitted. Bouch's 'Steam Retarder' was used, with the cigar shaped reversing screw and sector lever. The tenders at first were six wheeled S. & D. type carrying 2400 gals. of water and 6 tons of coal, but were later changed to standard Fletcher 2500 gal. type. The overall weight of engine and tender in working order was 54 tons 8 cwt. None of them lasted beyond 1909 (Figs.226-229).

Neither E. Craven nor Ken Hoole has much to say about these later 2-4-0 engines.

Fig.224 Photograph of No.1068 in original condition, as Fig.223

Fig.225 Photograph of No.1068 as rebuilt later

Fig.226 Photograph of No.1050 in original condition

Fig.227 Photograph of No.1050 as rebuilt later

Fig.228 Photograph of No.1035 in original condition

Fig.229 Photograph of No.1035 as rebuilt later

Data as early condition in general, and approximate.
n.k. = not known. n. = nominal. a = actual. Wheels: B = bogie; D = driving; L = leading.
Renumbering given separately in Appendix II.

No.	Name	Works No.	W.P. lb/in² (p.s.i.)	Boiler length Dia.	Surface Area F²	Cylinders Bore Stroke	Wheels Dia.	Wheelbase	Weight Tons Cwt.
160.	Brougham	1272.	120	10'-8" 4'-0" (11'-0": MacLean)	1048	16" 24"	B:3'-6" D:6'-0"	6'-5" + 6'-4½" + 7'-5"	48-14-2 total in W.O.
161.	Lowther	1273.	"	n.k. "	"	" "	"	" " "	" " "

Centre of bogie to driving axle 9'-7". No other boiler details, but MacLean states 176 iron tubes.
Tenders 6 x 3'-6" dia. wheels on 2 x 5'-0" base, 1500 gals. water, 3½ tons coal, weight 16 tons 14 cwt.

162.	Saltburn	1332.	120	10'-2" 4'-0"	1001 see notes*	16" 24"	B:3'-6" D:7'-0½"	6'-1" + 6'-6" + 7'-5"	52 3 total in W.O.
163.	Morecambe	1333.	"	" "	"	" "	"	" " "	" "
164.	Belfast	1334.	"	" "	"	" "	"	" " "	" "
165.	Keswick	1335.	"	" "	"	" "	"	" " "	" "

Bogie centre to driving axle 9'-0½". Firebox casing 4'-11" long; 174 x 2" i.d. tubes. Tenders as Nos.160/1.
(MacLean states 174 iron tubes 9'-8" long. Both MacLean & Colburn give the heating surface as 1053 Ft²)

Data as early condition in general, and approximate.
n.k. = not known. n. = nominal. a = actual. Wheels: B = bogie; D = driving; L = leading.
Renumbering given separately in Appendix II.

238.	unnamed	69.	140	11'-0" 4'-0"	c.1000*	17" 30"	B:3'-7" D:7'-1"	21'-0½" total 7'-3½" driving	63 16 total in W.O.

239, 240, 241; all as No.238 above. Works Nos.70, 71, 72.
Bogie wheel base 6'-6", pivot 5" forward of centre. 1667 x 1¾" tubes, 11'-3½" long b.t.p. (*MacLean gives heating surface as 1217 Ft²) Tenders 6 x 4'-0" dia. wheels on 2 x 5'-0" base, 1600 gals. water, 3½ tons coal, weight 21 tons 16 cwt.

1265.	unnamed	89.	140	11'-0" 4'-0"	1217	17" 30"	B:3'-7" D:7'-1"	as No.238.	as No.238.

1266, 1267, 1268, 1269, 1270: all as No.1265 above. Works Nos.90, 91, 92, 93, 94.
Number of tubes increased to 210, otherwsie all as Nos.238-241.

Fletcher rebuild into 2-4-0 form, in general;

			160	10'-3" 4'-3"	1097	17" 26"	L:4'-8" D:7'-1"	varied	c.39 7

Boilers were as fitted to Fletcher N.E.R. 901 class, with 206 x 1¾" o.d. tubes, 10'-7⅛" long b.t.p., copper firebox 5'-9½" high, 4'-8⅜" long, 3'-3" wide at top, 4'-9⅝" at bottom.

No.	Name	Works No.	W.P. lb/in² (p.s.i.)	Boiler length Dia.	Surface Area F²	Cylinders Bore Stroke	Wheels Dia.	Wheelbase	Weight Tons Cwt.
1071.	unnamed Tender weight 21 tons 8 cwt.	84.	Assumed as No.1068 class below.			17" 24"	L: n.k. D:6'-0"	n.k.	58 15
1068.	unnamed	105.	140	11'-2" 4'-0¾"	1091.2	1:17" 26" 2: " 24"	L:4'-8" D:6'-0½"	8'-6" + 8'-0"	54 8 total in W.O.
1050.	"	115.	"	" "	"	1:17" 26" 2: " 24"	"	" "	" "
1035.	"	123.	"	" "	"	17" 26"	"	" "	" "
1066.	"	124.	"	" "	"	" "	"	" "	" "
1062.	"	125.	"	" "	"	" "	"	" "	" "
1098.	"	126.	"	" "	"	" "	"	" "	" "

Copper firebox 5'-8" high, 4'-5" long, 3'-6" wide, casing 4'-10" long, 4'-0" wide, 158 x 2" o.d. tubes, 11'-5½" long b.t.p. Tenders 6 x 4'-0" dia. wheels on 2 x 5'-0" base, 2400 gals. water, 6 tons coal, weight 23 tons (later replaced by standard Fletcher 2500 gals. tenders).
Worsdell boilers fitted later similar to those fitted to Nos.1265-1270.

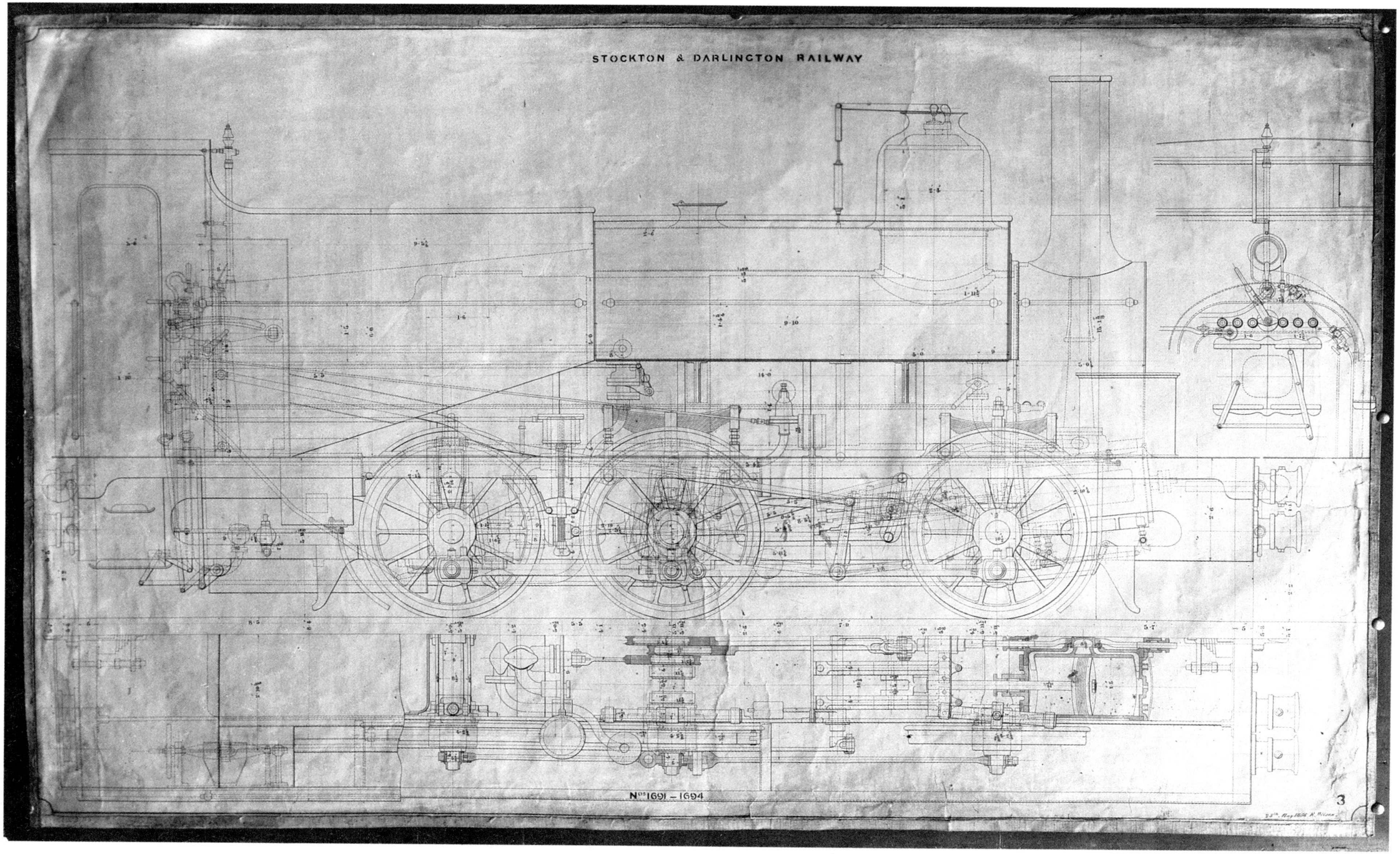

Fig.230 R. Stephenson & Co. Works drawing , 1886, of Nos.196-199, works Nos.1691-1694,0-6-0ST for Skinningrove zig-zag

Chapter 9: The Tank Engines and The Final Mineral Engines

The Tank Engines

The first four of these were designed by Bouch for the purpose of working as bankers on the Skinningrove mine branch, which had very heavy gradients between the ironstone mines (Loftus Mines) which were down on the south side of the Kilton Beck valley, and the main line which was close by but much higher. The gradients were very heavy, with a zig-zag system to enable the iron stone trains to be worked up and the empties down. The straight spur at the end of each zig-zag was about 500 feet long. There seems to have been some double track, but certainly some of the spurs were single, and the engines must have pushed and pulled alternately on each leg, the idea being that the tank engine was pushing on the last leg so that it could then run back down, while the train engine carried on to the main line.

In most major respects they were similar to the tender mineral engines, and their only advantage must have been the shorter length which would have enabled a few more trucks to be worked into and out of each leg of the zig-zag each time. They were built by Stephensons, and delivered in 1866, No.196 "Roseberry" and No.197 "Kildale" in November; No.198 "Whitby" and No.199 "Escomb" in December. They were cumbersome looking locomotives, as may be imagined, of long boiler type, with 4'-2 " diameter wheels on a 13'-0" wheelbase. The tanks did just meet across the top of the boiler, and they were designated 0-6-0ST, but with such a flat top they looked more like pannier tanks. The R.S. & Co. general arrangement (Fig.230) gives only a scrap view of the backhead as an end elevation of the drawing. According to Ken Taylor they were stationed at Carlin How shed when delivered.

The one illustrated by Inness (Fig.231) is the last one, No.199, and was fitted with an additional tank at the smokebox end, which makes it look even more peculiar. They were inside cylindered, of 17" bore x 26" stroke, the same as most of the later 0-6-0 tender engines, and the smaller wheels with a long crank throw will have given them a slightly higher tractive effort. Their boilers were also similar except for the number of tubes; 14'-0" long x 4'-0" inside diameter, with 138 x 2" tubes and a total heating surface of 1164 sq.ft., including firebox and midfeather, working at 130 p.s.i. The coal bunkers were side mounted and high, which restricted the firebox crown and so also the space available for the tubes, giving a small steam space. These bunkers carried $2^1/_2$ tons of coal, the side tanks carried 860 gals. and the front tank on No.199 held a further 540 gals. of water. The total weight given is 46 tons 10 cwt., with the inference that this included the front tank. The axle loading was excessive, as may be guessed, with consequent damage to the track, and eventually they were rebuilt as tender engines (Figs.232-235).

The suspension to the trailing axle was of transverse spring type as described earlier. The valve motion was of Stephenson link type, with the usual eccentric driven feed pumps, assisted by a No.6 Robinson & Gresham injector. Bouch's 'Steam Retarder', obviously of the band brake variety, was fitted to the centre (driven) and trailing wheels, and screw brakes were fitted as well to the centre and leading wheels. These brakes can be seen in the Inness drawing.

The cabs were very small and can hardly have given room to fire the boiler - very short handled shovels? (This is not entirely a joke; such short shovels are known to have been used in later industrial tank engines, although in this case there may have been more room than would appear). The steam dome was on the first ring, and without the tanks they looked a little like the Stephenson built 'Windsor' class (Figs.233 & 155) but the dome was not so high.

As tender engines they were quite satisfactory, and were reboilered from time to time as necessary; they were in service at various places, and ended up in 1908/9 for the first two, 1905 for the next, while No.199 lasted until 1911. They retained their original boilers until 1882 (No.196) and 1890 (No.197), the other two in between these dates. Again according to Ken Taylor they finished up, No.196 at Rosedale, No.197 at South Blyth, No.198 at Gateshead and No.199 at West Hartlepool.

No more tank engines were built until 1876, presumably when the earlier ones were taken off the job and rebuilt. The later ones were for the same purpose on the Skinningrove branch, but of much more conventional appearance. Probably the Bouch influence was minimal by then; he may have had some say in the design, but had been in ill health for some months before his death in January of the same year. They were all built at the North Road Works, and set to service all in the same year, No.1292 in September;

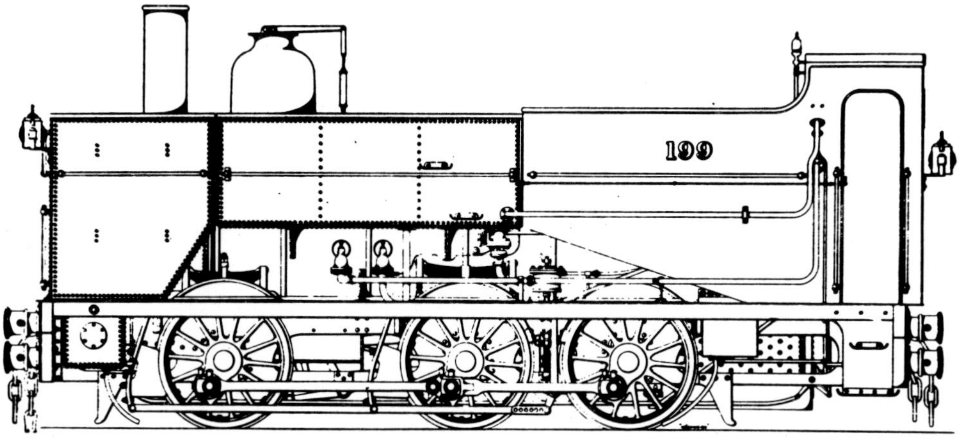

Fig.231 Drawing of 0-6-0ST No.199 "Escomb" as built, with additional forward tank

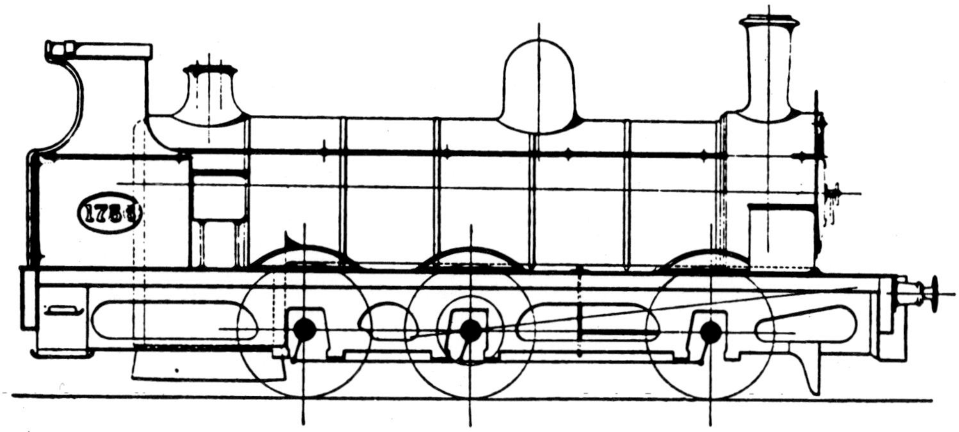

Fig.232 Drawing of No.1756, ex 0-6-0ST No.196 "Roseberry", rebuilt as a tender engine

Fig.233 Photograph of No.1197, ex 0-6-0ST No.197 "Kildale", as first rebuilt with original boiler

Fig.234 Photograph of No.1681, ex 0-6-6ST No.197 "Kildale", as rebuilt again later

Fig.235 Photograph of No.1686, ex 0-6-0ST No.198 "Whitby", rebuilt as a tender engine

No.1037 in October; No.1293 in November and No.1294 in December. As built they had a powerful chunky look, and in my opinion were of better appearance then than when rebuilt by the N.E.R. later with Worsdell boilers, and looking decidely 'Jintyish'; or do I mean J72ish? (Figs.236 & 237).

Originally they were fitted with 18" bore x 26" cylinders. The wheels were 4'-8½" diameter on a 16'-6" wheelbase. The boilers were 11'-2" long x 4'-0¾ diameter, with 158 x 2" tubes having a total heating surface of 1091.2 sq.ft., with a flat topped firebox, working at 140 p.s.i. As well as the usual Salter spring balance safety valves there seems to have been a further valve, probably Adams type, in a Beyer bonnet on the firebox. They were side tank engines with a conventional rear bunker; no capacities are given, but the working weight was 40 tons 4 cwt. They were fitted with a fairly conventional braking system by later standards, with smallish shoes working on the front faces of the tyres.

After rebuilding, No.1293 was fitted with 17" bore cylinders, and sold to the Woodlands Colliery Co. The others worked on, No.1037 until 1910 and the other two until 1911. One of them, No.1294, was involved in a spectacular derailment on the Skinningrove bank zig-zag in about 1905 (Fig.238).

One further tank engine is known of but without details; this was a small engine named "Sprite" which was used by the engineering department in the 1860's. Inness says that the railway fitted it with a new boiler, and that it was possibly purchased from Boultons by one of the contractors working on

Fig.236 Photograph of No.1293 in original condition

Fig.237 Photograph of No.1293 as rebuilt later

Fig.238 Photograph of No.1294 derailed on Skinningrove zig-zag, c.1905

the construction of some of the branch lines in the west, after which it was presumably sold to the railway for their own use. It may not have been built by Boultons; they did tend to specialize in buying, selling and exchanging small engines of this type. If it was of their make it was probably a geared drive engine of 0-4-0 type.

The Final 0-6-0's

Two more classes of 0-6-0 mineral engines remain to be described. The first of these was similar to earlier ones in many ways, being of long boiler type with inclined inside cylinders and with many usual S. & D.R. features, but the contracts to supply them were placed with outside engine builders who had never before supplied engines to this railway. Twenty engines were built, divided equally between Dübs & Co. of Glasgow and the Avonside Engine Co. at Bristol. They were delivered as follows; the first ten from Dübs, beginning with Nos.1271, 1272, 1273, 1274 and 1275 in May 1874, followed by Nos.1276, 1277, 1278, 1279 and 1280 in June. Nos.1281 and 1282 came from Avonside in November, Nos.1283, 1284, 1285 and 1286 in December, Nos.1287 and 1288 in February 1875 and the last two, Nos.1289 and 1290 in March.

They all had 5'-0½" diameter wheels on a wheelbase of 11'-10", 17" bore x 26" stroke cylinders, and boilers similar to the next four from the North Road Works, which I shall come to shortly. There were differences, mainly in the suspension details; the Dübs engines had the transverse spring arrangement to the trailing pair of wheels as described previously, while the Avonside engines had independent underhung springs to each trailing axlebox. The coupling rods had the same arrangement as referred to in chapter 7 for No.63 "Birkbeck and No.64 "Larchfield", in that the centre joint was actually on the crankpin of the centre (driven) pair of wheels, instead of this being a separate joint in the coupling rod adjacent to the crank bearing. Inness says that this arrangement was introduced on the North Eastern Railway by a Mr. John Stephenson, a former Locomotive Superintendent at York. The leading section of coupling rod had a forked bearing end and the trailing section a single end which worked in the fork, both carried on a long brass bush on the crankpin.

These engines were used on coke trains across the Pennines, over the Stainmore Summit to Tebay and Cockermouth for many years. Eventually they were reboilered under Worsdell's supervision and scattered all over the system, as usual. Nos.1271, 1274, 1287 and 1289 were stationed at Whitby, working to Loftus, Grosmont and Battersby. Nos.1282 and 1286 got up on the Rosedale branch and were fitted with the extended cabs on the tenders. No.1275, of course, was restored as far as possible to original condition for the 1925 Centenary Celebrations and took part in the procession. On its retirement it had covered 907,341 miles (Figs.239-244). This engine was in the North Road Station Museum at Darlington from 1975 until early 1989, together with the original No.1 "Locomotion" and the second No.25, Kitching's "Derwent", but No.1275 has now returned to the National Railway Museum at York (see also Fig.245).

A further four of the class were built at the North Road Works, similar to the Avonside ten in suspension details. These were Nos. 1067 and 1079, delivered in January 1875; No.1032 (Fig.246) in February, and No.1027 in March. The boilers had a length of 14'-0" x 4'-0¾" diameter, fitted with 158 x 2" tubes, with a total heating surface area of 1100 sq.ft. and working at 130 p.s.i. The tenders carried 2400 gals. of water and 6 tons of coal; they had six 4'-0" diameter wheels, the same diameter as the driving wheels of so many of the earlier Stockton & Darlington 0-6-0 mineral engines. The weights are not given. As preserved, No.1275 is said to weigh 35 tons, but this is probably a dry weight with empty boiler and tender. No.1032 was shown at the 1875 Jubilee exhibition; No.1079 lasted until 1915.

No.1275 has been refitted with the old type of S. & D.R. buffer beams intended to carry four buffers at each end, but without the lower pair at each end. Inness's drawing of the 226 class in 1870 shows all four buffers; the 1875 photograph of No.1032 shows only two at each end. It would be very interesting to know just when and why this practice was ceased on the mineral engines. I assume that originally the dumb buffers on chaldron wagons were lower than the buffers fitted to passenger coaches and the later trucks, so buffers of different heights were necessary on the locomotives, but why did they continue for so long? (See earlier remarks on these buffers in Chap.6; "Tyneside". Ken Hoole's notes confirm that chaldron wagons continued to be used until well into the 1870's and it is certain that they went on being used long after that. Many photographs of rebuilt and late 1860's 0-6-0 mineral engines show hanging lower dumb buffer beams.)

In the very last class of 0-6-0 mineral engines Bouch abandoned the short wheel base/long boiler principle. The wheels remained at 5'-0½" diameter but the wheelbase was increased to 16'-3". Some of the eight locomotives in this class had differences in the wheelbases, but no details are given. They were all built at the North Road Works, and delivered,

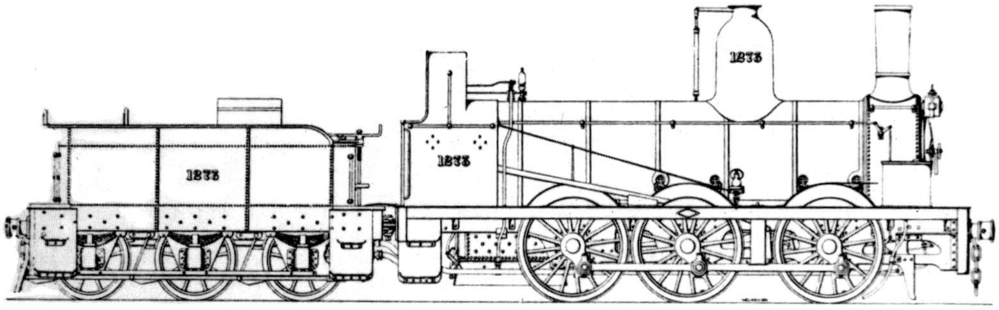

Fig.239 Drawing of No.1275 as built

Fig.240 Photograph of No.1275 as restored for the 1925 Centenary

Fig.241 Photograph of No.1275 as restored for the 1925 Centenary

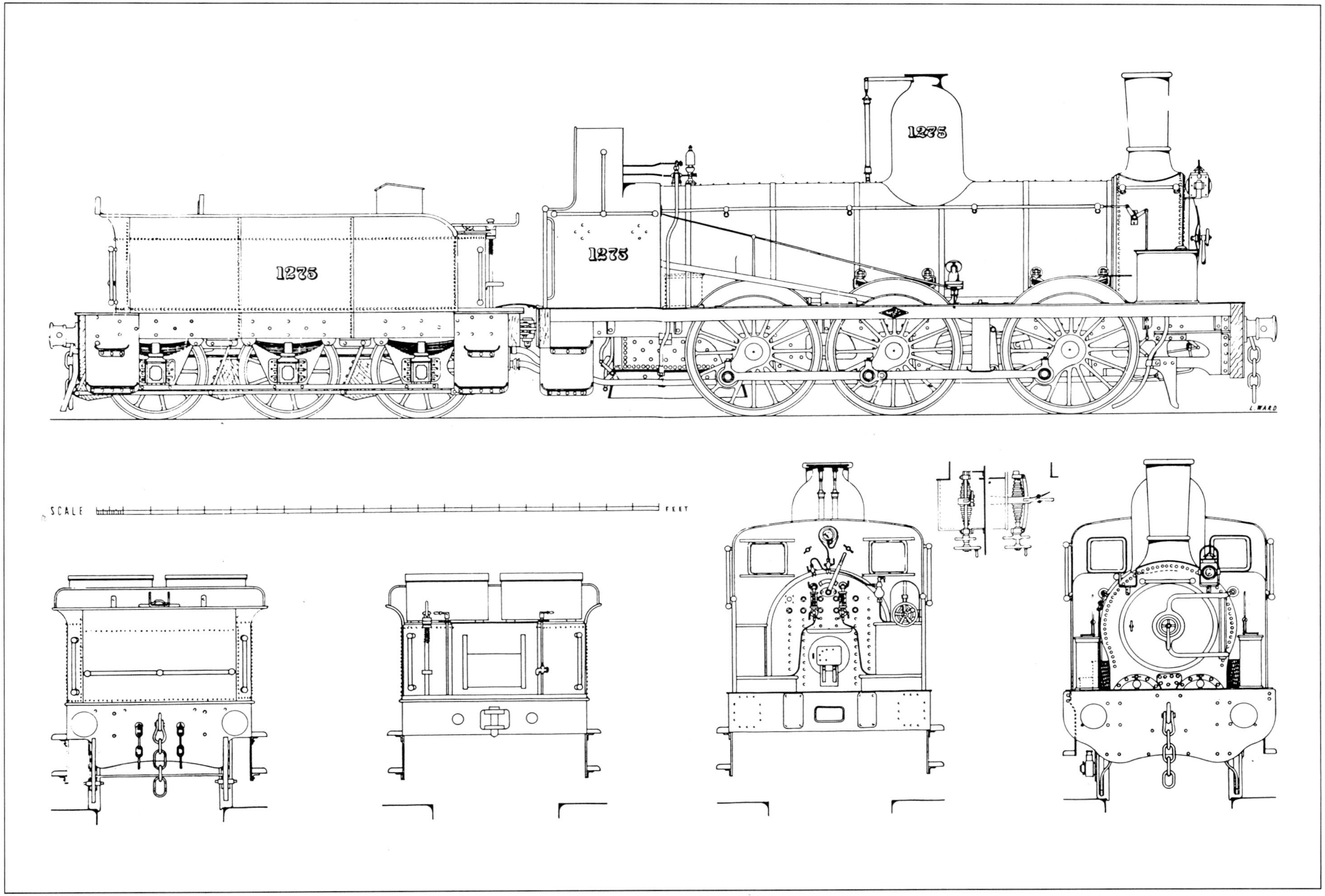

Fig.242 Arrangement drawing of No.1275 believed to show the engine as restored in 1925 and now preserved

No.1045 in February 1875; No.1038 in March; Nos.1052 and 1074 in May; No.1291 (Figs.248-250) in July; No.1063 in December; No.1070 (Fig.247) in January 1876, and No.1069 in February. Nos. 1038 and 1045 had taller steam dome casings than the others. The cylinders were of the usual 17″ bore x 26″ stroke.

The boiler dimensions are given as 11′-2″ long over the barrel x 4′-0¾″ diameter, with 158 x 2″ o.d. tubes, but the total heating surface is given as 1100 sq.ft., as much as for the 14′-0″ boilers. Here you can see that the original reason for the long boiler has gone. The working pressure was 130 p.s.i. Two No.8 Friedmann injectors were fitted instead of feed pumps, but most other fittings were the same as usual, including the 'Steam Retarder' and the cigar shaped reversing screw. The tenders were the same as for the long boiler engines above, and the total weight of engine and tender was 55 tons in working order. One of these engines, No.1038, lasted until 1925, and could have carried L.N.E.R. colours! Inness says that No.1291 only took seven days to erect, and was on show at the 1875 Jubilee celebrations.

This brings to an end my review of the locomotives of the Stockton & Darlington Railway, and those of the Darlington Section of the N.E.R. It was most certainly the close of an era and the end of a railway that was always different, but that led the way for others everywhere.

Fig.243 Photograph of No.1275 as rebuilt, before restoration

Fig.244 Photograph of No.1275 as rebuilt, before restoration

Fig.245 Photograph of No.1276 as rebuilt

Fig.246 Photogrpah of No.1032 in original condition

Fig.247 Photograph of No. 1070 in original condition

Fig.248 Photograph of No.1291 in original condition

Fig.249 Photograph of No.1291 in workshop grey, for photography

Fig.250 Photograph of 1291 as rebuilt later

Data as early condition in general, and approximate.
n.k. = not known. n. = nominal. a = actual.
Renumbering given separately in Appendix II.

No.	Name	Works No.	W.P. lb/in² (p.s.i.)	Boiler length Dia.	Surface Area F²	Cylinders Bore Stroke	Wheels Dia.	Wheelbase	Weight Tons Cwt.
196.	Roseberry	1691.	130	14'-0" 4'-0" i.d.	1164	17" 26"	4'-2¾"	7'-9" + 5'-3"	46 10 total in W.O.
197.	Kildale	1692.	"	" "	"	" "	"	" "	" "
198.	Whitby	1693.	"	" "	"	" "	"	" "	" "
199.	Escomb	1694.	"	" "	"	" "	"	" "	48 18

Copper firebox 5'-6" high, 4'-6⅛" long at top, 4'-7⅛" at bottom, 3'-5" wide, casing 5'-3" long, 4'-0" wide, mid-feather 4" wide x 2'-2" deep, 138 x 2" o.d. iron tubes, 14'-3⁷/16" long b.t.p.
Water capacity 860 gals. in saddle tank, 540 gals. in front tank, coal 2½ tons.

As rebuilt into mineral tender engines, at first they retained the same boilers; when again rebuilt with Worsdell boilers No.197 as N.E.R. No.1681 had the following dimensions:

			140	14'-0" 4'-3" i.d.	1321.75	" "	"	" "	50 11 total in W.O.

Copper firebox 3'-4"wide, 4'-0⅜" long, 5'-7" high, casing 4'-8¾" long, 4'-0" wide, 165 x 2" o.d. tubes, 14'-3⅜" long b.t.p. Tenders weight 15 tons 9 cwt., no other details.

1037.	unnamed	118.	140	11'-2" 4'-0¾" o.d.	1091.2	18" 26"	4'-8½"	8'-6" + 8'-0"	40 10 total in W.O.
1292.	"	117.	"	" "	"	" "	"	" "	" "
1293.	"	119.	"	" "	"	1:18" 26" 2:17" "	" "	" " " "	" " " "
1294.	"	121.	"	" "	"	18" 26"	"	" "	" "

These boilers are as originally fitted to the No.1068 class of 2-4-0 passenger engines. They had a copper firebox 5'-8" high, 4'-5" long, 3'-6" wide, casing 4'-10" long, 4'-0" wide, 158 x 2" o.d. tubes, 11'-5½" long b.t.p. Worsdell boilers were fitted in 1889-1891, again similar to those for the No.1068 class.

Data as early condition in general, and approximate.
n.k. = not known. n. = nominal. a = actual.

No.	Works No.	W.P. lb/in^2 (p.s.i.)	Boiler length	Boiler Dia. o.d.	Surface Area F^2	Cylinders Bore	Cylinders Stroke	Wheels Dia.	Wheelbase	Weight Tons Cwt.
1271.	704.	130	14'-0"	4'-0¾"	1100	17"	26"	5'-0½"	6'-7" + 5'-3"	n.k.
		(All similar to North Road built engines Nos.1027-1079 below)								
1272.	705.								" "	n.k.
1273.	706.								" "	n.k.
1274.	707.								" "	n.k.
1275.	708.								" "	n.k.
1276.	709.								" "	n.k.
1277.	710.								" "	n.k.
1278.	711.								" "	n.k.
1279.	712.								" "	n.k.
1280.	1003.								" "	n.k.
1281.	1003.	All as above.							" "	n.k.
1282.	1004.								" "	n.k.
1283.	1005.								" "	n.k.
1284.	1006.								" "	n.k.
1285.	1007.								" "	n.k.
1286.	1008.								" "	n.k.
1287.	1009.								" "	n.k.
1288.	1010.								" "	n.k.
1289.	1011.								" "	n.k.
1290.	1012.								" "	n.k.
1027.	99.	130	14'-0"	4'-0¾"	1100	17"	26"	5'-0½"	6'-5" + 5'-3"	n.k.
1032.	98.	"	"	"	"	"	"	"	" "	n.k.
1067.	96.	"	"	"	"	"	"	"	" "	n.k.
1079.	95.	"	"	"	"	"	"	"	" "	n.k.

Copper firebox 5'-8" high, 4'-0⅝" long, 3'-6" wide, casing 4'-8⅝" long, 4'-0" wide, 158 x 2" o.d. tubes, 14'-3½" long b.t.p. Tenders 6 x 4'-0" dia. wheels, 2400 gals. water, 6 tons coal, weight n.k.

Data as early condition in general, and approximate.
n.k. = not known. n. = nominal. a = actual.

No.	Works No.	W.P. lb/in² (p.s.i.)	Boiler length	Boiler Dia. o.d.	Surface Area F²	Cylinders Bore	Cylinders Stroke	Wheels Dia.	Wheelbase	Weight Tons Cwt.
1045.	100.	130	11'-2*	4'-0¾"	1100	17"	26"	5'-0½"	8'-3" + 8'-0"	55 total in W.O.
1038.	97.	"	"	"	"	"	"	"	" "	"
1052.	102.	"	"	"	"	"	"	"	" "	"
1074.	103	"	"	"	"	"	"	"	" "	"
1291.	104.	"	"	"	"	"	"	"	" "	"
1063.	113.	"	"	"	"	"	"	"	" "	"
1070.	114.	"	"	"	"	"	"	"	" "	"
1069.	116.	"	"	"	"	"	"	"	" "	"

Copper firebox 5'-8" high, 4'-5" long, 3'-6" wide, casing 4'-10" long, 4'-0" wide, 158 x 2" o.d. tubes, 11'-5⅛" long b.t.p. Tenders 6 x 4'-0" dia, wheels, 2400 gals. water, 6 tons coal, weight 23 tons.

Appendix I. Locomotive Names; Drivers' Names

The names of the Stockton & Darlington Railway locomotives are a very good subject for study in themselves; apparently simple but having all sorts of subtle references. Ken Hoole has written a couple of short articles on them in the N.E.R.A. magazine, "The North Eastern Express". Nos.51 & 52 of May & August 1973, which cover many of the more interesting points, but these were intended for members and people who know the area, and so I am going to try to give some further details which may be of use to those who do not know the background so well.

These names do not fall readily into classes, and in no case does one class of locomotive have a continuous run of names of one type; in several classes they start off in such a way, but change to another type of name in the middle. The spelling is not always the current one, and in some cases, such as "Gannymede", not the usual spelling at the time.

Other interpretations are, of course, always possible.

I have sorted them into rough groupings, but these do sometimes overlap, and I shall try to explain the background of the more difficult ones as I go. The groups are as follow:

1. Names of general meaning and frequent early usage, not falling into clear categories.

2. Astronomical and mythological names, seasons etc.

3. Birds and animals.

4. Royalty, royal palaces and castles, countries.

5. Placenames, lakes and rivers, which to some extent overlap with 6.

6. Names of personages connected with the railway, famous men and politicians, shareholders, landowners over whose estates the railway passed, and the residences of members of the management committee.

1. Includes Locomotion, Hope, Diligence, Black Diamond, Experiment, Rocket, Victory, Times, Dispatch, Leader, Beehive, Bee, Briton, Trader, Merchant, Ocean, Northumbrian, Alert, Active, Magnet, Enterprise, Arrow, Pilot, Swift, Driver, Commerce, Gem, Ruby, Dart, Alarm, Excelsior, Reliance, Iron Age, Lily, Director, Contractor, Miner, Union. Globe I have explained in Chap. 8. Driver may have been a well-known Darlington merchant.

2. North Star, Planet, Sun, Sunbeam, Gannymede, Meteor, Comet, Star, Orion, Uranus, Neptune, Mercury, Venus, Mars, Jupiter, Saturn, Priam, and one astronomer, Herschel. Alexander probably belongs in this group. Spring, Summer, Autumn, Winter.

3. Antelope, Unicorn, Elephant, Stag, Lion, Camel, Tiger, Gazelle, Panther, Zebra, Leopard, Fox, Mastiff, Eagle, Falcon, Hawk, Kestrel, Ostrich, Swallow, Lark. Some may like to include Bee here.

4. Royal George, Prince, Majestic, Coronation, William IVth., Adelaide (there was also a colliery of this name), Queen, Brunswick, Clarendon, Emperor, Prince of Wales, Victoria, Albert, Belfast, Windsor, Osborne, Balmoral, Edinburgh, Dublin, Princess, Alice, Helena. England, Ireland, Australia.

5. Placenames: these call for more explanation:
Bedlington; ironworks near Newcastle-upon-Tyne, owned by Michael Longridge, an associate and colleague of the Stephensons.
Stockton, Hopetown, Shildon, Tyneside, Manchester, London (all names indicating where the engine was built), Middlesbro', Huddersfield, Auckland, Stanhope, Redcar, Guisbro', Tow Law, Darlington, Yarm, Middleton (-in-Teesdale, or could have been Middleton St. George, where the Fighting Cocks inn and work camp were situated), Durham, Teesdale, Staindrop, Barnard Castle, Brough, Stanley (near Crook), York, Appleby, Kirkby Stephen, Kendal, Penrith, Saltburn, Morecambe, Keswick, Barrow, Acklam (Middlesbrough; Acklam Hall was the seat of the Hustler family), Gainford (near Darlington), Whitby, Carlton, Rokeby (near Barnard Castle), Wycliffe (near Barnard Castle), Cotherstone (ditto), all towns, villages, seaside resorts or districts served by the railway (not necessarily directly; see below)*. Some of them may also have had local collieries or ironstone mines.
Cleveland; maybe the district at the time, not the

present county borough, or possibly the Dukes of Cleveland, but then both father and son were opponents of the railway.
Normanby; village near Middlesbrough, ironstone mine and also Lord Normanby.
Eldon; one of the first collieries to be served by the railway (by a separate branch), also a village near Shildon. and Lord Eldon, who was one of the opponents of the railway at first, but gave way later to his financial benefit.
Deanery; a colliery near Shildon.
Kildale; village and dale south of Middlesbrough, also ironstone mine.
Skelton Castle, Witton Castle (also colliery), Raby Castle, all local, and Witton Park was the first starting point of the line. Skelton had ironstone mines.
Brougham (Castle), Lowther (Castle), both near Penrith. These two engines were named at the time when this line was completing.
Shotley; near Consett, probably also colliery.
Escomb, Wolsingham, both near Bishop Auckland. Escomb was also a colliery.
Lonsdale, Carlisle, probably the peers bearing those titles, and landowners over whose estates the line ran, maybe shareholders.
London (2nd.); probably just repeating the first engine's name.
Crossfell, source of both the Tees and the South Tyne, and near to the source of the Wear.
Tebay; junction with the west coast line south of Shap.
Clifton: junction with west coast line north of Shap, and giving access to the Cockermouth line.
Deepdale, Beelah (now Belah); viaducts on the line over Stainmore.
Ayton; village and station on the line south to Battersby Junction, but may have been named for Ayton Hall or House, possibly a director's residence, or for the Friends' School there. This was not an S. & D.R. line.
Marske; on the Redcar to Saltburn line, but may have been named for Marske Hall, as above, although this was a Pennyman residence. Cliffe House, Marske, was a Pease residence. There were Pease mines close by.
Hopetown; as mentioned above, but also part of Darlington, site of the Lister foundry and the Kitchings' early foundry and works, and site of the second North Road Station and other railway buildings, adjacent to North Road Works.
Conside; believed to be an early colloquialism for Consett.
Windermere, Ulleswater (sic); both lakes.
Tees, Tweed, Wear, Greta, Lune, Eden, Derwent; all rivers, although the Tweed seems to be somewhat out of orbit!

6. Tory, Whig, Lord Brougham, Earl Grey, Lord Durham, Wilberforce, Peel, Aberdeen, Baring (Alexander, mentioned under 2. above, could have been the first name of one of the Baring family of bankers), Zetland, Gladstone, The Duke (of Wellington); all politicians or famous men of the time, some of whom were connected with the Stockton & Darlington Railway bills. Lord Zetland may have had a closer connection.
Stephenson, Hackworth, Hawthorn, Oswald Gilkes, Edward Pease, Meynell, Hardinge (the Rev. Henry), John Wakefield, Graham, Stobart, John Dixon, Tufton, Birkbeck; all persons closely connected with the railway in one way or another. The 1868 stock list gives 'Birbeck' as the spelling of 'Birkbeck' but this may have been a misprint.
Miller may have been a person or may come under 1. above.

Residences of members of the Management Committee etc.
Woodlands; this house in Darlington was built by Botcherby & Son, builders, of Darlington, and it is probable that Robert Botcherby Jnr. lived there; he was an associate of Jonathan Backhouse and later became a director of the Great North of England Railway. It later became the property of Joseph Whitwell Pease, but the locomotive concerned had been built and named about three years earlier than this. The house came on to the market recently, and I believe is now the headquarters of a large firm.
Hutton (Hall), the later home of the same Joseph Whitwell Pease; also a station on the Guisborough line.
Nunthorpe (Hall); home of Isaac Wilson and later a Pease residence briefly ('Jack' Pease, later Baron Gainford, who largely rebuilt it). Also a station on the Guisborough line.
Lartington (Hall); Barnard castle, home of the Rev. T. Witham. There was also a station at Lartington.
Elton (Hall); between Stockton & Darlington, home of J. C. Hopkins.
Polam (Hall); Darlington, home of Jonathan Backhouse, and now Polam Hall School.
Hall Garth; Coatham Mundeville, Darlington, home of Henry Pascoe Smith.
Larchfield; a house in Darlington originally called "Paradise", home of Francis Mewburn, the company's solicitor.

Belle Vue, house in Darlington, home of John Dixon, and now owned by Ropners, shipowners and insurers.

Elmfield; a house in Northgate, Darlington, built by William Backhouse, and from 1855 the home of Alfred Kitching (listed as 'Elm Field').

East Mount; a house in Darlington, now gone, the home of John Pease.

Newlands; maybe a house in Darlington, or a hamlet near Braithwaite.

Southend; a house in Darlington, home of Joseph Pease, and now the Grange Hotel.

Brookside; a house in Northgate, Darlington, the home of Thomas MacNay, and also of H. G. Grievson at some time.

Pierremont; house and now a district of Darlington, home of Henry Pease (see notes below).

Malton; maybe The Lodge, Malton, home of W. C. Copperthwaite.

Long Hull; the name of the old Chaloner farmhouse residence, on their estates at Guisborough, rebuilt by Captain Chaloner in 1856, and then extended to form the present Gisborough Hall in 1907. The name is associated with the earlier name of a field on the estates, known as 'Long Howle'. (Sometimes shown erroneously as 'Long Hall'.)

Fryerage; probably The Friarage, Yarm, home of Thomas Meynell.

Etherley (House); home of W. C. Stobart, but also the Etherley collieries were two of the very early customers of the railway.

Further Notes

I am assured that among the Darlington railway cognoscenti the name of "Locomotion" was seldom used; the engine was always referred to as No.1. West calls it "Old No.1."! Likewise "Derwent" was always "The Derwent", in spite of the nameplate on the engine!

"Royal George" was certainly the first engine to be given a name; the others followed soon after, as mentioned in Chapter 1.

The names of "Belfast" and "Cotherstone" are difficult to explain. The latter place was not on the railway then, but was a popular residential area for many of the Darlington business men, especially the Quakers.

Pierremont is an especially interesting name. The house was stone built, by John Botcherby, c.1834, and was originally called "Pierpont". It became the home of Henry Pease in 1845, and he changed the name to Pierremont. There are now several streets bearing the name followed by Crescent, Drive, Gardens and Road in that district of Darlington.

Some of the omissions are surprising. There was an "Acklam", the home of the Hustlers, Lords of the Manor of Acklam and leading landowners in Middlesbrough long before it became a town, in fact the early Middlesbrough, farm and small priory, was in the parish of Acklam. But there is no "Ormesby", home of the Pennymans, the other major landowner of the area. There is still a Pennyman's Crossing on the Guisborough line near Middlesbrough. Neither is there an "Eston", although maybe there was intended to be an engine of this name (see below).

*Whitby, for example, was served directly by the Whitby & Pickering Railway, not by the S. & D.R., but was becoming a popular resort in the mid 1800's, as well as a busy port. After the North Yorkshire & Cleveland Railway joined up Picton and Grosmont in 1862 it was possible to reach Whitby from the old S. & D.R. system.

I am indebted to Ken Hoole for the following additional information; the following engines had originally been intended to receive the name in the second column, which in some cases was later used for another locomotive.

No.	Name.	Proposed Name.	Later used for:
2.	Graham	York	No.159, February 1862.
3.	Times	Utility	-
58.	Woodlands	Victoria	No.83, July 1854.
59.	Hallgarth	Albert	No.84, " "
61.	Star	Peel	No.72, October 1852.
63.	Birkbeck	Eston	-
64.	Larchfield	Perseverance	-
68.	Brunswick	Zetland	No.86, August 1864.

I have recently been given a list of S. & D.R. locomotive names and numbers, taken from a manuscript list in the Kew P.R.O., which does include several of these proposed names as if they had in fact been adopted. It would be very interesting to know the origin of this list, but unfortunately its provenance is not known to me. Its date is given as October 1850.

Driver's Names

The following list was sent to Ken Hoole in November 1964 by Mr. Herbert P. Bell, an alderman of Darlington. It should be noted, however, that these are only some of the drivers who had charge of the engines named at a particular time, and I have added others known to me.

Name of Engine	Name of Driver
Locomotion	James Stephenson
Hope	James Stephenson
Black Diamond	Robert Murray
Diligence	George Jennison
Royal George	William Gowland
Experiment	Michael Law
Rocket	James Stephenson
Victory	John Mounfrey (Mountrey?)
Globe	John Morgan
Planet	Joseph Gladdon
North Star	George Newcombe
Majestic	William Baxter
Coronation	Willam Gladstone
William the Fourth	John Morgan
Northumbrian	George Newcombe
Director	Henry Lanchester
Lord Brougham	Ralph White
Shildon	John Holme
Darlington	Michael Law
Adelaide	Richard Parkinson
Earl Grey	Thomas Lanchester
Lord Durham	George Jennison
Wilberforce	Edward Corner
Magnet	John Mounfrey
Enterprise	William Gaffins
Arrow	Edward Corner
Swift	Edward Corner
Sunbeam	Edward Corner
Queen	Edward Corner
Raby Castle	- Appleby
Additions:	
No.1	George Graham
"	John Cree (killed by the explosion)
No.5 (1st.)	James Stephenson (at the explosion)
Experiment	Jas. Pickering
Globe	Johny Morgan
	George Muken
	Ralph Wade
Derwent	Jacob Dobson (age 90 in 1924)
Raby Castle	Matthew Appleton

Appendix II. Renumbering out of Class

The following engines, all of them but two 0-6-0 mineral locomotives, were given numbers which did not fall into the N.E.R. 1xxx plus the original S. & D. No. class, at a date or dates from 1890 on. The exceptions were No.163 "Morecambe", a passenger 4-4-0 and later a 2-4-0 rebuild, and also No.71, a late 2-4-0. This list may not be complete.

S&D No.	Name	NER No.	1st. Change Date	1st. Change No.	2nd. Change Date	2nd. Change No.	3rd. Change Date	3rd. Change No.
73.	Aberdeen	1073.	Jan.1890	1845.	Jan.1894	1727.	Jul.1899	2266.
75.	Baring	1075.	" "	1846.				
84.	Albert	1084.	Jan.1892	1708.	Jan.1894	1666.		
85.	Hardinge	1085.	" "	1950.	Jan.1897	1694.		
90.	Merchant	1090.	Jan.1890	1835.	Jan.1894	1749.	Mar.1899	2261.
95.	Elephant	1095.	Jul.1892	1709.	" "	1685.		
103.	Darlington	1103.	Jan.1892	1951.	" "	1724.	Jul.1899	2263.
104.	Durham.	1104.	Jan.1890	1836.	" "	1742.	Mar.1899	2259.
105.	Teesdale	1105.	May 1889	1793.				
122.	Kestrel	1122.	Jul.1891.	1916.				
126.	John Wakefield	1126.	May 1899	1789.				
140.	Penrith	1140.	Jan.1892	1952.	Jan.1894	1695.		
142.	Wycliffe	1142.	" "	1953.	" "	1696.		
143.	Windermere	1143.	Jul.1892	1711.	" "	1725.	Jul.1899	2264.
144.	Ulleswater	1144.	Jan.1892	1715.	" "	1680.		
145.	Panther	1145.	May 1899	1788.				
151.	Mercury	1151.	Jul.1892	1720.	Jan.1894	1667.		
152.	Venus	1152.	Jan.1892	1954.				
155.	Saturn	1155.	Jul.1892	1728.	Jan.1894	1747.	Mar.1899.	2260.
157.	Planet	1157.	" "	1955.	" "	1697.		
159.	York	1159.	Jul.1904	1674.				
163.	Morecambe	1163.	n.k.	1736.				
172.	Barrow	1172.	Jul.1904	1676.				
187.	Iron Age	1187.	Jul.1891	1917.	Jan.1894	1743.		
188.	Lily	1188.	" "	1918.				
196.	Roseberry	1196.	Nov.1892	1743.	Jan.1894.	1756.		
197.	Kildale	1197.	" "	1744.	" "	1681.		
198.	Whitby	1198.	" "	1745.	" "	1686.		
199.	Escomb	1199.	" "	1747.	" "	1738.		
71.	unnamed	1071.	Jul.1891	1915.				

NER No.	= S&D No.	NER No.	= S&D No.	NER No.	= S&D No.	NER No.	= S&D No.	NER No.	= S&D No.
1666.	84.	1708.	84.	1835.	90.	1915.	71.	2259.	104.
1667.	151.	1709.	95.	1836.	104.	1916.	122.	2260.	155.
1674.	159.	1711.	143.	1845.	73.	1917.	187.	2261.	90.
1676.	172.	1715.	144.	1846.	75.	1918.	188.	2263.	103.
1680.	144.	1720.	151.			1950.	85.	2264.	143.
1681.	197.	1724.	103.			1951.	103.	2266.	73.
1685.	95.	1725.	143.			1952.	140.		
1686.	198.	1727.	73.			1953.	142.		
1694.	85.	1728.	155.			1954.	152.		
1695.	140.	1736.	163.			1955.	157.		
1696.	142.	1738.	199.						
1697.	157.	1742.	104.						
		1743.	187. & 196.						
		1744.	197.						
		1745.	198.						
		1747.	155. & 199.						
		1749.	90.						
		1756.	196.						
		1788.	145.						
		1789.	126.						
		1793.	105.						

No.28 "Ocean" took the place of No.28 "Tees", but in 1843 "Ocean" was renumbered 13 to fill a blank, and in May 1845 the number 28 was given to "Conside".

Author's note: see text regarding use of the number 12 for "Ocean" and others.

Appendix III. Theodore West Sketch Sheets and Clement E. Stretton's Drawings

Theodore West Plates III & IV.

Containing sketches of the following Stockton & Darlington Railway Locomotives:

III.

Locomotion
Experiment
Royal George
Rocket
Globe
Planet
Wilberforce
Coronation
Swift
Comet (incorrect)
Arrow
London
Middlesbro'
Raby Castle
Pilot
Dart
Manchester
Stockton
Trader
Tyneside
Meteor
Sun
Gannymede
Active
Miner
Commerce
Rokeby
Priam
Woodlands
Cleveland

IV.

Hackworth
Peel
Lonsdale
The Duke
Ayton
Edward Pease
Panther
Brougham
Saltburn
Contractor
Roseberry class 0-6-0ST
No.1221.
No.1238.
No.1269.
No.1035.
No.1045.
No.1292.

C. E. Stretton Drawings:

No.1	"Locomotion"
No.5	"Stockton" (Chittapratt)
No.6	"Experiment"
No.5	"Royal George"
No.9	"Globe"
No. 13	"Coronation
No. 26	"Swift"
No. 27	"Arrow"
No. 43	"Sunbeam"
No.165	"Keswick"

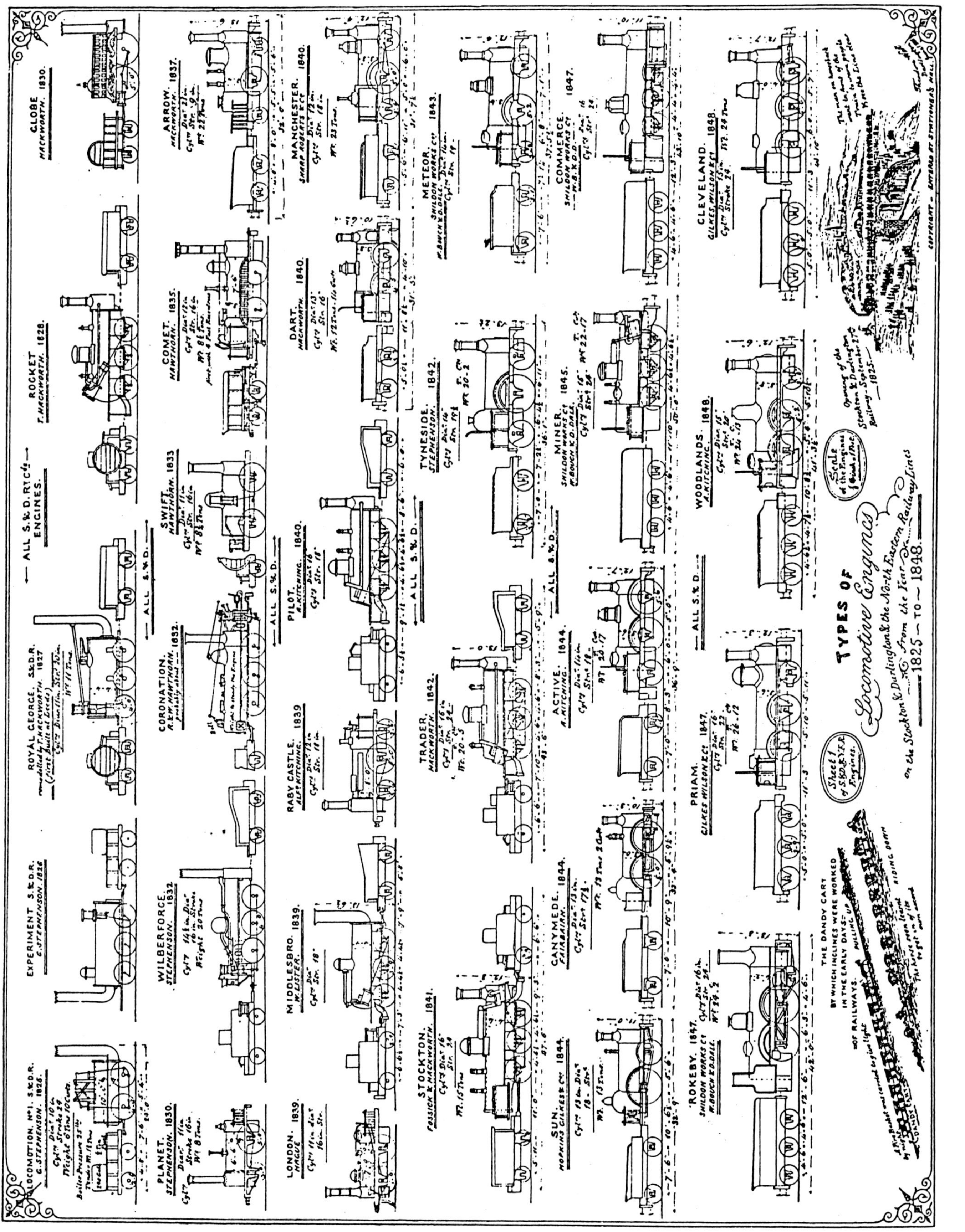
TYPES OF
Locomotive Engines
on the Stockton & Darlington & the North Eastern Railway Lines
from the Year
1825 — TO — 1848.
Sheet 1 of S&D RY ER Engines.
Scale of the Engines ¼ inch = 1 Foot.
ALL S.& D. RY Cos ENGINES.
LOCOMOTION. No 1. S.& D.R. G. STEPHENSON. 1825.
EXPERIMENT S.& D.R. G. STEPHENSON. 1826
ROYAL GEORGE. S.& D.R. remodelled by T. HACKWORTH. 1827 (first built at Leeds)
ROCKET T. HACKWORTH. 1828.
GLOBE HACKWORTH. 1830.
PLANET. STEPHENSON. 1830.
WILBERFORCE. STEPHENSON. 1832
CORONATION. R. & W. HAWTHORN. probably about 1832.
SWIFT. HAWTHORN. 1833
COMET. HAWTHORN. 1835.
ARROW. HACKWORTH. 1837.
LONDON. HAGUE 1839.
MIDDLESBRO. W. LISTER. 1839.
RABY CASTLE. ALF. KITCHING. 1839
PILOT. A. KITCHING. 1840.
DART. HACKWORTH. 1840.
MANCHESTER. SHARP ROBERTS & Co 1840.
STOCKTON. FOSSICK & HACKWORTH. 1841.
TRADER. HACKWORTH. 1842.
TYNESIDE. STEPHENSON. 1842.
METEOR. SHILDON WORKS Co W. BOUCH & D. DALE. 1843.
SUN. HOPKINS GILKES & Co 1844.
GANYMEDE. FAIRBAIRN. 1844.
ACTIVE. A. KITCHING. 1844.
MINER. SHILDON WORKS Co W. BOUCH & D. DALE. 1845.
COMMERCE. SHILDON WORKS Co W.B. & D.D. 1847.
ROKEBY. 1847. SHILDON WORKS Co W. BOUCH & D. DALE.
PRIAM. GILKES WILSON & Co 1847.
WOODLANDS. A. KITCHING. 1848.
CLEVELAND. GILKES, WILSON & Co 1848.
ALL S.& D.
THE DANDY CART BY WHICH INCLINES WERE WORKED IN THE EARLY DAYS OF RAILWAYS.
PULLING UP
RIDING DOWN
DANDY CART
Opening of the Stockton & Darlington Railway — September 27th 1825.
The man on horseback went in front of the Train to warn people & see the Line clear
COPYRIGHT — ENTERED AT STATIONER'S HALL.

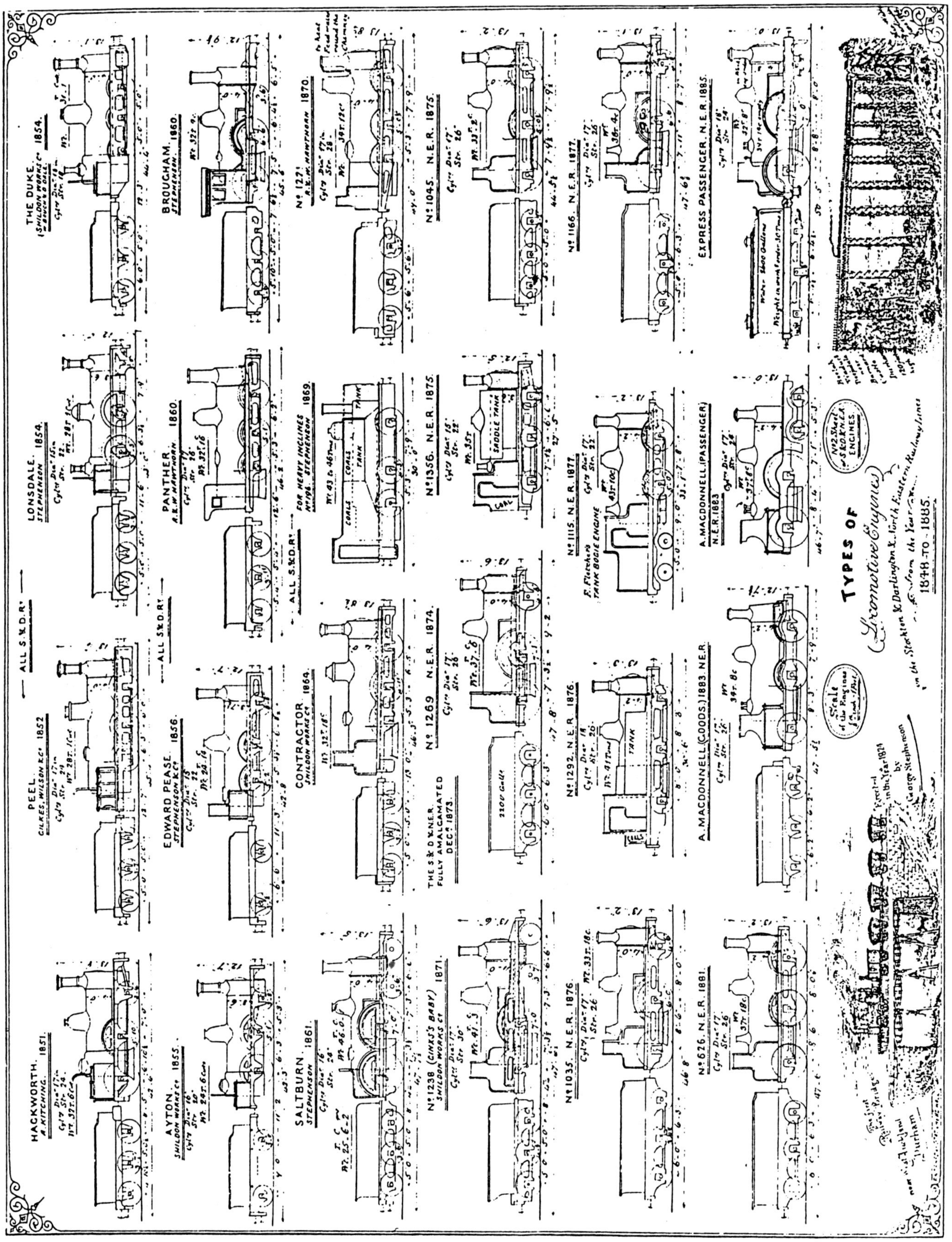
TYPES OF
Locomotive Engines
on the Stockton & Darlington & North Eastern Railway Lines
from the Year 1848 to 1885.
HACKWORTH. 1851.
AYTON. 1855.
SALTBURN. 1861.
Nº 1238. (GINKS'S BABY) 1871.
Nº 1035. N.E.R. 1876.
Nº 626. N.E.R. 1881.
PEEL. 1852
EDWARD PEASE 1856.
CONTRACTOR 1864.
Nº 1269 N.E.R. 1874.
Nº 1292. N.E.R. 1876.
A. MACDONNELL (GOODS) 1883. N.E.R.
LONSDALE. 1854.
PANTHER. 1860.
FOR HEAVY INCLINES 1869.
Nº 1356. N.E.R. 1875.
Nº 1115. N.E.R. 1877.
A. MACDONNELL (PASSENGER) N.E.R. 1883
THE DUKE. 1854.
BROUGHAM. 1860.
Nº 1271 1870.
Nº 1045. N.E.R. 1875.
Nº 1166. N.E.R. 1877.
EXPRESS PASSENGER. N.E.R. 1885.
ALL S.&D.Rʸ
THE S&D & N.E.R. FULLY AMALGAMATED DEC: 1873.

Copyright

EARLY LOCOMOTIVE ENGINES OF THE STOCKTON AND DARLINGTON RAILWAY COMPANY 1825 TO 1862

PARTICULARS AND DIMENSIONS OF ENGINES SHOWN

Ry Coy's Number	Name of Engine	Name of Maker	Maker's Number	Date when delivered	Cylinders	Boiler Barrel	Tubes	Fire-box Outside	Grate Area	Heating Surface	Wheels	Wheel Base	Total Length of Engine-frame over Buffer-beams	Intermediate Lengths of Frames	Weight of Engine in working order without Tender	Tenders	Remarks
1	Locomotion	R. Stephenson & Co.	3	Sept. 1825	10 24	40100 —	1 24	— 2040	8	Fire-tube 60	4 4 40 —	5 4	130	Over all	6 10	[illegible]	Vertical cylinders. 2 Blast pipes.
5	Stockton	R. Wilson & Co.		1826	6 18	44130 —	1 24	— 2040		Fire-tube 81.6	4 Nil 40 —	5 9	160	0 — 53 — 5050		[illegible]	4 cylinders. Axles independent. Cranks at right angles.
6	Experiment	R. Stephenson & Co.	9	October 1826	9 24	40100 —	2 18	— 30 9	12.	2 Fire-tubes 94	6 6 40 —	8 6	15 5	0 — 3143 — 4330	7 0	1	Outside inclined cylinders. 2 blast pipes.
5	Royal George	Rebuilt by T. Hackworth at Shildon Works.	1	„ 1827	11 20	44130 —	2 [illegible]	— 2140	5.6	Return tube 141	6 6 40 —	9 6	16 2	0 — 4049 — 4926	11 0	[illegible]	Return tube boiler. Vertical cylinders. single blast pipe.
9	Globe	Designed by Hackworth built by Stephenson.	30	1830	11 16	30100 —	1	—		Fire-tube	4 4 50 —	5 3	15 5	0191924 53 — 44		1	Horizontal inside cylinders under firebox. Copper dome.
13	Coronation	Designed by Hackworth built by Hawthorn.	6	1831	14 16	[illegible]	[illegible]	Fire-tube 9 02 [illegible]	[illegible]	[illegible]	6 6 40 4 [illegible]	9 0	20 6	0 5620 — 616 — 4636	11 15	[illegible]	Vertical outside cylinders. Small tubes in flue.
26	Swift	Ditto	[illegible]	March 1836	11 16	[illegible]	[illegible]	492 543	8.4	217 395 2565	4 4 40 40	6 5	[illegible] 7	0 [illegible] — 3234	8 15	1	Vertical cylinders between wheels driving an auxiliary shaft.
27	Arrow	Soho Works Shildon		1837	22 9	[illegible]	[illegible]	2736 42	9.13	[illegible]	6 Nil 5 03 0	[illegible]	[illegible] 7	[illegible] — 823	22 0	[illegible]	Horizontal inside cylinders under smoke box. Single driver. Outside frames.
43	Sunbeam	Designed by Hackworth built by Hawthorn.		1837	12 18	[illegible]	[illegible]	[illegible]	[illegible]	[illegible]	4 Nil 5 04 0	5 6	15 2	[illegible] — 16	10 0	[illegible]	Horizontal inside cylinders under smoke box. Outside frames.
165	Keswick	R. Stephenson & Co.	[illegible]	March 1862	16 24	40 06 [illegible]	[illegible]	[illegible]	[illegible]	950 103 [illegible]	8 4 7 0 3 6	[illegible]	[illegible]	[illegible] — 7342		[illegible]	Modified bogie engine. Outside cylinders, feed water heating tank under foot plate.

PARTICULARS AND DIMENSIONS OF ENGINES SHOWN

Ry Co's Number	Name of Engine	Name of Maker	Maker's Number	Date when delivered	Cylinders: Diameter (ins)	Cylinders: Stroke (ins)	Boiler Barrel: Diameter (F. I.)	Boiler Barrel: Length (F. I.)	Boiler Barrel: Length between Tube-Plates (F. I.)	Tubes: Number	Tubes: Diameter Outside (ins)	Fire-box Outside: Length (F. I.)	Fire-box Outside: Width (F. I.)	Fire-box Outside: Height above bars (F. I.)	Grate Area (Sq. feet)	Heating Surface: Tubes (Sq. feet)	Heating Surface: Fire-box (Sq. feet)	Heating Surface: Total (Sq. feet)
1	Locomotion	R. Stephenson & Co	3	Septr 1825	10	24	4·0	10·0	—	1	24	—	2·0	1·0	8	Fire-tube		60
5	Stockton	R. Wilson & Co		1826	6	18	4·4	13·0	—	1	24	—	2·0	1·0		Fire-tube		81.6
6	Experiment	R. Stephenson & Co	9	October 1826	9	24	4·0	10·0	—	2	18	—	3·0	·9	12.	2 Fire-tubes		94
5	Royal George	Rebuilt by T. Hackworth at Shildon Works.	1	" 1827	11	20	4·4	13·0	—	2	25/18	—	2·1	1·0	5.6	Return tube		141
9	Globe	Designed by Hackworth built by Stephenson.	30	1830	11	16	3·0	10·0	—	1		—				Fire-tube		
13	Coronation	Designed by Hackworth built by Hawthorn.	6	1831	14½	16	3·8	13·0	4·0	106	1¾	Fire-tube 9·0	2·5	1·5	10.87	191.22	71.10	265.32
26	Swift	Ditto	199	March 1836	11	16	2·11 wide × 4·0 high	9·9	5·0	102	1⅝	4·9	2·5	1·3	8.4	217.	39.5	256.5
27	Arrow	Soho Works Shildon		1837	22	·9	3·4	7·9	8·2	133	1⅝	2·7½	3·6	4·2	9.19	461.9	57.1	519.0
13	Sunbeam	Designed by Hackworth built by Hawthorn.		1837	12	18	3·2	8·0	8·4	104	1¾	3·5½	2·10	3·2	9.46	396.9	44.8	441.7
163	Keswick	R. Stephenson & Co	1335	March 1862	16	24	4·0	0·6	0·6	198	1¾	5·6	3·10	5·10	12.75	950.	103	1053.

Ry Co's Number	Name of Engine	Wheels: Number	Wheels: Number Coupled	Wheels: Diameter of large (F. I.)	Wheels: Diameter of small (F. I.)	Wheel Base (F. I.)	Total Length of Engine-frame over Buffer-beam (F. I.)	Front of Buffer-beam (F. I.)	Front of Smoke-box (F. I.)	Back of Smoke-box (F. I.)	Leading Axle (F. I.)	Driving Axle (F. I.)	Front F.B. Casing (F. I.)	Back F.B. Casing (F. I.)	Trailing Axle (F. I.)	Back of Frame (F. I.)	Weight of Engine in working order without Tender (Tons)	(Cwts)	Tenders	Remarks
1	Locomotion	4	4	4·0	—	5·4	13·0	← Over all →									6	10	1 tender 4 Wheels 240 galls	2 Vertical cylinders. 2 Blast pipes.
5	Stockton	4	Nil	4·0	—	5·9	16·0	0	–	–	5·3	–	–	–	5·9	5·0			1 Tender 4 Wheels	4 Cylinders. Axles independent. Cranks at right angles.
6	Experiment	6	6	4·0	—	8·6	15·5	0	–	–	3·1	4·3	–	–	4·3	3·10	7	0	1 4 Wheels	Outside inclined cylinders. 2 blast pipes.
5	Royal George	6	6	4·0	—	9·6	16·2	0	–	–	4·0	4·9	–	–	4·9	2·8	11	0	2 tenders one at each end 4 Wheels	Return tube boiler. Vertical cylinders. single blast pipe.
9	Globe	4	4	5·0	—	5·3	15·5	0	1·9	1·9	2·4	5·3	–	–	–	4·4			4 Wheels 1	Horizontal inside cylinders under firebox. Copper dome.
13	Coronation	6	6	4·0	4·0	9·0	20·6	0	5·6	2·0	·6	1·6	–	–	4·6	3·6	11	15	2 Tenders one at each end. Four Wheels	Vertical Outside cylinders. Small tubes in flue.
26	Swift	4	4	4·0	4·0	6·5	13·7	0	1·1	1·3	1·7½	3·2½	–	–	3·2	3·4½	8	15	4 Wheels 1	Vertical cylinders between axles driving an auxiliary shaft.
27	Arrow	6	Nil	5·0	3·0	10·11	16·7	0	1·0	1·9	·8	5·6	1·6	3·3	·8	2·3	22	0	4 Wheels 815 Gall. Water. Screw Brake.	Horizontal inside Cylinders under smoke box. Single driver. Outside frames.
13	Sunbeam	4	Nil	5·0	4·0	5·6	15·2	0	1·0	1·10	1·0	5·6	1·6	2·10	—	1·6	10	0	4 Wheels Iron tank	Horizontal inside Cylinders. under smoke box. Outside frames.
163	Keswick	8	4	7·0½	3·6	20·0½	26·9½	0	2·7½	1·8	2·7½	1·8¾	6·7¾	–	7·3½	4·2			6 Wheels 25 tons 6½ Cwts.	4 Coupled bogie engine. Outside cylinders. feed water heating tank under foot plate.

Appendix IV. Numerical and Alphabetical Lists of Locomotives

List of Locomotives in Numerical Order

See also remarks on numbering in Chapter 1. under 'General Note'.

Almost all locomotives still in use when the Stockton & Darlington Railway merged with the N.E.R. in July 1863 were eventually renumbered into the N.E.R. 1xxx (1001) class by adding their original S. & D.R. number to 1000, but some were also renumbered again later; see Appendix II. In the following list I have only used the 1xxx number where it was given to the locomotive when built.

Classes of 0-6-0 mineral engines in this list are more extended than may be considered strictly accurate; for instance, the 'Peel' class has been used to cover most of the small wheel mineral engines up to 1866, although wheel sizes did vary slightly, cylinder bore and stroke also varied, and so did the wheelbase, but they form a convenient class in themselves. The 5'-0" diameter wheeled engines have been split into more classes than this. Later unnamed 0-6-0 mineral engines in general, whether built by Shildon, the North Road Works or Hopkins, Gilkes, have been classed as N.R.W.min. since the numbers were in most cases not consecutive. N.R.W. works Nos. given are to some extent based on conjecture.

The following abbreviations have been used:

Robert Stephenson & Co.	-	R.S.& Co.
W. & A. Kitching or A. Kitching	-	Kitchings
North Road Works, Darlington	-	N.R.W.
Timothy Hackworth	-	Hackworth
Fossick & Hackworth	-	F. & H.
R. & W. Hawthorn, Newcastle	-	Hawthorn
Stephenson/Hawthorn/Shildon	-	S/H/Sh. ('Majestic' class)
Shildon Works Co.	-	Shildon W.
Soho Works, Shildon	-	Soho W.
Gilkes, Wilson, Middlesbrough	-	Gilkes, W.
Hopkins, Gilkes, ,,	-	Hopkins, G.
Neasham & Welsh	-	N. & W.
Wm. Lister	-	Wm. Lister
Hartlepool Iron Co.	-	Hart.I.Co.
John Hague, London	-	J. Hague
Fairbairns, Manchester	-	Fairbairns
Edward Bury, Clarence Foundry, Liverpool	-	E. Bury
Sharp, Roberts, Manchester	-	Sharp, R.
Kirtley, Warrington	-	Kirtley
Benjamin Hick & Son, Bolton	-	B. Hick
West Hartlepool Harbour & Railway Co.	-	W.H.H.R.
Dübs & Co., Glasgow	-	Dübs & Co.
Avonside Engine Co., Bristol	-	Avonside
Long Wheel Base 0-6-0	-	Long w.b.
Not known	-	n.k.
Nominal	-	n.
Disposed of	-	Disp.

No.	Name	Class	Works No.	Wheel Arr't	Designer	Builder	Year Rec'd	Year Disp'd	Page No.	Fig. No.
1.a.	Locomotion	Eponymous	3.	0-4-0	R. S. & Co.	R. S. & Co.	1825	1850 (sold, now preserved)	28	14/18
b.	unnamed	N.R.W.min.	26.	0-6-0	W. Bouch	N.R.W.	1868	1907	150	177
2.a.	Bedlington ? ?	Locomotion	4.	0-4-0	R. S. & Co.	R. S. & Co.	1825	1827?	39	
b.	Hope as rebuilt	" Enterprise	4. 1.	0-4-0 0-6-0	" Hackworth?	" Kitchings	1825/6 1835	 1840's	28 40	19/20 52
c.	Bedlington (later No.12.)	Majestic		0-6-0	Hackworth	S/H/Sh.	1831	1839	60	
d.	Graham	Times	22.	0-6-0	W. Bouch	Kitchings	1849	1865	127	
e.	unnamed	N.R.W.min.	9.	0-6-0	"	N.R.W./Sh.	1865	1901.	149	
None	"Chittapratt"	Individual		0-4-0	R. Wilson	R. Wilson	1825	1827	34	21/3
3.a.	Black Diamond as rebuilt	Locomotion	5.	0-4-0 0-6-0	R. S. & Co.	R. S. & Co. Wm.Lister	1826 1830	 1848?	38 40	19
b.	Times	eponymous	21.	0-6-0	W. Bouch	Kitchings	1849	1865	127	
c.	unnamed	N.R.W.min.	10.	0-6-0	"	N.R.W./Sh.	1865	1908	149	
4.a.	Diligence as rebuilt	Locomotion	6.	0-4-0 0-6-0	R. S. & Co.	R. S. & Co. Hackworth	1826 1834	1841 (sold)	38 40	
b.	Stockton	Tory (variant?)		0-6-0	Hackworth?	F. & H.	1841	1870	73	56/7
c.	unnamed	N.R.W.min.	44.	0-6-0	W. Bouch	N.R.W.	1870	1909	150	182
5.a.	unnamed	Locomotion	7.	0-4-0	R. S. & Co.	R. S. & Co.	1827	1828 (exploded)	28	
b.	Royal George	individual		0-6-0	Hackworth	Shildon W.	1827	1840 (sold)	45	26/30
c.	Hopetown	Tory	10.	0-6-0	Hackworth	Kitchings	1841	1869	75	
d.	unnamed	N.R.W.min.	36.	0-6-0	W. Bouch	N.R.W.	1869	1908	150	
6.a.	Experiment	individual	8.	0-4-0/ 0-6-0	R. S. & Co.	R. S. & Co.	1827/8	1839 (sold)	47	31/3
b.	Dispatch	Tory		0-6-0	Hackworth	Soho W.	1839	1869	75	
c.	unnamed	N.R.W.min.	30.	0-6-0	W. Bouch	N.R.W.	1869	1907	150	178
-	Trial engine	individual	n.k.	0-4-0	R. S. & Co.	R. S. & Co.	1827	1827	50	
-	Quickspeed	-	18?	0-4-0	R. S. & Co.		1829?		52	34

No.	Name	Class	Works No.	Wheel Arr't	Designer	Builder	Year Rec'd	Year Disp'd	Page No.	Fig. No.
7.a.	Experiment	(changed almost immediately to No.6.)								
b.	Rocket	individual	15.	0-6-0	Hack./R.S.	R. S. & Co.	1829	1840? (sold)	52	35/6
c.	Prince	Tory		0-6-0	Hackworth	Shildon W.	1843	1869	75	
d.	unnamed	N.R.W.min.	32.	0-6-0	W. Bouch	N.R.W.	1869	1909	150	179
8.a.	Victory	individual		0-6-0	Hackworth	Shildon W.	1829	1841 (sold)	53	37
b.	Leader (rebuild of No.25 "Enterprise")	Tory		0-6-0	Hackworth	Soho W.	1842	1869	71	60
c.	unnamed	N.R.W.min.	28.	0-6-0	W. Bouch	N.R.W.	1869	1908	150	
9.a.	Globe	individual	29.	0-4-0	Hackworth	R. S. & Co.	1830	1838 (exploded)	54	38/9
b.	Middlesbro'	Tory		0-6-0	Hackworth	Wm. Lister	1839	1865	75	59
c.	unnamed	Eskdale	219.	0-6-0	W. Bouch	Hopkins, G.	1865	1903	149	167
10.a.	Planet	Planet	30.	2-2-0	R. S. & Co.	R. S. & Co.	1830	c.1839 (sold)	57	40/1
b.	Auckland	Tory		0-6-0	Hackworth	Soho W.	1839	c.1869	75	
c.	unnamed	N.R.W.min.	41.	0-6-0	W. Bouch	N.R.W.	1870	1907	150	
11.a.	North Star (Stat. Eng. from Dec. 1837)	Planet	31.	2-2-0	R. S. & Co.	R. S. & Co.	1830/1	1842	57	40/1
b.	Beehive/ Bee	Enterprise/ Tory		0-6-0	Hackworth	Soho W. Shildon W.	1837	Rebuilt 1869	1842 70	
c.	unnamed	N.R.W.min.	37.	0-6-0	W. Bouch	N.R.W.	1869	1909	150	180
12.a	Majestic (see No.2.c.)	eponymous		0-6-0	Hackworth	S/H/Sh.	1831	1839	60	42/3
b.	Briton	Enterprise		0-6-0	Hackworth	Shildon W.	1837	1842?	70	
c.	Trader (rebuild of No.12 "Briton")	Tory		0-6-0	Hackworth	Soho W.	1842	1868	71	
d.	unnamed	N.R.W.min.	24.	0-6-0	W. Bouch	N.R.W.	1868	1910	150	176
13.a.	Coronation	Majestic	123.	0-6-0	Hackworth	S/H/Sh.	1831	1841	60	42/3
b.	Ocean (see No.28 below)									
c.	unnamed	N.R.W.min.	42.	0-6-0	W. Bouch	N.R.W.	1870	1907	150	

No.	Name	Class	Works No.	Wheel Arr't	Designer	Builder	Year Rec'd	Year Disp'd	Page No.	Fig. No.
14.a.	William IVth.	Majestic.		0-6-0	Hackworth	S/H/Sh.	1831	1839 (sold)	60	42/3
b.	Tees	Enterprise/ Tory	3.	0-6-0	Hackworth	Kitchings/ Shildon W.	1837 1840?	Rebuilt " 2nd. 1871?	1840? 1848 70	67
c.	unnamed	N.R.W.min.	57.	0-6-0	W. Bouch	N.R.W.	1871	1910	154	
15.a	Northumbrian	Majestic		0-6-0	Hackworth	S/H/Sh.	1831	1838	60	42/3
b.	Tory	eponymous		0-6-0	Hackworth	Soho W.	1838	1869	73	
c.	unnamed	N.R.W.min.	33.	0-6-0	W. Bouch	N.R.W.	1869	1910	150	
16.a.	Director	eponymous	1(2nd.)	0-6-0	Hack./St.	R. S. & Co.	1832	1845 (sold)	63	44/8
b.	Stanhope	Tory	14.	0-6-0	Hackworth	Kitchings	1845	1870	75	62
c.	unnamed	N.R.W.min.	45.	0-6-0	W. Bouch	N.R.W.	1870	1906	150	
17.a.	Lord Brougham	Majestic		0-6-0	Hackworth	S/H/Sh.	1831	1838 (scrapped)	60	42/3
b.	Whig	Tory	6.	0-6-0	Hackworth	Kitchings	1839	1870	73	
c.	unnamed	N.R.W.min.	46.	0-6-0	W. Bouch	N.R.W.	1870	1908	150	
18.a.	Shildon	Majestic		0-6-0	Hackworth	S/H/Sh.	1831	1839 (sold)	60	42/3
b.	Etherley	Tory		0-6-0	Hackworth	Wm. Lister	1840	1869	75	
c.	unnamed	N.R.W.min.	38	0-6-0	W. Bouch	N.R.W.	1869	1906	150	181
19.a.	Darlington	Director	127	0-6-0	Hack./St.	Hawthorn	1832	1856 (sold)	63	44/8
b.	Tweed	individual		0-6-0	J. Carson	W.H.H.R.	1865	1877	148	163
20.a.	Adelaide (Stat. Eng. from	Director c.1855)	2(2nd.)	0-6-0	Hack./St.	R. S. & Co.	1832	1860's (sold)	63	44/8
b.	Normanby			0-6-0	J. Carson	W.H.H.R.	1865	1877	148	164
21.a.	Earl Grey	Director	142.	0-6-0	Hack./St.	Hawthorn	1832	1858	63	44/8
b.	Cleveland			0-6-0	J. Carson	W.H.H.R.	1865	1877	148	164
22.a.	Lord Durham	Director	3(2nd.)	0-6-0	Hack./St.	R. S. & Co.	1832	1848	63	44/8
b.	Alert	Tory	18	0-6-0	Hackworth	Kitchings	1848	1869	74	
c.	unnamed	N.R.W.min.	39	0-6-0	W. Bouch	N.R.W.	1869	1907	150	

No.	Name	Class	Works No.	Wheel Arr't	Designer	Builder	Year Rec'd	Year Disp'd	Page No.	Fig. No.
23.a.	Wilberforce	Director	150.	0-6-0	Hack./St.	Hawthorn	1832/3	1858+?	63	44/9
b.	unnamed	Eskdale	220.	0-6-0	W. Bouch	Hopkins, G.	1865	1907	149	167
24.a.	Magnet	individual		0-6-0	Hackworth	Soho W.	1835	n.k. (Rebuilt 1850?)	66	50/1
b.	Skelton Castle	Tory	13.	0-6-0	Hackworth	Kitchings	1846	1870	75	
c.	unnamed	N.R.W.min.	43.	0-6-0	W. Bouch	N.R.W.	1870	1907	150	
25.a.	Enterprise	eponymous	2.	0-6-0	Hackworth	Kitchings	1835	1841?	68	52
b.	Derwent	Tory	12.	0-6-0	Hackworth	Kitchings	1845	1868 (sold, now preserved)	75	63/5
c.	unnamed	N.R.W.min.	29.	0-6-0	W. Bouch	N.R.W.	1869	1907	150#	
26.a.	Arrow* (later No.51)	individual		2-2-2	Hackworth	Soho W.	1837	1864	87	68/9
b.	Pilot	Tory	8.	0-6-0	Hackworth	Kitchings	1840	1868/9 (sold)	75	61/6
c.	unnamed	N.R.W.min.	27.	0-6-0	W. Bouch	N.R.W.	1868	1907	150	
27.a.	Swift* (renumbered?)	individual	199.	0-4-0	Hawthorn	Hawthorn	1836	1839 (sold)	88	70/2
	(*the numbers of these two engines are reversed in some lists)									
b.	Witton Castle	Tory		0-6-0	Hackworth	N. & W.	1840	1867/75?	75	
c.	(1027) unnamed	1271	99.	0-6-0	W. Bouch	N.R.W.	1875	1910	196	
28.a.	Ocean (later No.13, then 12 - see notes in text)	Tory		0-6-0	Hackworth	Wm. Lister	1840	1870	75	
b.	Sunbeam (later No.43 - see notes in text)	individual	227/8?	2-2-0	Hawthorn	Hawthorn	1837	1863	90	73
c.	Conside	individual		0-6-0	Hart.I.Co.	Hart.I.Co.	1845	c.1858	121	113/4
d.	unnamed	Eskdale	221.	0-6-0	W. Bouch	Hopkins, G.	1866	1908	149	167
29.a.	Queen (later No.40	individual	5.	0-4-0/ 0-4-2	Kitchings	Kitchings	1837	1866?	90	74
b.	Miner	eponymous		0-6-0	Hack/Bouch	Shildon W.	1845	1870	81	
c.	unnamed	N.R.W.min.	53.	0-6-0	W. Bouch	N.R.W.	1870	1907	154	185
30.a.	Raby Castle (later No.49)	individual	7.	2-2-2	Kitchings	Kitchings	1839	1858	91	75/6

No.	Name	Class	Works No.	Wheel Arr't	Designer	Builder	Year Rec'd	Year Disp'd	Page No.	Fig. No.
30.b.	Wear	Miner		0-6-0	Hack/Bouch	Shildon W.	1845	1871	81	
c.	unnamed	N.R.W.min.	55.	0-6-0	W. Bouch	N.R.W.	1871	1909	154	185
31.a.	Redcar	Miner		0-6-0	Hack/Bouch	Shildon W.	1845	1869	81	
b.	unnamed	N.R.W.min.	31.	0-6-0	W. Bouch	N.R.W.	1869	1906	150	
32.a.	Eldon	Miner		0-6-0	Hack/Bouch	Shildon W.	1846	1875	81	
b.	(1032) unnamed	1271	98.	0-6-0	W. Bouch	N.R.W.	1875	1907	196	246
33.	Shildon	Miner		0-6-0	Hack/Bouch	Shildon W.	1846	1877	81	64
34.a.	Driver	"		"	" "	" "	1846	1870	81	
b.	unnamed	N.R.W.min.	54.	0-6-0	W. Bouch	N.R.W.	1870	1908	150	184
35.a.	Commerce	eponymous		0-6-0	W. Bouch	Shildon W.	1847	1876	125	115/7
b.	(1035) unnamed	'Gamecock'	123.	2-4-0	W. Bouch	N.R.W.	1876	1909	187	228/9
36.a.	Guisbro'	Commerce		0-6-0	W. Bouch	Shildon W	1847	1870	125	
b.	unnamed	N.R.W.min.	48.	0-6-0	W Bouch	N.R.W.	1870	1908	150	
37.a.	Gem	Commerce		0-6-0	W. Bouch	Shildon W.	1847	1876	125	
b.	(1037) unnamed	1037	118.	0-6-0T	W. Bouch	N.R.W.	1876	1910	194	
38.a.	Rokeby (Later No.39)	eponymous		2-4-0	W. Bouch	Shildon W.	1847	1875	102	89/92
b.	(1038) unnamed	Long w.b.	97.	0-6-0	W. Bouch	N.R.W.	1876	1925	199	
39.a.	Ruby	Rokeby		2-4-0	W. Bouch	Shildon W.	1847	1871	102	
b.	unnamed	N.R.W.min.	59?	0-6-0	W. Bouch	N.R.W.	1875	1907	154	
40.a.	Queen (see No.29a above)									
b.	unnamed	Eskdale	222.	0-6-0	W. Bouch	Hopkins, G.	1866	1908	149	167
41.	Dart	individual		0-4-0	Hackworth	Soho W.	1840	1880	94	78/80
42.a.	London	individual		0-4-0	J. Hague?	J. Hague?	1839	c.1859	92	77
b.	unnamed	Eskdale	223.	0-6-0	W. Bouch	Hopkins, G.	1866	1906	149	167
43.a.	Sunbeam (see No.28a above)									
b.	unnamed	Eskdale	224.	0-6-0	W. Bouch	Hopkins, G.	1866	1905	149	167
44.a.	Sun	Bury type		2-2-0	E. Bury	Fairbairns	1844	c.1869	99	86

No.	Name	Class	Works No.	Wheel Arr't	Designer	Builder	Year Rec'd	Year Disp'd	Page No.	Fig. No.
44.b.	unnamed	N.R.W.min.	25.	0-6-0	W. Bouch	N.R.W.	1868	1910	150	
45.a.	Gannymede	Bury type		0-4-0	E. Bury	Fairbairns	1844	1875	100	
b.	(1045) unnamed	Long w.b.		0-6-0	W. Bouch	N.R.W.	1875	1922	199	
46.a.	Antelope	Bury type		2-2-0	E. Bury	E. Bury	1844	c.1868	100	
b.	unnamed	Eskdale	225.	0-6-0	W. Bouch	Hopkins, G.	1866	1908	149	167
47.a.	Unicorn	Bury type		2-2-0	E. Bury	E. Bury	1844	c.1868	100	
b.	unnamed	N.R.W.min.	51.	0-6-0	W. Bouch	N.R.W.	1870	1907	150	
48.a.	Active	individual	11.	0-4-2	Kitchings	Kitchings	1844	1871?	101	88
B.	unnamed	N.R.W.min.	56.	0-6-0	W. Bouch	N.R.W.	1871	1909	154	
49.a.	Raby Castle (see No.30a above)									
b.	unnamed	Eskdale	226.	0-6-0	W. Bouch	Hopkins, G.	1866	1909	149	167
50.a.	Meteor	individual		2-2-2	W. Bouch	Shildon W.	1842	1875	98	84/5
b.	unnamed	'Gamecock'	115.	2-4-0	W. Bouch	N.R.W.	1875	1908	187	226/7
51.a.	Arrow (see No.26a above)									
b.	unnamed	Eskdale	227.	0-6-0	W. Bouch	Hopkins, G.	1867	1906	149	167
52.a.	Comet	individual		2-2-2	Kirtley	Kirtley	1840/1	1875	94	81
b.	(1052)unnamed	Long.w.b.		0-6-0	W. Bouch	N.R.W.	1875	1905	199	
53.a.	Manchester	individual	94	2-2-2	Sharp, R.	Sharp, R.	1840	1867	96	82
b.	unnamed	N.R.W.min. (as 2nd. "Uranus")	18.	0-6-0	W. Bouch	N.R.W.	1867	1909	150	
54.a.	Tyneside	individual	354.	2-2-2	R. S. & Co.	R. S. & Co.	1842	1870	96	83
b.	unnamed	N.R.W.min.	50.	0-6-0	W. Bouch	N.R.W.	1870	1907	150	
55.	Wolsingham	individual	17.	0-4-2	Kitchings	Kitchings	1847	1879	101	
56.a.	Shotley*	eponymous	30.	0-6-0	W. Bouch	Gilkes, W.	1852	1866	130	
b.	unnamed	N.R.W.min.	12?	0-6-0	W. Bouch	N.R.W.	1866	1909	150	
57.a.	Tow Law*	Shotley	31.	0-6-0	W. Bouch	Gilkes, W.	1853	1866	130	
b.	unnamed	N.R.W.min.		0-6-0	W. Bouch	N.R.W.	1866	1904	150	
	(*the names & numbers of these two engines are reversed in some lists)									

No.	Name	Class	Works No.	Wheel Arr't	Designer	Builder	Year Rec'd	Year Disp'd	Page No.	Fig. No.
58.	Woodlands	eponymous	19.	2-4-0	W. Bouch	Kitchings	1848	1877	107	95/6
59.	Hallgarth	Woodlands	20.	2-4-0	W. Bouch	Kitchings	1848	1881	107	"
60.a.	Cleveland	eponymous	16.	0-6-0	J. Graham?	Gilkes, W.	1848	1870	126	119/ 120
b.	unnamed	N.R.W.min.	52.	0-6-0	W. Bouch	N.R.W.	1870	1907	150	
61.a.	Star	Cleveland	17.	0-6-0	as No.60	Gilkes, W.	1849	1870	126	
b.	unnamed	N.R.W.min.	47.	0-6-0	W. Bouch	N.R.W.	1870	1908	150	
62.a.	Southend	Cleveland	18.	0-6-0	as No.60	Gilkes, W.	1849	1876	126	
b.	(1062)unnamed	'Gamecock'	125.	2-4-0	W. Bouch	N.R.W.	1876	1907.	187	
63.a.	Birkbeck	eponymous		0-6-0	W. Bouch	Shildon W.	1849	1875	126	118
b.	(1063)unnamed	Long w.b.		0-6-0	W. Bouch	N.R.W.	1875	1910	199	
64.a.	Larchfield	Birkbeck		0-6-0	W. Bouch	Shildon W.	1849	1873	126	
b.	unnamed (possible, ex J.H.P. list)		85?	0-6-0	W. Bouch	N.R.W.	1873			
65.a.	Newmarket(?)/ Stephenson	Priam	10.	2-4-0	R.S.& Co.?	Gilkes, W.	1850?	1871	103	
b.	unnamed (as No.64b)			0-6-0	W. Bouch	N.R.W.	1871			
66.a.	Priam	eponymous	8.	2-4-0	as No.65a	Gilkes, W.	1847	1876	103	93/4
b.	(1066)unnamed	'Gamecock'	124.	2-4-0	W. Bouch	N.R.W.	1875	1909	187	
67.a.	Orion	Priam	9.	2-4-0	as No.65a	Gilkes, W.	1848	1875	103	93
b.	(1067)unnamed	1271	96.	0-6-0	W. Bouch	N.R.W.	1875	1908	196	
68.a.	Brunswick	Priam	11.	2-4-0	as No.65a	Gilkes, W.	1850	1875	103	
b.	(1068)unnamed	'Gamecock'	105.	2-4-0	W. Bouch	N.R.W.	1875	1909	187	223/5
69.a.	Clarendon	Priam	12.	2-4-0	as No.65a	Gilkes, W.	1850	1876	103	
b.	(1069)unnamed	Long w.b.		0-6-0	W. Bouch	N.R.W.	1876	1915	199	
70.a.	Alarm	Priam	13.	2-4-0	as No.65a	Gilkes, W.	1851	1875	103	
b.	(1070)unnamed	Long w.b.		0-6-0	W. Bouch	N.R.W.	1876	1921	199	247
71.a.	Hackworth (but see remarks in text)	individual	23.	2-4-0	W. Bouch	Kitchings	1851	1873?	107	97/9
b.	(1071)unnamed	individual	84.	2-4-0	W. Bouch	N.R.W.	1873?	1901	187	99

No.	Name	Class	Works No.	Wheel Arr't	Designer	Builder	Year Rec'd	Year Disp'd	Page No.	Fig. No.
72.	Peel	eponymous	33.	0-6-0	W. Bouch	Gilkes, W.	1852	1889	130	121/4
73.	Aberdeen	Peel	34.	0-6-0	"	"	1852	1903	130	125
74.a.	Emperor	Peel	39.	0-6-0	"	"	1853	1875	130	
b.	(1074)unnamed	Long w.b.		0-6-0	W. Bouch	N.R.W.	1875	1911	199	
75.	Baring	Peel	40	0-6-0	W. Bouch	Gilkes, W.	1853	1890	130	
76.	Prince of Wales	Peel	46.	0-6-0	"	"	1854	1881	130	
77.	Alexander	Peel	47.	0-6-0	"	"	1854	1905	130	
78.a.	Lonsdale	s/h	700?	2-4-0	R. S. & Co.	R. S. & Co.	1854	1870	109	100
b.	unnamed	N.R.W.min.	49.	0-6-0	W. Bouch	N.R.W.	1870	1906	150	183
79.a.	Carlisle	s/h	701?	2-4-0	R. S. & Co.	R. S. & Co.	1854	1875	109	100
b.	(1079)unnamed	1271	95.	0-6-0	W. Bouch	N.R.W.	1875	1915	196	
80.	The Duke	eponymous		0-6-0	W. Bouch	Shildon W.	1854	1882	130	128
81.a.	Miller	s/h	532.	0-6-0	n.k.	Hawthorn	1854	1867	130	
b.	unnamed	Eskdale	228.	0-6-0	W. Bouch	Hopkins, G.	1867	1909	149	167
82.	Hawthorn	s/h	533.	0-6-0	n.k.	Hawthorn	1854	1881	130	127
83.	Victoria	Peel	51.	0-6-0	W. Bouch	Gilkes. W	1854	1893	130	
84.	Albert	Peel	52.	0-6-0	"	"	1854	1894	130	126
85.	Hardinge	Peel	54.	0-6-0	"	"	1854	1903	130	
None	unnamed			0-4-2	Fairbairn	Fairbairn	1854	1854?	109	
86.	Zetland	s/h		0-6-0	n.k.	Hawthorn	1854	1881	135	
87.	Fryerage	Bury type		0-4-0	E. Bury	E. Bury	1854	1882	101	
88.a.	Deanery	" "		0-4-0	"	"	1854	1871	101	
b.	unnamed	N.R.W.min.	58.	0-6-0	W. Bouch	N.R.W.	1871	1912	154	191
89.	Huddersfield	Bury type		0-4-0	E. Bury	E. Bury	1854	1877	101	87
90.	Merchant	Peel	55.	0-6-0	W. Bouch	Gilkes, W.	1855	1881	135	
91.	Australia	Peel	59.	0-6-0	"	"	1855	1902	135	
92.	Yarm	Peel	60.	0-6-0	"	"	1855	1883	135	

No.	Name	Class	Works No.	Wheel Arr't	Designer	Builder	Year Rec'd	Year Disp'd	Page No.	Fig. No.
93.a.	Uranus	s/h		2-2-2	B. Hick	B. Hick	1855	1867	109	101
b.	Uranus	N.R.W.min.	20.	0-6-0	W. Bouch	N.R.W.	1867	1909	150	174/5
94.a.	Neptune	s/h		2-2-2	B. Hick	B. Hick	1855	1867	109	101
b.	Neptune	N.R.W.min.	21.	0-6-0	W. Bouch	N.R.W.	1867	1909	150	
95.	Elephant	Peel	61.	0-6-0	W. Bouch	Gilkes, W.	1855	1903	135	
96.	Middleton	Peel	62.	0-6-0	"	"	1855	1894	135	
97.	Gainford	Peel	63.	0-6-0	"	"	1855	1882	135	
98.a.	Pierremont	eponymous	24.	2-4-0	W. Bouch	Kitchings	1855	1876	111	102/4
b.	(1098)unnamed	'Gamecock'		2-4-0	W. Bouch	N.R.W.	1876	1905	187	
99.	Ayton	Pierremont		2-4-0	W. Bouch	Shildon W.	1855	1882	111	108
100.	Stobart	Pierremont		2-4-0	W. Bouch	Shildon W.	1855	1887	112	
101.	Marske	Pierremont	25.	2-4-0	"	Kitchings	1855	1882	111	103/4
102.	Cotherstone	Peel	64.	0-6-0	W. Bouch	Gilkes, W.	1855	1885	135	
103.	Darlington	Peel	68.	0-6-0	"	"	1855	1905	135	
104.	Durham	Peel	69.	0-6-0	"	"	1855	1902	135	132/3
105.	Teesdale	Peel	70.	0-6-0	"	"	1855	1891	135	
106.	Staindrop	Peel	71.	0-6-0	"	"	1856	1885	135	
107.	Barnard Castle	Peel	72.	0-6-0	"	"	1856	1888	135	130
108.	Eagle	Peel	959.	0-6-0	W. Bouch	Hawthorn	1856	1912	135	131
109.	Falcon	Peel	960.	0-6-0	"	"	1856	1903	135	131
110.	Hawk	Peel	961.	0-6-0	"	"	1856	1903.	135	131
111.	Stag	Peel	962.	0-6-0	"	"	1856	1904	135	131
112.	Lion	Peel	963.	0-6-0	"	"	1856	1902	135	134
113.	Camel	Peel	964.	0-6-0	"	"	1856	1900	135	131
114.	Edward Pease	eponymous	1071.	2-4-0	Bouch/Steph.	R. S. & Co.	1856	1877	112	105
115.	Meynell	Ed. Pease	1072.	2-4-0	" "	"	1856	1877	112	106/7
116.	Lartington	Woodlands	74.	2-4-0	W. Bouch	Gilkes, W.	1856	1879	112	
117.	Nunthorpe	Woodlands	75.	2-4-0	"	"	1856	1880	112	109/ 110

No.	Name	Class	Works No.	Wheel Arr't	Designer	Builder	Year Rec'd	Year Disp'd	Page No.	Fig. No.
118.	Elm Field	Woodlands	26.	2-4-0	W. Bouch	Kitchings	1857	1880	112	111
119.	Greta	Peel	82.	0-6-0	W. Bouch	Gilkes, W.	1856	1882	135	
120.	Brough	Peel	83.	0-6-0	"	"	1856	1888	135	
121.	Tiger	Peel	84.	0-6-0	"	"	1857	1880	135	
122.	Kestrel	Peel	85.	0-6-0	"	"	1857	1892	135	135
123.	Belle Vue	Peel	88.	0-6-0	"	"	1857	1888	135	
124.	Brookside	Peel	89.	0-6-0	"	"	1857	1889	135	
125.	Gazelle	Peel	90.	0-6-0	"	"	1858	1879	135	136
126.	John Wakefield	Peel	91.	0-6-0	"	"	1858	1909	135	
127.	Polam	Peel	92.	0-6-0	"	"	1858	1889	135	
128.	East Mount	Peel	93.	0-6-0	W. Bouch	Gilkes, W.	1858	1903	135	
129.	Stanley	Peel	1026.	0-6-0	W. Bouch	Hawthorn	1858	1909	136	137
130.	Hutton	Peel	1027.	0-6-0	"	"	1858	1905	136	138
131.	Long Hull	Peel	1028.	0-6-0	"	"	1858	1905	136	
132.	Appleby	Peel	27.	0-6-0	W. Bouch	Kitchings	1858	1903	136	
133.	Kirkby Stephen	Peel	28.	0-6-0	"	"	1858	1880	136	
134.	Kendal	Peel	95.	0-6-0	W. Bouch	Gilkes, W.	1858	1895	136	
135.	Eden	The Duke		0-6-0	W. Bouch	Shildon W.	1858	1882	130	129
136.	Tebay	Peel	1185.	0-6-0	W. Bouch	R. S. & Co.	1858	1907	136	141
137.	Deepdale	Peel	1186.	0-6-0	"	"	1858	1887	136	141
138.	Beelah	Peel	1187.	0-6-0	"	"	1858	1902	136	139
139.	Crossfell	Peel	96.	0-6-0	W. Bouch	Gilkes, W.	1859	1904	136	140
140.	Penrith	Peel	97.	0-6-0	"	"	1859	1901	136	
141.	Excelsior	Peel (mod.)	98.	0-6-0	"	"	1859	1885	136	
142.	Wycliffe	Peel	99.	0-6-0	"	"	1859	1903	136	
143.	Windermere	Peel	100.	0-6-0	"	"	1859	1902	136	142
144.	Ulleswater	Peel	104.	0-6-0	"	"	1860	1904	136	143
145.	Panther	eponymous	1074.	0-6-0	W. Bouch	Hawthorn	1860	1889	136	144/6

No.	Name	Class	Works No.	Wheel Arr't	Designer	Builder	Year Rec'd	Year Disp'd	Page No.	Fig. No.
146.	Ostrich	Panther	1075.	0-6-0	W. Bouch	Hawthorn	1860	1905	136	144/6
147.	Leopard	Panther	1076.	0-6-0	"	"	1860	1906	136	"
148.	Zebra	Panther	1077.	0-6-0	"	"	1860	1903	136	144/8
149.	Fox	Panther	1078.	0-6-0	"	"	1860	1904	136	"
150.	Mastiff	Panther	1079.	0-6-0	"	"	1860	1903	136	144/7
151.	Mercury	Panther	110.	0-6-0	W. Bouch	Gilkes, W.	1860	1902	136	
152.	Venus	Panther	111.	0-6-0	"	"	1860	1889	136	
153.	Mars	Panther	112.	0-6-0	"	"	1860	1892	136	
154.	Jupiter	Panther	113.	0-6-0	"	"	1860	1880	136	
155.	Saturn	Panther	114.	0-6-0	"	"	1861	1903	136	149
156.	Herschel	Panther	115.	0-6-0	W. Bouch	Gilkes, W.	1861	1880	136	
157.	Planet	Panther	119.	0-6-0	"	"	1861	1902	136	
158.	Lune	Panther	120.	0-6-0	"	"	1861	1880	136	
159.	York	Panther	125.	0-6-0	"	"	1862	1905	136	
160.	Brougham	eponymous	1272.	4-4-0	Bouch/Steph.	R. S. & Co.	1860	1888	175	196/7 & 202
161.	Lowther	Brougham	1273.	4-4-0	" "	"	1860	1888	175	198-202
162.	Saltburn	eponymous	1332.	4-4-0	W. Bouch	R. S. & Co.	1862	1880	177	203/5 & 209
163.	Morecambe	Saltburn	1333.	4-4-0	"	"	1862	1888	177	206
164.	Belfast	Saltburn	1334.	4-4-0	"	"	1862	1882	177	207
165.	Keswick	Saltburn	1335.	4-4-0	"	"	1862	1886	177	208
166.	Oswald Gilkes	Woodlands	29.	2-4-0	W. Bouch	Kitchings	1860	1877	116	112
167.	Newlands	Panther	139.	0-6-0	W. Bouch	Gilkes, W.	1862	1889	143	
168.	Clifton	Panther	140.	0-6-0	"	"	1862	1891	143	
169.	Tufton	Panther	147.	0-6-0	"	"	1863	1895	143	150
170.	Reliance	Panther	158.	0-6-0	"	"	1864	1905	143	151
171.	Gladstone	Panther	3.	0-6-0	W. Bouch	Shildon W.	1863	1889	143	153
172.	Barrow	Panther	4.	0-6-0	"	"	1863	1905	143	153

No.	Name	Class	Works No.	Wheel Arr't	Designer	Builder	Year Rec'd	Year Disp'd	Page No.	Fig. No.
173.	London	Panther	5.	0-6-0	W. Bouch	Shildon W.	1864	1904	143	153
174	John Dixon	Panther	6.	0-6-0	"	"	1864	1908	143	153
175.	Contractor	individual	1.	0-6-0	W. Bouch	N.R.W.	1864	1908	143	154
176.	Windsor	eponymous	1571.	0-6-0	W. Bouch	R. S. & Co.	1865	1907	144	155/6
177.	Osborne	Windsor	1572.	0-6-0	"	"	1865	1907	144	155/8
178.	Balmoral	Windsor	1573.	0-6-0	"	"	1865	1907	144	155/6
179.	Edinburgh	Windsor	1574.	0-6-0	"	"	1865	1907	144	155/6
180.	Dublin	Windsor	1575.	0-6-0	"	"	1865	1910	144	155/6
181.	Malton	Windsor	1576.	0-6-0	"	"	1865	1910	144	155/6
182.	Elton	Windsor	1577.	0-6-0	"	"	1865	1908	144	155/9
183.	Acklam	Windsor	1578.	0-6-0	"	"	1866	1909	144	155/6
184.	Lark	Windsor	1579.	0-6-0	W. Bouch	R. S. & Co.	1866	1906	144	155/6
185.	Swallow	Windsor	1580.	0-6-0	"	"	1866	1907	144	155/6
186.	Union	Windsor	2.	0-6-0	W. Bouch	N.R.W.	1865	1903	144	
187.	Iron Age	Panther	186.	0-6-0	W. Bouch	Gilkes, W.	1864	1902	143	
188.	Lily	Panther	187.	0-6-0	"	"	1865	1893/4	143	
189.	Spring	Panther	7.	0-6-0	W. Bouch	Shildon W.	1865	1906	144	
190.	Summer	Panther	8.	0-6-0	"	"	1865	1908	144	161/2
191.	Autumn	Windsor	11.	0-6-0	W. Bouch	N.R.W.	1866	1909	144	160
192.	Winter	Peel	14.	0-6-0	W. Bouch	N.R.W./ Shildon	1866	1908	149	165
193.	Princess	Peel	15.	0-6-0	"	"	1866	1909	149	
194.	Alice	Peel	16.	0-6-0	"	"	1866	1905	149	166
195.	Helena	Peel	17.	0-6-0	"	"	1866	1910	149	
196.	Roseberry	eponymous	1691.	0-6-0ST	W. Bouch	R. S. & Co.	1866	1909	192	230/2
197.	Kildale	Roseberry	1692.	0-6-0ST	"	"	1866	1909	192	230/4
198.	Whitby	Roseberry	1693.	0-6-0ST	"	"	1866	1906	192	230/5
199.	Escomb	Roseberry	1694.	0-6-0ST	"	"	1866	1911	192	231

No.	Name	Class	Works No.	Wheel Arr't	Designer	Builder	Year Rec'd	Year Disp'd	Page No.	Fig. No.
200.	Eskdale	eponymous	229.	0-6-0	W. Bouch	Hopkins, G.	1867	1905	149	167
201.	Carlton	Eskdale	230.	0-6-0	"	"	1867	1908	149	167
202.	Ireland	N.R.W.min.	22.	0-6-0	W. Bouch	Shildon W.	1867	1905	150	
203.	England	N.R.W.min.	23.	0-6-0	"	"	1867	1909	150	
204.	unnamed	Eskdale	248.	0-6-0	W. Bouch	Hopkins, G.	1867	1908	149	167
205.	"	Eskdale	249.	0-6-0	W. Bouch	"	1867	1909	149	167
206.	"	Eskdale	250.	0-6-0	W. Bouch	"	1867	1906	149	167
207.	"	Enlarged N.R.W.min.	1410.	0-6-0	W. Bouch	Hawthorn	1867	1906	150	168
208.	"	" " "	1411.	0-6-0	"	"	1868	1905	150	
209.	"	" " "	1412.	0-6-0	"	"	1868	1906	150	
210.	"	" " "	1413.	0-6-0	"	"	1868	1906	150	169
211.	"	" " "	1414.	0-6-0	W. Bouch	Hawthorn	1868	1907	150	
212.	"	" " "	1415.	0-6-0	"	"	1868	1908	150	
213.	"	" " "	1416	0-6-0	"	"	1868	1910	150	
214.	"	" " "	1417.	0-6-0	"	"	1868	1908	150	170
215.	"	" " "	1418.	0-6-0	"	"	1868	1907	150	
216.	"	" " "	1419.	0-6-0	"	"	1868	1910	150	
217.	"	" " "	1420.	0-6-0	"	"	1868	1906	150	171
218.	"	" " "	1421	0-6-0	"	"	1868	1903	150	
219.	"	" " "	34.	0-6-0	W. Bouch	N.R.W.	1869	1909	150	
220.	"	" " "	35.	0-6-0	"	"	1869	1907	150	
221.	"	" " "	1486.	0-6-0	W. Bouch	Hawthorn	1870	1908	150	172
222.	"	" " "	1487.	0-6-0	"	"	1870	1909	150	173
223.	"	" " "	1488.	0-6-0	"	"	1870	1906	150	
224.	"	" " "	1489.	0-6-0	"	"	1870	1908	150	
225.	"	" " "	1490.	0-6-0	"	"	1870	1907	150	
226.	"	" " "	271.	0-6-0	W. Bouch	Hopkins, G.	1870	1907	154	
227.	"	" " "	272.	0-6-0	"	"	1870	1907	154	186

No.	Class	Works No.	Wheel Arr't	Designer	Builder	Year Rec'd	Year Disp'd	Page No.	Fig. No.
228.	Enlarged N.R.W.min.	273.	0-6-0	W. Bouch	Hopkins, G.	1871	1909	154	
229.	" "	274.	0-6-0	"	"	1871	1909	154	187/8
230.	" "	275.	0-6-0	"	"	1871	1909	154	189
231.	" "	62.	0-6-0	W. Bouch	N.R.W.	1870	1907	154	190
232.	" "	63.	0-6-0	"	"	1870.	1906	154	190
233.	" "	64.	0-6-0	"	"	1870	1912	154	190
234.	"	65.	0-6-0	"	"	1872	1906	154	190
235.	" "	66.	0-6-0	"	"	1872	1906	154	190
236.	" "	67.	0-6-0	"	"	1872	1906	154	190
237.	" "	68.	0-6-0	"	"	1872	1907	154	190
238.	'Ginx's Babies'	69.	4-4-0/ 2-4-0	W. Bouch	N.R.W.	1871	1909	178	210/2
239.	" "	70.	" "	W. Bouch	N.R.W.	1872	1910	178	
240.	" "	71.	" "	"	"	1872	1910	178	213/5
241.	" "	72.	" "	"	"	1872	1909	178	216
242.	N.R.W.min.	73.	0-6-0	W. Bouch	N.R.W.	1872	1909	154	190
243.	"	74.	0-6-0	"	"	1873	1908	154	190
244.	"	75.	0-6-0	"	"	1873	1908	154	190
245.	"	76.	0-6-0	"	"	1873	1907	154	190
246.	"	293.	0-6-0	W. Bouch	Hopkins, G.	1872	1910	154	190
247.	"	294.	0-6-0	"	"	1873	1907	154	190
248.	"	295.	0-6-0	"	"	1873	1912	154	190
249.	"	296.	0-6-0	"	"	1873	1912	154	190/2
250.	"	77.	0-6-0	W. Bouch	N.R.W.	1873	1912	154	190
251.	"	78.	0-6-0	"	"	1873	1907	154	190
252.	"	79.	0-6-0	"	"	1873	1912	154	190
253.	"	80.	0-6-0	"	"	1873	1907	154	190/3
254.	"	81.	0-6-0	"	"	1873	1908	154	190

No.	Class	Works No.	Wheel Arr't	Designer	Builder	Year Rec'd	Year Disp'd	Page No.	Fig. No.
255.	N.R.W.min.	82.	0-6-0	W. Bouch	N.R.W.	1873	1921	154	190
256.	"	83.	0-6-0	"	"	1873	1906	154	190/4
257.	"	327.	0-6-0	W. Bouch	Hopkins, G.	1873	1910	154	190
258.	"	328.	0-6-0	"	"	1873	1906	154	190
259.	"	329.	0-6-0	"	"	1874	1915	154	190
260.	"	330.	0-6-0	"	"	1874	1906	154	190
261.	"	87.	0-6-0	W. Bouch	N.R.W.	1874	1907	154	190
262.	"	349.	0-6-0	W. Bouch	Hopkins, G	1875	1907	154	190
263.	"	350.	0-6-0	"	"	1875	1917	154	190
264.	"	351.	0-6-0	"	"	1875	1907	154	190
1265.	'Ginx's Babies'	89.	4-4-0/ 2-4-0	W. Bouch	N.R.W.	1874	1914	178	
1266.	" "	90.	" "	"	"	1874	1914	178	
1267.	" "	91.	" "	"	"	1874	1914	178	
1268.	" "	92.	" "	"	"	1874	1914	178	217/8
1269.	" "	93.	" "	"	"	1874	1914	178	219/ 220
1270.	" "	94.	" "	"	"	1874	1914	178	221/2
1271.	eponymous	704.	0-6-0	W. Bouch	Dübs & Co.	1874	1912	196	
1272.	1271.	705.	0-6-0	"	"	1874	1915	196	
1273.	1271.	706.	0-6-0	"	"	1874	1915	196	
1274.	1271.	707.	0-6-0	"	"	1874	1913	196	
1275.	1271.	708.	0-6-0	"	"	1874	1925 (preserved)	196	239-244
1276.	1271.	709.	0-6-0	"	"	1874	1907	196	245
1277.	1271.	710.	0-6-0	"	"	1874	1914	196	
1278.	1271.	711.	0-6-0	"	"	1874	1908	196	
1279.	1271.	712.	0-6-0	"	"	1874	1907	196	
1280.	1271.	713.	0-6-0	"	"	1874	1909	196	

No.	Class	Works No.	Wheel Arr't	Designer	Builder	Year Rec'd	Year Disp'd	Page No.	Fig. No.
1281.	1271.	1003.	0-6-0	W. Bouch	Avonside	1874	1915	196	
1282.	1271.	1004.	0-6-0	"	"	1874	1917	196	
1283.	1271.	1005.	0-6-0	"	"	1874	1907	196	
1284.	1271.	1006.	0-6-0	"	"	1874	1908	196	
1285.	1271.	1007.	0-6-0	"	"	1874	1908	196	
1286.	1271.	1008.	0-6-0	"	"	1874	1921	196	
1287.	1271.	1009.	0-6-0	"	"	1875	1921	196	
1288.	1271.	1010.	0-6-0	"	"	1875	1907	196	
1289.	1271.	1011.	0-6-0	"	"	1875	1921	196	
1290.	1271.	1012.	0-6-0	"	"	1875	1915	196	
1291.	Long w.b.		0-6-0	W. Bouch	N.R.W.	1875	1909	197	248-250
1292.	1037.	117.	0-6-0T	W. Bouch	N.R.W.	1876	1910	194	
1293.	1037.	119.	0-6-0T	"	"	1876	n.k. (sold)	194	236/7
1294.	1037.	121.	0-6-0T	"	"	1876	1911	194	238

Alphabetical List of Named Locomotives

Name	No.	Type	Date	Name	No.	Type	Date
Aberdeen	73	0-6-0	1852	Alice	194	0-6-0	1866
Acklam	183	0-6-0	1866	Antelope	46	2-2-0	1844
Active	48	0-4-2	1844	Appleby	132	0-6-0	1858
Adelaide	20	0-6-0	1832	Arrow	26	2-2-2	1837
Alarm	70	2-4-0	1851	Auckland	10	0-6-0	1839
Albert	84	0-6-0	1854	Australia	91	0-6-0	1855
Alert	22	0-6-0	1848	Autumn	191	0-6-0	1866
Alexander	77	0-6-0	1854	Ayton	99	2-4-0	1855
Balmoral	178	0-6-0	1865	Belle Vue	123	0-6-0	1857
Baring	75	0-6-0	1853	Birkbeck	63	0-6-0	1849
Barnard Castle	107	0-6-0	1856	Black Diamond	3	0-4-0	1826
Barrow	172	0-6-0	1863	" "	3	0-6-0	1830
Bedlington	2	0-4-0	1825	Briton	12	0-6-0	1837
"	2	0-6-0	1831	Brookside	124	0-6-0	1857
Beehive/Bee	11	0-6-0	1837/42	Brough	120	0-6-0	1856
Beelah	138	0-6-0	1858	Brougham	160	4-4-0	1860
Belfast	164	4-4-0	1862	Brunswick	68	2-4-0	1850
Camel	113	0-6-0	1856	Comet	52	2-2-2	1840/1
Carlisle	79	2-4-0	1854	Commerce	35	0-6-0	1847
Carlton	201	0-6-0	1867	Coronation	13	0-6-0	1831
Chittapratt	3rd	0-4-0	1825	Conside	28	0-6-0	1845
Clarendon	69	2-4-0	1850	Contractor	175	0-6-0	1864
Cleveland	60	0-6-0	1848	Cotherstone	102	0-6-0	1855
"	21	0-6-0	1865	Crossfell	139	0-6-0	1859
Clifton	168	0-6-0	1862				
Darlington	19	0-6-0	1832	Diligence	4	0-6-0	1834
"	103	0-6-0	1855	Director	16	0-6-0	1832
Dart	41	0-4-0	1840	Dispatch	6	0-6-0	1839
Deanery	88	0-4-0	1854	Driver	34	0-6-0	1846
Deepdale	137	0-6-0	1858	Dublin	180	0-6-0	1865
Derwent	25	0-6-0	1845	The Duke	80	0-6-0	1854
Diligence	4	0-4-0	1826	Durham	104	0-6-0	1855
Eagle	108	0-6-0	1856	Emperor	74	0-6-0	1853
Earl Grey	21	0-6-0	1832	England	203	0-6-0	1867
East Mount	128	0-6-0	1858	Enterprise	25	0-6-0	1835
Eden	135	0-6-0	1858	Escomb	199	0-6-0ST	1866
Edinburgh	179	0-6-0	1865	Eskdale	200	0-6-0	1867
Edward Pease	114	2-4-0	1856	Etherley	18	0-6-0	1840
Eldon	32	0-6-0	1846	Excelsior	141	0-6-0	1859
Elephant	95	0-6-0	1855	Experiment	6	0-4-0	1827
Elm Field	118	2-4-0	1857	"	6	0-6-0	1828
Elton	182	0-6-0	1865				
Falcon	109	0-6-0	1856	Fryerage	87	0-4-0	1854
Fox	149	0-6-0	1860				

Name	No.	Type	Date	Name	No.	Type	Date
Gainford	97	0-6-0	1855	Globe	9	0-4-0	1830
Gannymede	45	0-4-0	1844	Graham	2	0-6-0	1849
Gazelle	125	0-6-0	1858	Greta	119	0-6-0	1856
Gem	37	0-6-0	1847	Guisbro'	36	0-6-0	1847
Gladstone	171	0-6-0	1863				
Hackworth	71	2-4-0	1851	Herschel	156	0-6-0	1861
Hallgarth	59	2-4-0	1848	Hope	2	0-4-0	1825
Hardinge	85	0-6-0	1854	"	2	0-6-0	1835
Hawk	110	0-6-0	1856	Hopetown	5	0-6-0	1841
Hawthorn	82	0-6-0	1854	Huddersfield	89	0-4-0	1854
Helena	195	0-6-0	1866	Hutton	130	0-6-0	1858
Iron Age	187	0-6-0	1864	John Dixon	174	0-6-0	1864
Ireland	202	0-6-0	1867	John Wakefield	126	0-6-0	1858
				Jupiter	154	0-6-0	1860
Kendal	134	0-6-0	1858	Kildale	197	0-6-0ST	1866
Kestrel	122	0-6-0	1857	Kirkby Stephen	133	0-6-0	1858
Keswick	165	4-4-0	1862				
Larchfield	64	0-6-0	1849	London	42	0-4-0	1839
Lark	184	0-6-0	1866	"	173	0-6-0	1864
Lartington	116	2-4-0	1856	Long Hull	131	0-6-0	1858
Leader	8	0-6-0	1842	Lonsdale	78	2-4-0	1854
Leopard	147	0-6-0	1860	Lord Brougham	17	0-6-0	1831
Lily	188	0-6-0	1865	Lord Durham	22	0-6-0	1832
Lion	112	0-6-0	1856	Lowther	161	4-4-0	1860
Locomotion	1	0-4-0	1825	Lune	158	0-6-0	1861
Magnet	24	0-6-0	1835	Mercury	151	0-6-0	1860
Majestic	12	0-6-0	1831	Meteor	50	2-2-2	1842
Malton	181	0-6-0	1865	Meynell	115	2-4-0	1856
Manchester	53	2-2-2	1840	Middlesbro'	9	0-6-0	1839
Mars	153	0-6-0	1860	Middleton	96	0-6-0	1855
Marske	101	2-4-0	1855	Miller	81	0-6-0	1854
Mastiff	150	0-6-0	1860	Miner	29	0-6-0	1845
Merchant	90	0-6-0	1855	Morecambe	163	4-4-0	1862
Newlands	167	0-6-0	1862	Normanby	20	0-6-0	1865
Newmarket?	65	2-4-0	1850	North Star	11	2-2-0	1830
Neptune	94	2-2-2	1855	Northumbrian	15	0-6-0	1831
"	94	0-6-0	1867	Nunthorpe	117	2-4-0	1856
Ocean	28/13/12	0-6-0	1840	Ostrich	146	0-6-0	1860
Orion	67	2-4-0	1848	Oswald Gilkes	166	2-4-0	1860
Osborne	177	0-6-0	1865				
Panther	145	0-6-0	1860	Planet	157	0-6-0	1861
Peel	72	0-6-0	1852	Polam	127	0-6-0	1858
Penrith	140	0-6-0	1859	Priam	66	2-4-0	1847
Pierremont	98	2-4-0	1855	Prince of Wales	76	0-6-0	1854
Pilot	26	0-6-0	1840	Prince	7	0-6-0	1843
Planet	10	2-2-0	1830	Princess	193	0-6-0	1866
Queen	29	0-4-0/2	1837				

Name	No.	Type	Date
Raby Castle	30	2-2-2	1839
Redcar	31	0-6-0	1845
Reliance	170	0-6-0	1864
Rocket	7	0-6-0	1829
Rokeby	38	2-4-0	1847
Roseberry	196	0-6-0ST	1866
Royal George	5	0-6-0	1827
Ruby	39	2-4-0	1847
Saltburn	162	4-4-0	1862
Saturn	155	0-6-0	1861
Shildon	18	0-6-0	1831
"	33	0-6-0	1846
Shotley	56	0-6-0	1852
Skelton Castle	24	0-6-0	1846
Southend	62	0-6-0	1849
Spring	189	0-6-0	1865
Stag	111	0-6-0	1856
Staindrop	106	0-6-0	1856
Stanhope	16	0-6-0	1845
Stanley	129	0-6-0	1858
Star	61	0-6-0	1849
Stephenson	65	2-4-0	1850
Stobart	100	2-4-0	1855
Stockton	4	0-6-0	1841
Summer	190	0-6-0	1865
Sun	44	2-2-0	1844
Sunbeam	28	2-2-0	1837
Swallow	185	0-6-0	1866
Swift	27	0-4-0	1836
Tebay	136	0-6-0	1858
Tees	14	0-6-0	1837
Teesdale	105	0-6-0	1855
Tiger	121	0-6-0	1857
Times	3	0-6-0	1849
Tory	15	0-6-0	1838
Tow Law	57	0-6-0	1853
Trader	12	0-6-0	1842
Tufton	169	0-6-0	1863
Tweed	19	0-6-0	1865
Tyneside	54	2-2-2	1842
Ulleswater	144	0-6-0	1860
Unicorn	47	2-2-0	1844
Union	186	0-6-0	1865
Uranus	93	2-2-2	1855
"	93	0-6-0	1867
Venus	152	0-6-0	1860
Victoria	83	0-6-0	1854
Victory	8	0-6-0	1829
Wear	30	0-6-0	1845
Whig	17	0-6-0	1839
Whitby	198	0-6-0ST	1866
Wilberforce	23	0-6-0	1832/3
William IVth.	14	0-6-0	1831
Windermere	143	0-6-0	1859
Windsor	176	0-6-0	1865
Winter	192	0-6-0	1866
Witton Castle	27	0-6-0	1840
Wolsingham	55	0-4-2	1847
Woodlands	58	2-4-0	1848
Wycliffe	142	0-6-0	1859
York	159	0-6-0	1862
Yarm	92	0-6-0	1855
Zebra	148	0-6-0	1860
Zetland	86	0-6-0	1854

Appendix V. List of Lines & Opening Dates

Date of Act	Date of Opening	Line & Company	Remarks
19-4-1821	27-9-1825	Etherley, Witton Park Colliery, to Cottage Row, Stockton-on-Tees. 1st. S. & D.R. Act.	Witton Park Colliery to New Shildon, later abandoned after opening of alternative lines. At the time of Inness's series Albert Hill Junction to Oak Tree Junction & Bowesfield Junction to Cottage Row were in use for goods only.
19-4-1821	27-9-1825	Brusselton North Bankfoot to West Auckland. 1st. S.&D.R.Act.	Mineral line, later abandoned.
19-4-1821	27-9-1825	Darlington Junction to Darlington Depot (now Westbrook area) 1st. S. & D.R. Act.	Mineral line, later abandoned.
19-4-1821	17-10-1825	Urlay Nook to Yarm Depot. 1st. S. & D.R. Act.	Abandoned 1871.
19-4-1821	10-7-1827	Shildon Junction to Black Boy Colliery. 1st. S. & D.R. Act.	Mineral line, later abandoned.
23-5-1823	27-10-1829	Albert Hill Junction (see text) to Croft Wharf. 2nd. S.&D.R. Act.	Mineral line, most of it sold to the G.N. of E.R. 1836.
19-4-1821	1-5-1830	West Auckland to Evenwood. 1st. S. & D.R. Act.	West Auckland to Spring Gardens Junction. Opened for passenger traffic 1-2-1863.
17-5-1824	1-5-1830	Evenwood Junction to Cockfield Fell. 3rd. S. & D.R. Act.	Mineral line.
17-5-1824	2-10-1830	Cockfield Fell to Haggerleases Lane. 3rd. S. & D.R. Act.	Mineral line.
23-5-1828	27-12-1830	Bowesfield Junction to Middlesbrough Old Town (Newport). 4th. S&DR Act.	At the time of Inness's series was in use for goods only.
15-7-1837	19-4-1842	Shildon Junction to South Church (east of Bishop Auckland). Shildon Tunnel Co.	Worked by S. & D.R. Leased 1-10-1847.
No Act.	12-5-1842	Middlesbrough Old Town Junction to Middlesbrough Dock. Owners of the Middlesbrough Estate.	amalgamated 1849.
Not known.	c.1842	Junction north of Shildon Tunnel to Adelaide Colliery.	Not shown on 1861-63 O.S. map.
15-7-1837	8-11-1843	South Church to Crook Old Station. Bishop Auckland & Weardale Railway.	Worked by S. & D.R. Leased 1-10-1847.

Date of Act	Date of Opening	Line & Company	Remarks
No Act.	16-5-1845	Crook Old Station to Waskerley Junction. Weardale Extension Railway.	See earlier notes, Chap.1. Worked by S. & D.R. Incorporated by Act of 29-9-1847. Leased 1-10-1847. Opened for passenger traffic 1-4-1846. Crook Junction to Sunniside abandoned on opening of deviation line 1867.
No Act.	16-5-1845	Stanhope Lime Kilns to Carr House. Originally Stanhope & Tyne Railway. Wear & Derwent Railway from 1842. Wear & Derwent Junction Railway from early 1845.	See earlier notes, Chap.1. Leased early 1845. Purchased by S.&D.R. on the date given. Incorporated 29-9-1847. Opened for passenger traffic: Waskerley Junction to Cold Rowley 1-4-1846; Waskerley Junction to Whitehall Junction abandoned on opening of deviation line in 1859.
Not known	25-12-1845	Spur to lime kilns near South Church from Black Boy line.	Not shown on 1861-63 O.S. map.
21-7-1845	4-6-1846	Middlesbrough Dock Junction to Redcar Old Station. Middlesbrough and Redcar Railway.	Worked by S. & D.R., leased 1-10-1847. Redcar Junction to Redcar Old Station later left derelict, new station opened 1861 on Saltburn line.
31-7-1845	3-8-1847	Witton Junction to Frosterley. Wear Valley Railway.	Worked by S. & D.R., leased 1-8-1847. Mineral line.
31-7-1847	3-8-1847	Bishopley Junction to Bishopley Kilns. Wear Valley Railway.	All as above.
Not known.	c.1850	Connecting line from junction on Black Boy line to Shildon/South Church line just north of tunnel.	Probably also joined the private line to Eldon Colliery.
Not known.	c.1853	Allen's Curve; from junction on S. & D.R. to junction on Leeds Northern line north of Yarm.	Gave access to new Yarm depot and Leeds Northern line to Stockton. Original Yarm spur was not used afterwards but some of the buildings were not abandoned until 1871.
17-6-1852	11-11-1853	Guisborough Junction, at Middlesbrough, to Guisborough. Middlesbrough & Guisborough Ry.	Worked by S. & D.R. opened for passenger traffic 25-2-1854.
17-6-1852	11-11-1853	Low Cross, Guisborough, to Cod Hill. Middlesbrough & Guisborough Railway.	Worked by S. & D.R. Mineral line to Pease mines.
4-7-1854	8-7-1856	Hopetown Junction, Darlington, Barnard Castle Old Station. Darlington & Barnard Castle Ry.	Worked by S. & D.R.

Date of Act	Date of Opening	Line & Company	Remarks
3-7-1854	13-9-1856	Shildon Tunnel North Junction to West Auckland. S. & D.R. Act.	Mineral line. Fieldon Bridge to West Auckland opened for passenger traffic 1-2-1863.
16-7-1855	19-10-1858	Crook to Hedley Hill Junction. S. & D.R. Act.	Mineral line. Stanley Colliery to Hedley Hill Junction later abandoned.
16-7-1855	4-7-1859	Burnhill Junction to Whitehall Junction. S. & D.R. Act.	Waskerley deviation line.
Not known	before 1860	Parkgate Junction to Horton Road Junction	Known as Fighting Cocks curve. Apparently not used for regular services.
Not known	c.1861	Haugton Road Junction to Allen Street Junction.	Known as Ironstone curve.
Not known	Later	Albert Hill Junction to Allen Street Junction.	No records.
16-7-1855	8-8-1861	Charity Junction to Stooperdale Junction, Darlington. S.&D.R.Act.	Stooperdale curve. Inness gives 1859.
17-7-1857	26-3-1861	Barnard Castle Junction to Barras. South Durham & Lancashire Union Railway.	Worked by S. & D.R., opened for passenger traffic 8-8-1861.
17-7-1857	4-7-1861	Barras to Tebay. S.D. & L.U.R.	As above.
23-7-1858	17-8-1861	Redcar Junction to Saltburn. S. & D.R. Act.	
21-5-1858	12-4-1862	Kirkby Stephen to Clifton Junction Eden Valley Railway.	Worked by S. & D.R., opened for passenger traffic 7-6-1852.
28-6-1861	3-4-1862	Frosterley to Newlandside. Frosterley & Stanhope Junction Ry.	Worked by S. & D.R., opened for pass'r traffic 22-10-1862.
28-6-1861	22-10-1862	Newlandside to Stanhope Old Station. F. & S.J.R.	Worked by S. & D.R.
13-8-1859	1862	Bowesfield Junction to Hartburn Junction, Stockton. S.&D.R. Act.	Joining S. & D.R. to N.E.R. (old Leeds Northern) line to north.
17-7-1857	1-2-1863	Spring Gardens Junction to Barnard Castle Junction. S.D. & L.U.R.	Part of this, up to Cockfield Fell, had already been covered in the 3rd. Act.
23-7-1858	1-2-1863	Bishop Auckland to Fielden Bridge Junction (near West Auckland).	Linking line. Opened for passenger traffic 1-8-63.
Not known.	1-5-1863	Hartburn Junction to Bowesfield Junction.	Connecting curve between S. & D.R. main line to Stockton and Middlesbrough and the N.E.R. (old Leeds Northern) line north.

Date of Act	Date of Opening	Line & Company	Remarks
		Changes under N.E.R.	
Not known	5-9-1886	Junction north of Clifton to junction on Penrith to Keswick line.	Redhill curve, giving access north to Keswick.
1862	10-4-1867	Crook to Tow Law.	Deviation line; Sunniside Incline then abandoned except for access to Pease controlled pits.
1883	1-7-1887	Polam Junction to Oak Tree Junction.	Deviation line, including new station at Dinsdale. Old S.&D. R. main line through Fighting Cocks no longer used for pass. traffic, but survived for minerals, goods, and to serve several factories etc. until 1964 and very little of the original line now survives, although most of Fighting Cocks Station is still there, a private house.
		Other Lines involved and Later Additions of Importance	
		North Yorkshire & Cleveland Railway	
2-7-1855	3-3-1857	Picton to Stokesley, and Whorlton branch.	Whorlton branch was mineral line.
	1-2-1858	Stokesley to Ingleby (Greenhow).	
	6-4-1858	Ingleby to Kildale.	Ingleby Mining Co. private line bought by N.E.R. almost immediately.
8-8-1859		Taken over by N.E.R.	
	1-4-1861	Kildale to Castleton.	
	1-6-1864	Battersby Junction to Nunthorpe Junction.	Joining old Middlesbrough & Guisborough Railway.
	2-10-1865	Castleton to Grosmont.	Joining old Whitby & Pickering Railway.
		Ingleby Mining Co.	
	6-4-1858	Battersby Junction to Ingleby (Rudd Scar) Mine.	Bought almost immediately by N.Y. & C.R. Line to Ingleby J'n rebuilt to standard gauge.
		Rosedale Branch	
8-8-1859		Taken over by N.E.R.	Continued line to Bankfoot.
	27-3-1861	Ingleby to Rosedale West.	
	18-8-1865	Blakey Junction to Rosedale East.	

Date of Act	Date of Opening	Line & Company	Remarks
		Tees Valley Railway	
19-6-1865	13-5-1868	Tees Valley Junction to Middleton-in-Teesdale.	
19-6-1882		Taken over by N.E.R.	
		Cleveland Railway	
23-7-1858	23-11-1861	Cargo Fleet Jetty to Slapewath.	Crossed Middlesbrough & Guisborough Railway.
	1862	Slapewath to Boosbeck.	
	c.1862	Short connecting line to M. & G.R. (Chaloner Branch.)	Guisborough Junction.
	23-2-1865	Boosbeck to Brotton.	
	21-4-1865	Brotton to Skinningrove, and the Skinningrove (Loftus Mines) zigzag.	Opened for passenger traffic 1-11-1875.
5-7-1865		Taken over by N.E.R.	Line from Normanby Mines to junction with old M. & G.R. abandoned.
	27-5-1867	Skinningrove to Loftus.	
	1-6-1872	Saltburn West Junction to North Skelton Junction and on to join Brotton line from Boosbeck.	Opened for passenger traffic 1-4-1875.
	Jan. 1873	Priestcroft Junction to North Skelton Junction.	
	1-11-1878	Guisborough to Brotton opened for passenger traffic.	
		Whitby, Redcar & Middlesbrough Union Railway	
16-7-1866		Taken over by N.E.R. before any part opened.	
19-7-1875	3-12-1883	Whitby to Loftus.	
		Wear Valley Extension Railway	
20-6-1892		Taken over by N.E.R. before completion.	
31-7-1894	21-10-1895	Frosterley to Wearhead, including new Stanhope Station.	Old station believed used as goods depot for a while.

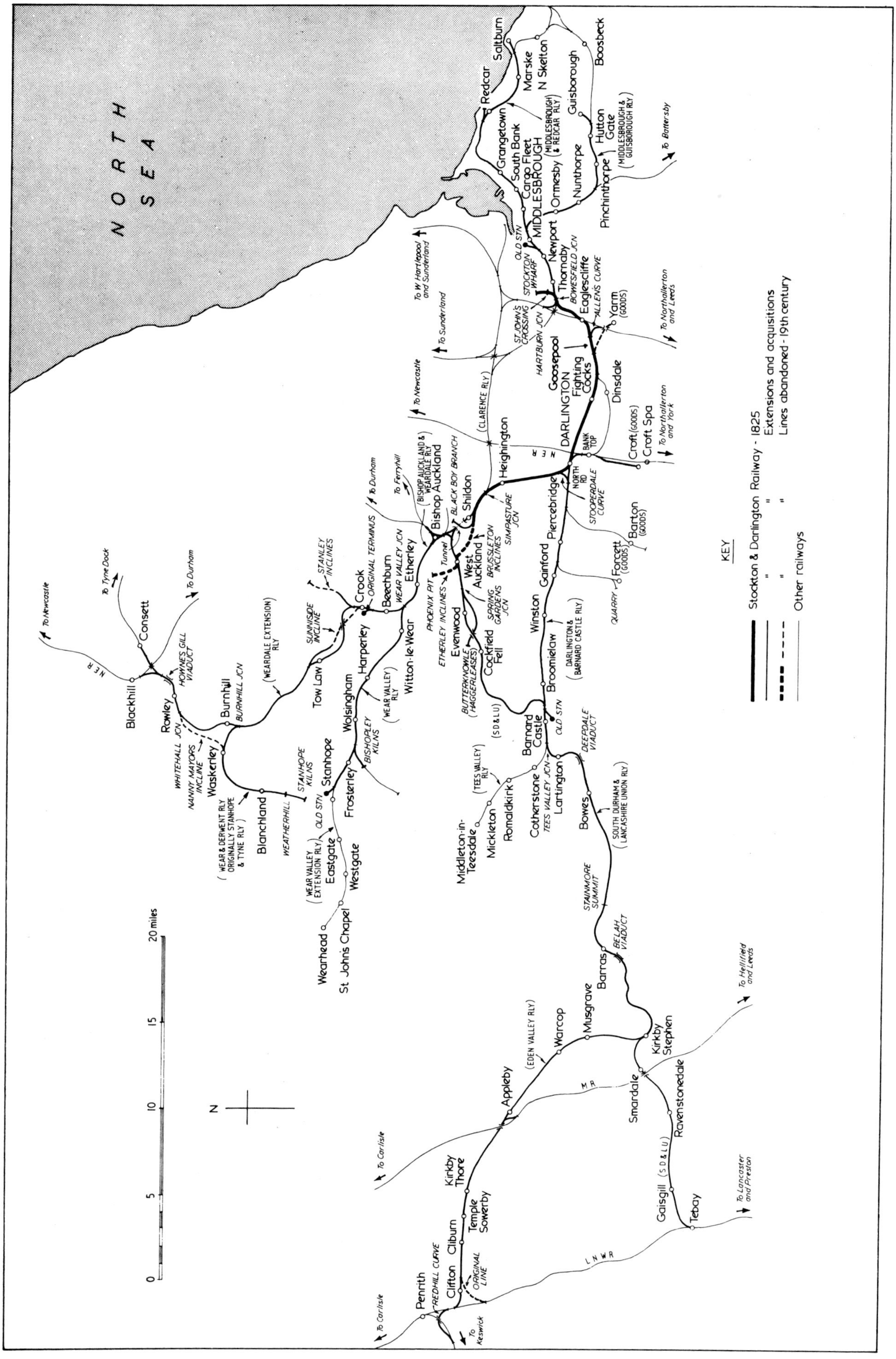
NORTH SEA
Saltburn
Redcar
Marske
N Skelton
Boosbeck
Guisborough
Hutton Gate
(MIDDLESBROUGH & GUISBOROUGH RLY)
(MIDDLESBROUGH & REDCAR RLY)
To Battersby
Grangetown
South Bank
Cargo Fleet
MIDDLESBROUGH
Ormesby
Nunthorpe
Pinchinthorpe
OLD STN
Newport
BOWESFIELD JCN
Thornaby
STOCKTON WHARF
Eaglescliffe
ALLENS CURVE
Yarm (GOODS)
To Northallerton and Leeds
To W Hartlepool and Sunderland
To Sunderland
ST JOHN'S CROSSING
HARTBURN JCN
Goosepool
Fighting Cocks
Dinsdale
To Northallerton and York
To Newcastle
(CLARENCE RLY)
DARLINGTON
BANK TOP
Croft (GOODS)
Croft Spa
Heighington
N E R
NORTH RD
STOOPERDALE CURVE
Barton (GOODS)
BLACK BOY BRANCH
(BISHOP AUCKLAND & WEARDALE RLY)
Bishop Auckland
To Shildon
SIMPASTURE JCN
Piercebridge
Gainford
Forcett (GOODS)
QUARRY
To Ferryhill
To Durham
STANLEY INCLINES
ORIGINAL TERMINUS
WEAR VALLEY JCN
Beechburn
Etherley
Crook
Tunnel
West Auckland
BRUSSLETON INCLINES
PHOENIX PIT
ETHERLEY INCLINES
SPRING GARDENS JCN
Winston
Evenwood
Cockfield Fell
(DARLINGTON & BARNARD CASTLE RLY)
Broomielaw
BUTTERKNOWLE (HAGGERLEASES)
(S D & L U)
Barnard Castle
OLD STN
DEEPDALE VIADUCT
SUNNISIDE INCLINE
Tow Law
Harperley
Witton-le-Wear
Wolsingham
(WEAR VALLEY RLY)
BISHOPLEY KILNS
(WEARDALE EXTENSION RLY)
To Tyne Dock
To Durham
To Newcastle
Consett
N E R
HOWNES GILL VIADUCT
Blackhill
Rowley
WHITEHALL JCN
NANNY MAYORS INCLINE
Burnhill
BURNHILL JCN
Waskerley
STANHOPE KILNS
Stanhope
Frosterley
OLD STN
Blanchland
WEATHERHILL
(WEAR & DERWENT RLY ORIGINALLY STANHOPE & TYNE RLY)
(WEAR VALLEY EXTENSION RLY)
Eastgate
Westgate
Wearhead
St John's Chapel
(TEES VALLEY RLY)
Romaldkirk
Mickleton
Middleton-in-Teesdale
Cotherstone
TEES VALLEY JCN
Lartington
Bowes
(SOUTH DURHAM & LANCASHIRE UNION RLY)
STAINMORE SUMMIT
BELAH VIADUCT
Barras
Musgrave
Warcop
(EDEN VALLEY RLY)
Kirkby Stephen
To Hellifield and Leeds
M R
Appleby
Smardale
Ravenstonedale
(S D & L U)
Gaisgill
Tebay
To Lancaster and Preston
To Carlisle
Kirkby Thore
Temple Sowerby
Cliburn
Clifton
ORIGINAL LINE
REDHILL CURVE
Penrith
To Keswick
To Carlisle
L N W R
0 5 10 15 20 miles
N
KEY
Stockton & Darlington Railway - 1825
" " " Extensions and acquisitions
" " " Lines abandoned - 19th century
Other railways

Acknowledgements and Attributions

This book is very largely based on a series of 38 illustrated articles by R. H. Inness published anonymously in "The Locomotive" magazine between February 1925 and August 1930, but additional information has been gathered from many sources, and a list of the books consulted is appended following the text.

However, particular acknowledgement is due to the following:- the late Mr. E. A. Forward, of the Science Museum, South Kensington, and his articles in the series "Links in the History of the Locomotive", published in the "Engineer" magazine in 1940, 1941 and 1943; the late Mr. C. F. Dendy-Marshall and his books published under the banner of the Locomotive Publishing Co.; the late Mr. J. G. H. Warren and his excellent book on the Stephensons' family business; the late Mr. Robert Young of the Hackworth family; the Seguin family of France, and the various manuscript notebooks consulted by courtesy of various institutions, listed later.

Also, of course, Mr. Theodore West, because in a roundabout way he started me off on this fascinating exploration, and the late Mr. Ken Hoole for pointing me in the right direction and for putting up with a flow of questions afterwards. His recent death is a tragedy for all railway historians.

The late Mr. Brian Reed, whose monograph "Loco Profile No.25" on "Locomotion" No.1 is probably the best and most comprehensive work on that engine, deserves a lot of credit, and Mr. Alan Prior, whose drawings of 19th. century locomotives have helped so much is also to be thanked, especially for the jacket design using his paintings.

Mr. Tony Hall-Patch, lately of the Science Museum, for reading and criticism, and for the Introduction.

In May 1993 a binder containing 120 pages of typed notes, forming the draft of a substantial book, came to light in the Ken Hoole Study Centre, North Road Station Railway Museum, Darlington. This solved several puzzles regarding numbering of engines, and gave much additional information. Ken Taylor then found and loaned to me a binder full of m.s. notes by the late Mr. E. Craven, which were obviously the basis of the above typescript. These contain a comprehensive summary of the various relevant Committee Minutes wherever they refer to the locomotives. Sometime later Colin Foster found and loaned to me a further file containing more original m.s. notes by E. Craven and many by Ken Hoole, all of which has helped to fill out my earlier material.

The Kitching and Whessoe papers at the Durham County Records Office have also been of great help.

There are many others - some reference sources I have not yet managed to sight at all. Most of them are long out of print but worth looking for. Mr. John H. Proud's library has been at my disposal and I am greatly obliged to him in this and other ways. Mr. Joe Clarke, historian of my own old Institution and, at the time, an official of the Newcomen Society, has been more than kind in suppling me with photocopies of pages from old periodicals which I could not have managed to get so easily myself.

If I have neglected to acknowledge or thank anyone whose work I have used I apologize; it will be due more to being unable to trace them than from deliberate omission. If any such care to contact me I will try to make amends.

Special thanks are due to the following for their help and encouragement:

The North Eastern Railway Association
The Ken Hoole Trust
My son, Thomas F. Pearce
The late Ken Hoole, N.E.R.A.,
John H. Proud, N.E.R.A. (also for proof-reading)
John M. Fleming, N.E.R.A.
Joe F. Clarke, Marine Historian.
Tom Eden, Darlington R.P.S. & Friends of the N.R.S.M.
Jane Hackworth-Young.
M. Pierre Seguin.
Tony Hall-Patch, lately of the Science Museum.
David Jenkinson.
Peter Holmes.
David Baxter.
Ray Gordon, Shildon.
The late Alan Russell, Yarm.
John Liffen, Science Museum, South Kensington.
Alan Pearce, Hackworth Museum, Shildon.
Paul Talbot, Tyne & Wear Museum Service,
Maurice Kirby, Lancaster University.
Reg. Sowler, N.E.R.A., and his son Nigel.
Miss Muriel Ablett and Mrs. Bee Robertson for proof reading.
My son-in-law, Bill Colombi, for electronic aid & advice.
My dear Wife, for living with it and encouraging it for more than seven years!

Acknowledgement for use of text and illustrations:

Ian Allan Ltd. (including the Locomotive Publishing Co. and Real Photographs Ltd.: these Photographs now all transferred to the N.R.M.)
Ken Hoole, N.E.R.A.
John H. Proud, N.E.R.A.
John M. Fleming, N.E.R.A.
Ken Taylor, N.E.R.A.
Colin Foster, N.E.R.A.

The Science Museum, South Kensington (John Liffen).
The Public Record Office, Kew.
Jane Hackworth-Young and the Hackworth Society.
The Hackworth Museum, Soho Cottages, Shildon (Mr. Alan Pearce)
The Stephenson Railway Museum, North Shields (Mr. Paul Talbot at the time)
The University of Newcastle-upon-Tyne.
The "Engineer" magazine.
David & Charles.
The Cleveland Institution of Engineers.
The Newcomen Society.
The University of London (Goldsmiths' Library)
The Institution of Mechanical Engineers.
David Baxter/Moorland Publishing Co.
The N.E.R.A. Library.
Alan Prior.
Dr. P. J. Holmes.
Mrs. J. B. Reed and Profile Books Ltd.
The British Railway Journal (Mr.Paul Karau)
The National Railway Museum Library (Mr. C. P. Atkins)
North Road Station Railway Museum, Darlington (Mr. Steven Dyke)
The Ken Hoole Study Centre, Darlington (Mrs. Ann Wilson)
Darlington Branch Library, Reference Library and Local History Section (Mrs. C. Copeland)
Durham County Records Office (Miss J. Gill)
Stockton Reference Library (Mrs. Joyce Chesney)
Nunthorpe Branch Library (Miss Judith Hanson)
Darlington Railway Preservation Society.
The "George & Dragon" Inn, Yarm (Mr. Chris. Neil in 1993)
Dr. Alan Earnshaw
The late Mr. Laurie Ward, N.E.R.A.

Sources of Illustrations

I have tried to trace the copyright holder of all those items still covered, but not always with success. Will any person whose illustrations I have used without permission please accept that it was not intentional, and contact me if required?
In many cases the same illustration has been used in several publications, or, for instance, is the property of the Science Museum originally, in which case I have tried to give both.
Instead of giving the attribution in the caption to each illustration, I am listing them below with the number of illustrations and the Fig. Nos. concerned

Author and author's collection: 64 items, Fig. Nos.11, 12, 13, 23, 51, 59, 70, 78, 86, 91, 93, 99, 103, 108, 115, 123, 124, 126, 127, 130, 131, 132, 133, 135, 136, 138, 142, 143, 144, 146, 149, 150, 151, 156, 165, 169, 174, 175, 176, 177, 178, 180, 184, 185, 192, 195, 196, 208, 209, 212, 213, 216, 218, 219, 220, 221, 224, 228, 229, 235, 240, 244, 245, 246

W. H. Inness and "Locomotive Magazine": 29 items, Fig. Nos.9, 10, 25, 83, 89, 90, 92, 100, 101, 102, 112, 113, 118, 121, 122, 128, 145, 148, 153, 163, 164, 167, 187, 190, 210, 211, 231, 232, 239.

"Locomotive Magazine": 25 items, Fig. Nos.54, 125, 134, 140, 141, 147, 152, 158, 162, 168, 173, 188, 189, 191, 193, 215, 225, 227, 234, 236, 237, 243, 248, 250, and portrait of William Bouch.

"Engineer Magazine" (Ahrons, Dunlop, Forward, Graham): 19 items, Fig. Nos.6, 7, 18, 19, 20, 22, 31, 32, 47, 54, 58, 68, 73, 74, 85, 87, 116, 120, and Graham Nov.1840 list.

Robert Young, "Timothy Hackworth & the Locomotive": 4 items, Fig. Nos.8, 39, 69 and portrait of Timothy Hackworth (original now in the Soho Cottage Hackworth Railway Museum, Shildon).

A. L. Deghilage, "Origine de la Locomotive", and Robert Young: 1 item, Fig. No.71.

E. L. Ahrons, "The British Railway Steam Locomotive - 1825-1925": 2 items, Fig. No.3, and drawing of Bouch's Blast Pipe (originally published in "Engineering")

J. G. H. Warren, "A Century of Locomotive Building by Robert Stephenson & Co. - 1823- 1923": 11 items, Fig. Nos.5, 16, 34, 35, 36, 37, 38, 41, 44, 45, 203.

Zerah Colburn, "Locomotive Engineering & the Mechanism of Railways": 3 items, Fig. Nos.48, 204 (ex Warren), 205.

Theodore West, Sketch Sheets: 10 items, Fig. Nos.33, 43, 56, 77, 82, 88, 97, 119 and two reduced size sketch sheets (Appendix III). Some of these items are from the Ken Hoole Study Centre full size sheets.

National Railway Museum: 11 items, Fig. Nos.61, 64, 80, 84, 94, 95, 107, 117, 154, 166, 249 (from author's copy negatives.

Locomotive Publishing Co., Real Photographs Ltd., (but now in the National Railway Museum): 9 items, Fig. Nos.60, 105, 129, 157, 160, 183, 197, 207, 226.

Ken Hoole collection: 7 items, Fig. Nos.130, 161, 181, 182, 206, 238, 241.

Laurie Ward, N.E.R.A. and Laurie Ward collection: 8 items, Fig. Nos.57, 79, 141, 155, 201, 202, 230, 242.

Kenneth L. Taylor collection: 9 items, Fig. Nos.62, 106, 111, 159, 170, 171, 179, 186, 233.

Alan Prior, "19th. Century Railway Drawings in 4 mm. Scale": 4 items, Fig. Nos.14, 30, 49, 65.

Samuel Smiles, "Story of the Life of George Stephenson": 2 items, portrait of George Stephenson, and Forth Street Works, Newcastle-upon-Tyne.

Northern Echo: 2 items, Fig. No.63 and photograph of Soho Works, Shildon, in 1925.

Tyne & Wear Museum Service and University of Newcastle-upon-Tyne: 1 item, Fig. No.15.

Newcomen Society: 2 items, Fig. Nos.21, 24.

J. U. Rastrick, 1829 notebooks: University of London (Goldsmiths') Library, 2 items, Fig. Nos.26, 29.

Institution of Mechanical Engineers: 1 item, Fig. No.46.

Durham County Library Services, Darlington Branch Library: 1 item, Fig. No.50.

Durham County Record Office and Whessoe Engineering: 3 items, Fig. Nos.52, 76, 104.

Public Archives, Canada: 1 item, Fig. No.53.

Michael Grocock collection: 3 items, Fig. Nos.109, 110, 247.

J. S. MacLean, "The Locomotives of the North Eastern Railway, 1841-1922": by permission of Mr. J. M. Flemimg. 5 items, Fig. Nos.72, 194, 217, 223, 226.

Ken Hoole Study Centre, Darlington Railway Museum, North Road Station:3 items, Fig. Nos.66, 67, 198.

W. W. Tomlinson, "North Eastern Railway": 1 item, Fig. No.75.

"The Model Engineer", the late Dr. A. T. Thompson, Darlington, and an unknown Patient, 1 item, Fig. No.81.

John H. Proud collection: 2 items, Fig. Nos.137, 214.

"British Railway Journal" and Jim Stephens: 2 items, Fig. Nos.199, 200.

John W. Ward collection: 1 item, Fig. No.222.

E. Craven, notes and sketches: 1 item, Fig. No.114.

Alan Earnshaw collection: 1 item, Fig. No.139

David Williamson collection: 1 item, Fig. No.172.

P.R.O. Kew: 2 items, Fig. Nos.96, 98.

Science Museum, South Kensington, & Clement E. Stretton: 6 items, Fig. Nos.17, 27, 28, 42, Sketch sheet and list, Appendix III.

Nicholas Wood: 2 items, Fig. Nos.4, 40.

Comte de Pambour: 1 item, 1833 list of locomotives.

Author's tracing from Canadian photograph: 1 item, Fig. No.55.

Alan Prior: Dust Cover, Conjectural paintings of Nos.1 & 2 as built.

Bibliography

Many of the small older books listed may be considered trivial and of little interest historically, but have helped to show attitudes of the time. All information is of some value, even if only negative.

R. H. Inness (unattributed): "Locomotive History of the Stockton & Darlington Railway, 1825-1876"; "The Locomotive" Magazine, 1925-1930.

J. G. H. Warren: "A Century of Locomotive Building by Robert Stephenson & Co., 1823-1923", Andrew Reid & Co. Ltd., 1923, & David & Charles reprint, 1970.

J. G. H. Warren: Lecture to the Institute of Locomotive Engineers, at the Mechanics' Institute, Darlington, 1925.

Nicholas Wood: "A Practical Treatise on Rail-Roads" etc.; Knight & Lacey, London, 1825.

J. Adamson: "Sketches of our Information as to Railroads"; Newcastle-upon-Tyne, 1826.

J. Adamson: "An Account of the Stockton & Darlington Railway", 1826.

Comte de Pambour: "Practical Treatise on Locomotive Engines"; John Weale, London, 1835 & 1840.

D. Lardner: "The Steam Engine"; Taylor & Walton, London, 1836.

Thomas Tredgold: "The Steam Engine"; John Weale, London, 1838. (an earlier edition is better if obtainable)

Francis Whishaw; "Railways of Great Britain & Ireland"; 1842, David & Charles reprint, 1969.

Zerah Colburn: "Locomotive Engineering & the Mechanism of Railways"; Collins, Glasgow, 1864 & 1871.

John Dixon et al.: Stock Valuation Lists, S. & D.R., 1838-1844, M.S., P.R.O. Kew.

John Dixon: Stock list from manuscript book, 1849 & later; Science Museum.

North Eastern Railway: A List of Locomotive Engines on the North Eastern Railway; Stock Valuation List, 1868.

Timothy Hackworth: Notebook, 1827-8; M.S., Science Museum.

John Graham: Notebooks (4 vols.) 1831-1845; M.S., Science Museum.

George Graham: Reminiscences (recorded by H. Oxtoby c.1896-7); M.S., P.R.O. Kew & typed version, Stockton-on-Tees Reference Library.

Robert Young: "Timothy Hackworth & the Locomotive"; Locomotive Publishing Co., London 1923, and reprint by the Shildon Jubilee Committee, 1975.

John Wesley Hackworth: "A Chapter in the History of Railway Locomotion"; Practical Mechanics' Journal c.1875 and reprints.

E. L. Ahrons: "The British Steam Locomotive, 1825-1925"; The Locomotive Publishing Co., London, 1927.

C. F. Dendy-Marshall:

"Two Essays in Early Locomotive History", 1928.

"Early British Locomotives", 1939.

"A History of Railway Locomotives down to the Year 1831", 1953.

Locomotive Publishing Co., London.

"A History of British Railways down to the Year 1830", 1938.

Oxford University Press, London.

W. W. Tomlinson: "The North Eastern Railway - Its Rise and Development"; 1914; David & Charles reprint, 1967.

Joseph Tomlinson: Presidential Address to the Institution of Mechanical Engineers, May 1890; Proceedings of the I.Mech.E.

J. S. MacLean: "The Locomotives of the North Eastern Railway, 1841-1922"; R. Robinson & Co. Ltd., Newcastle-upon-Tyne, 1923.

J. S. Jeans: "Jubilee Memorial of the Railway System; a History of the Stockton & Darlington Railway"; Longmans, Green and Co., 1875, and Frank Graham reprint, 1974.

J. U. Rastrick: Notebooks, 1829, M.S., University of London (Goldsmiths') Library (extracts).

E. A. Forward: "Links in the History of the Locomotive"; six parts, the "Engineer" magazine, 1940, 1941 and 1943.

Earlier & unattributed series; "Links in the History of the Locomotive", four parts, the "Engineer" magazine, 1879.

E. A. Forward: "Land Transport No.3, Railway Locomotives & Rolling Stock, Part I, Historical Review", Science Museum, 1947 reprint.

E. A. Forward: "Chapman's Locomotives, 1812-1815, Some Facts and Some Speculations"; Paper, 1951, Newcomen Society.

J. Dunlop: "The Development of Locomotive Valve Gear", part III; the "Engineer" magazine, 1920.

Ferdinand Achard & Laurent Seguin: "British Railways of 1825, as seen by Marc Seguin"; Paper, 1927, the Transactions of the Newcomen Society.

C. von Oeynhausen & H. von Dechen: "Wrought Iron Railway at Darlington in the County of Durham": translation from the original, the Newcomen Society, 1827.

C. von Oeynhausen & H. von Dechen: "Railways in England, 1826 & 1827"; W. Heffer & Sons, Cambridge, for the Newcomen Society, 1971.

Theodore West: "Typical Engines on the Stockton & Darlington and North Eastern Railways"; Paper, 1886, Transactions of the Cleveland Institution of Engineers.

Theodore West: "Development of the Locomotive"; Wm. Dresser & Sons, Darlington, c.1900.

Theodore West: Journal, Railway Jubilee, "The Railway 'do' at Darlington"; M.S., September 27th. 1875.

Brian Reed: "Loco Profile No.25, 'Locomotion'"; Profile Books, 1975.

Alan Prior: "19th. Century Railway Drawings in 4 mm. Scale"; David & Charles, 1983.

Charles E. Lee: "The Evolution of Railways"; The Railway Gazette and the Permanent Way Institution, 1937 & 1943.

K. Hoole:

"The Stockton & Darlington Railway - A History"; David & Charles, 1975.

"A Regional History of the Railways of Great Britain; Volume 4, The North East"; David & Charles, reprint, 1985.

"Stockton & Darlington Railway"; Dalesman Press, 1974.

"Forgotten Railways: Vol.1, North East England"; David & Charles, 1973, 1984.

"North Road Locomotive Works"; Roundhouse Books. 1967.

"North Eastern Locomotive Sheds"; David & Charles, 1972.

"An Illustrated History of N.E.R. Locomotives"; O.P.C., 1988.

"Railways in Cleveland", Dalesman Books, 1971.

"The Stainmore Railway"; Dalesman Press, 1973.

"The Railways of East Durham"; Dalesman Books, 1985.

R. A. Cook & K. Hoole: "North Eastern Railway - Historical Maps"; Railway & Canal Historical Society, 1975.

C. J. Allen: "The North Eastern Railway"; Ian Allan, 1964.

O. S. Nock: "Locomotives of the North Eastern Railway"; Ian Allan, 1954.

Randall Davies: "The Railway Centenary - A Retrospect"; L. & N.E.Ry., 1925.

H. G. Lewin: "Early British Railways"; Locomotive Publishing Co., 1925.

J. B. Snell: "Early Railways"; George Weidenfeld & Nicolson Ltd., 1964.

J. B. Snell: "Railways; Mechanical Engineering"; Arrow Press, 1973.

L. T. C. Rolt: "Great Engineers"; G. Bell & Sons, London, 1962.

L. T. C. Rolt: "George & Robert Stephenson - The Railway Revolution"; Penguin Books, 1984; Longman, 1960.

Hunter Davies: "George Stephenson"; Quartet Books, 1977.

W. O. Skeat: "George Stephenson - The Engineer and His Letters"; The Institution of Mechanical Engineers, London, 1973.

Samuel Smiles: "Story of the Life of George Stephenson"; John Murray, 1860 (abridged edition).

Samuel Smiles: "The Lives of the Engineers - Vol.3."; (1862); David & Charles reprint, 1968.

J. F. Layson: "Memorable Engineers - VI, The Stephensons and Other Railway Pioneers"; John G. Murdoch, London.

Ruth Maxwell: "George Stephenson"; George E. Harrap & Co. Ltd., 1920.

H. C. Knight: "The Rocket - The Story of the Stephensons, Father & Son"; Nelson, 1886.

Henry Frith: "The Flying Horse"; Griffith Farran Brown & Co. Ltd., 1893.

J. Hamilton Fyfe: "Triumphs of Invention and Discovery"; Nelson, 1867.

Brian Haresnape: "Railway Design since 1830", Vol.1, 1830-1914; Ian Allan, 1968.

Bryan Morgan: "Early Trains"; Hampton House, 1977.

C. Hamilton Ellis: "The Lore of the Train"; Crescent Books, 1973.

C. E. Stretton: "The Locomotive Engine & its Development"; Crosby Lockwood & Son, 5th. edition, 1896.

Railway Centenary Procession & Exhibition, 1925; L. & N.E.Ry.

Railway Centenary Supplement to the "Locomotive" magazine, 1925.

Catalogue of the Collection of Railway Relics and Modern Stock, at Faverdale, Darlington, in connection with the Railway Centenary Celebrations, July 1925: London & North Eastern Railway Company.

"The Railway Centenary, 1825-1925"; Stockton-on-Tees Railway Centenary Committee, 1925.

Northern Echo Railway Centenary Supplement, 1925 (1975 facsimile edition)

R. W. Kidner: "A Short History of Mechanical Traction & Travel, Part 3; The Early History of the Railway Locomotive, 1804-1879"; The Press at Coombelands Ltd., 1946.

C. A. McDougall: "The Stockton & Darlington Railway - 1821-1963"; Durham County Libraries, 1975.

P. W. B. Semmens: "Stockton & Darlington - One Hundred & Fifty Years of British Railways"; New English Library, 1975.

P. W. B. Semmens: "Exploring the Stockton & Darlington Railway"; Frank Graham, 1975.

Maurice W. Kirby: "The Origins of Railway Enterprise - The Stockton & Darlington Railway, 1821-1863"; Cambridge University Press, 1993.

Stockton & Darlington Railway: Study Folder by the Witton Park Working Party, 1975.

J. C. Tyson, General Editor: Archive Teaching Units; University of Newcastle-upon-Tyne;

No.3, "Railways in the Making", 1969,

No.11, "The Stockton & Darlington Railway, 1825", 1975.

N. Moorsom: "The Stockton & Darlington Railway - The Foundation of Middlesbrough"; J. G. Peckston Ltd., Middlesbrough, 1975.

Robert Corkin: "Shildon, Cradle of the Railways"; Frank Graham, 1977.

Philip R. B. Brooks: "William Hedley, Locomotive Pioneer"; Tyne and Wear Industrial Monuments Trust, Newcastle-upon-Tyne, 1980.

L. G. Charlton: "The First Locomotive Engineers"; Frank Graham, 1974.

Colin E. Mountford & L. G. Charlton: "Industrial Locomotives of Durham"; Handbook L, Industrial Railway Society, 1977.

Local History Guides No.3 - Stockton & Darlington Railway; Durham County Libraries, 1983.

F. Satow, M. G. Satow & L. S. Wilson: "Locomotion"; The Locomotion Trust, Beamish, 1976.

T. C. Eden: Darlington Railway Preservation Society Leaflets;

"Enterprise"; September, 1985.

"Locomotion & Derwent"; November, 1986, with later notes.

George Ottley: "Bibliography of British Railway History"; H.M.S.O./ Science Museum, 1966 and 1983.

J. F. Clarke: "Power on Land & Sea"; a History of R. & W. Hawthorn Leslie; published by the firm, 1979.

Carol Fox: "Norton's Iron Road"; Norton Heritage Group, 1982.

The History of Whessoe: Whessoe p.l.c., Darlington.

Dennis W. Hockin: "Whessoe"; The Pentland Press, 1994.

Mary Williams: "The Pottery that Began Middlesbrough"; C Books, Redcar, 1985.

George Turnbull: Account of the Drops used for the Shipment of Coals at Middlesbro' on Tees, with a Short Description of the Town and Port of Middlesbrough; Paper, The Institution of Civil Engineers, 1847, Thomas Telford Publications.

P. J. Holmes: "The Stockton & Darlington Railway, 1825-1975"; First Avenue, Glasgow, 1975.

P. J. Holmes: "Passenger Coaches of the Stockton & Darlington Railway"; Paper (M.S.), January 1992.

James W. Lowe: "British Steam Locomotive Builders"; Guild Publishing, London, (1975) 1989.

Bertram Baxter, ed. David Baxter: "British

Locomotive Catalogue, 1825-1923"; Vol.5A., North Eastern Railway, Hull & Barnsley Railway, Moorland Publishing Co., 1986.

W. MacNay: The MacNay Papers, collection, Shildon Town Council.

The "Railway Engineer" magazine, 1899.

W. Huby: "Recollections of a Branch Line Station in the 1890's"; undated, published in five parts in the "North Eastern Express", 1962-3.

Many other articles in the "North Eastern Express" over many years.

T. E. Rounthwaite: "The Railways of Weardale"; R.C.T.S., 1965.

C. Whittle: "The Railways of Consett and North-West Durham"; David & Charles, 1971.

Col. H. C. B. Rogers: "Turnpike to Iron Road"; Seeley, Service & Co. Ltd., London, 1961.

G. Drysdale Dempsey: "A Rudimentary Treatise on the Locomotive Engine in all its Phases"; John Weale, London, 1857, and Kingsmead Reprints.

P. J. G. Ransome: "The Archaeology of Railways"; World's Work Ltd., 1981.

P. J. G. Ransome: "The Archaeology of the Transport Revolution, 1750-1850"; World's Work Ltd., 1984.

Alan Russell: "Yarm & the First Railway"; Yarm Civic Society, 1990.

Alan Russell: "The Locomotives of Timothy & Thomas Hackworth"; Alan Russell, 1992.

Alan Russell: "Braddyll, South Hetton Coal Company", Alan Russell, Hackworth Locomotive Register, 1993.

K. E. Hartley: "The Easingwold Railway"; The Oakwood Press, 1970.

Longstaffe: "The History and Antiquities of the Parish of Darlington in the Bishopric", 1854 (extract).

Thomas Richmond: "Local Records of Stockton", 1868 (extract).

Contemporary Newspapers: "Sunderland Herald", 26th. January 1838 (extract).

John A. Elliott: "Private Owner Wagons of the North East, Vol.1, The Chaldrons"; Chilton Iron Works, 1994.

Colin Walker: "Happy Return - The Rail 150 Steam Cavalcade"; Pendyke Publications, 1975.

Stockton & Darlington Railway Rail 150 Exhibition Steam Cavalcade, Souvenir Guide; Joint Committee, 1975.

Stockton & Darlington Railway 1825-1975 150th. Anniversary; Stockton's Week, Souvenir Programme, Stockton-on-Tees Borough Council, 1975.

"Cavalcade Reflections"; British Rail Eastern Region, 1975.

I have not yet myself read the following, except in quoted extract:

M. Heavisides: "History of the First Public Railway"; Ottley, 1912. (interesting, but not significant re locomotives)

A. L. Deghilage: "Origine de la Locomotive"; Paris, 1886.

Thomas Tredgold: "A Practical Treatise on Railroads & Carriages"; 1835.

Theodore West: "An Outline of the Growth of the Locomotive Engine, and a Few Early Railway Carriages"; also known as "The Development of the Locomotive"; 1885. (This consists mainly of the eight or so large sheets of sketches referred to in the text)

A Short History of R. & W. Hawthorn Leslie & Co. Ltd.; Hawthorn Leslie, Newcastle-upon-Tyne, c.1910.

Ronald Embleton: "The Stockton & Darlington Railway; 1825-1975"; Frank Graham, 1975.

The Committee Minutes of the Stockton & Darlington Railway - now in the Public Record Office, Kew (except in the Craven notes).

John Dixon's 1849 manuscript book, in full (in private hands)

List of Subscribers

Without whom this book would never have been published

1. John H. Proud.
2. Norman Moorsom.
3. Anthony Wickens.
4. John Tuffnell.
5. Colin B. Foster.
6. Paul Talbot.
7. Dr. Peter J. Holmes.
8. D. J. Mawson.
9. John Keavy.
10. Ian Bean.
11. Per Boyen.
12. Richard Lacey.
13. A. J. Higgitt.
14. D. Corfield.
15. R. Hetherington.
16. G. R. Foster.
17. Devon Library Services, Railway Studies Collection.
18. Keith Dear.
19. Michael J. Denholme.
20. Dr. David C. Robinson.
21. John R. Strong.
22. S. J. Hoyle.
23. Dr. Mike Cleverley.
24. John Edgington.
25. P. Robinson.
26. P. Robinson.
27. Peter J. Hogg.
28. Tim Barmby.
29. Don Wilcock.
30. D. R. Wilcock.
31. R. V. Webster.
32. Henry Pearce.
33. John Addyman.
34. Peter Elliott.
35. J. Roebuck.
36. Mike Grocock.
37. Alan R. D. Orr.
38. Peter Everton.
39. Rev. Alan Cliff.
40. K. R. McCall.
41. M. M. Bradbury.
42. Norman R. Gard.
43. A. J. Scott.
44. Les. Cairnes.
45. Les. Cairnes.
46. Alan Pearson.
47. Ted Tomiak.
48 David J. Williamson.
49. D. Howlett.
50. R. Chipchase.
51. Phil. Johnson.
52. J. F. Sedgwick.
53. Stephen Duffell.
54. Dennis Birtle.
55. J. B. Dawson.
56. M. J. Middleditch.
57. R. A. Panting.
58. D. A. Cox.
59. S. C. Walker.
60. P. Fletcher.
61. Robin Barnes.
62. Ken Fox.
63. F. G. Fletcher.
64. S. G. Dyke.
65. North Road Station Railway Museum.
66. Dick Webber.
67. John L. Spalding.
68. John Sloane.
69. Raymond Gordon.
70. D. A. Digby.
71. Kenneth L. Taylor.
72. John B. Hamilton.
73. P. S. Eilertsen.
74. R. H. Burdon.
75. Malcolm Shelmerdine.
76. Michael G. South.
77. Prof. J. C. Mann.
78. Colin J. Bullock.
79. J. Thompson.
80. J. R. Bushby.
81. G. W. Green.
82. John Moore.
83. David Ward.
84. L. H. Buchanan.
85. Dr. M. Andrews.
86. Michael Southcott.
87. Roy Bealby.
88. C. Lemmon.
89. David Morton.
90. J. Hancock (Biddulph).
91. Eric Peacock.
92. D. Thompson.
93. Peter Jesperson.
94. Bernard Cunningham.
95. P. Walker.
96. I. E. Burrows.
97. Neil A. MacKillop.
98. Paul. A. Dixon.
99. Cdr. H. W. Polhill.
100. G. D. Scurr.
101. B. P. Wolstenholme.
102. Ian Thomson.
103. Peter Semmens.
104. D. A. B. Brown.
105. Rev. David Littlfair.
106. Dr. W. T. W Cory.
107. Prof. David M. Knight.
108. E. P. Cremin.
109. John Liffen.

110. Dennis Goode.
111. M. A. Lane.
112. W. G. Lloyd.
113. John Glithero.
114. Alan Halfpenny.
115. David Dowdell.
116. F. W. Terry.
117. John G. Godfrey.
118. Capt. & Mrs. W. L. Lowndes.
119. Anthony D. Sutherland.
120. Michael R. Bailey.
121. Grant F. Milne.
122. C. Biggs.
123. L. P. Austin.
124. W. A. Shickle.
125. Miss P. Reed.
126. Dr. J. G. Rye.
127. Vincent Bennett.
128. J. M. Firth.
129. Murray Lewis.
130. J. Tomlinson.
131. W. G. Davenport.
132. J. Baxendale.
133. Miss Sally Adamson.
134. F. Vincent Brett.
135. N. Bamforth.
136. J. Finlay.
137. Alan Pearce.
138. J. B. Rowley
139. T. Hutchinson.
140. Robert Slegers.
141. C. H. Shepherd.
142. David Glanville.
143. T. Robinson.
144. Alan Robinson.
145. A. Betteney.
146. Peter Barton.
147. Andrew Frith.
148. Don Ballard.
149. A. F. Young.
150. D. J. Twibell.
151. P. R. Cowper.
152. Guy R. Sloman.
153. J. W. Baskett.
154. Graham A. Carpenter.
155. Reg. Arnell.
156. K. G. Watson.
157. Harry Paar.
158. R. Tarpey.
159. G. Toole.
160. Ken Plant.
161. Grahame Boyes.
162. Tony Swift.
163. Walter Nunn.
164. Tom Toward, Shildon Town Council.
165. Bob Robinson.
166. Brian Williams.
167. Michael Seymour.
168. Andrew J. Rice.
169. Rev. Graham Woolfenden.
170. Keith Wooley.
171. Dr. M. R. D. Wood.
172. Paul Spencer.
173. Alan Bowman.
174. Alan Bowman.
175. A. H. Peck.
176. Dr. P. J. Rodgers.
177. Tom Quinn.
178. Timothy P. Straw.
179. John Harrison.
180. Alwyn Boulby.
181. Colin Binnie.
182. Graham Stowell.
183. Kenneth A. Bell.
184. Barry Lamb.
185. Reg. Sowler.
186. Jim Brodie.
187. G. Eric Bailey.
188. Thomas Colombi.
189. George Colombi.
190. Bill & Alison Colombi.
191. Denham & Diana Cook.
192. Eric & Mary Lockwood.
193. Martin N. Hale.
194. Eric Wood.
195. Philip Rand.
196. Linda A. Rand.
197. John & Glynis Wenyon
198. Brian Collinson.
199. Tom Bain.
200. Alan Powell.
201. Geoff. Briggs.
202. Joe F. Clarke.
203. Peter Morris.
204. Richard Pepper.
205. Dave Lund.
206. John Hancock (Belper).
207. David S. Cooke.
208. Richard Simpson.
209. Betty & Alan Bolton.
210. Tony Pearson.
211. Rev. Robert Bull.
212. Bob Grey.
213. John Wigston.
214. Clare & Gerry Pierson.
215. Bill J. Cummins.
216. W. S. Wolfe.
217. D. R. Spence.
218. Halcyone Purvis.
219. Bee & Dick Robertson.
220. George Heron.
221. Dirk Eric Servaege.

No further names listed.